EMC MADE SIMPLE ®

Printed Circuit Board and System Design

Mark I. Montrose
Montrose Compliance Services, Inc.

EMC MADE SIMPLE ®

Printed Circuit Board and System Design

Mark I. Montrose
Montrose Compliance Services, Inc.

MONTROSE COMPLIANCE SERVICES, INC

Montrose Compliance Services, Inc.
2353 Mission Glen Dr.
Santa Clara, CA 95051-1214 USA
Office (408) 247-5715, fax (408) 247-5715
E-mail: mark@montrosecompliance.com

Printed in the United States of America
10 9 8 7 6 5 4 3 2 1

Library of Congress Cataloging-in-Publication Data:

Montrose, Mark I.
EMC Made Simple-Printed Circuit Board and System Design / Mark I. Montrose
p. cm.
Includes bibliographical references and index
ISBN 978-0-9891032-0-6 (paper)
ISBN 978-0-9891032-1-3 (electronic)
1. Electromagnetic compatibility 2. Printed circuit board and system design

Printed in the United States of America

Table of Contents

Table of Contents v
Preface xvii
Acknowledgment xxi

1 EMC or Maxwell Made Simple 1
1.1 Time Domain vs. Frequency Domain 2
1.2 History of Electromagnetics (Made Simple) 2
1.3 Theory of Electromagnetics (Maxwell Made Simple) 9
1.4 Antenna Definitions Related to Field Propagation from Source 12
1.4.1 Reactive Near Field Region 12
1.4.2 Radiating Near-Field (Fresnel) Region 13
1.4.3 Far Field (Fraunhofer) Region 13
1.5 Relationship Between Electric and Magnetic Sources 13
1.6 Electromagnetic Fields Represented as Antenna Elements 17
1.6.1 Conductive pathways 19
1.7 Maxwell Simplified-Further Still (Conversion Frequency to Time Domain) 19
1.8 Concept of Flux Cancellation (Flux Minimization) 24
1.9 Skin Effect and Lead Inductance 24
1.10 What are Common-Mode and Differential-Mode Currents? (Made Simple) 27
1.10.1 Differential-Mode Current Description 29
1.10.2 Differential-Mode Radiated Emission Equations 30
1.10.3 Common-Mode Current Description 32
1.10.4 Common-Mode Radiated Emission Equations 35
1.10.5 How Common-Mode Current Drives I/O Cables 36
1.10.6 Conversion Between Differential-Mode and Common-Mode Currents 38
1.11 Antenna Efficiency 39
1.12 Fundamental Principals and Concepts for Suppression of RF Energy 40
1.12.1 Fundamental Principles of EMI Suppression 40
1.12.2 Fundamental Concepts of EMI Suppression 40
1.13 Hidden Schematic or Parasitics of Passive Components 41
1.13.1 Wires, Printed Circuit Board Traces and Transmission Lines 43
1.13.2 Resistors 45
1.13.3 Capacitors 46
1.13.4 Inductors 47
1.13.5 Transformers 48

2 Transmission Line Theory Made Simple 51
2.1 The Definition of Signal Integrity 51
2.2 Primary Concerns Related to High-Speed Signal Integrity Problems 53
2.3 Defining Transmission Line Structures 55
2.4 Types of Transmission Lines 56
2.4.1 Coaxial cable 56
2.4.2 Microstrip 56
2.4.3 Embedded microstrip 56
2.4.4 Stripline (Single and Dual) 57
2.4.5 Balanced lines 57
2.4.6 Lecher lines 59
2.4.7 Single-wire 59

2.4.8 Waveguide 59
2.4.9 Optical fiber 60
2.5 Description of a Typical Transmission Line System 60
2.6 Transmission Line Structures in a Printed Circuit Board 61
2.6.1 Lossless Transmission Line 61
2.6.2 Lossy Transmission Line 63
2.7 Transmission Line Effects on Signal Propagation 64
2.7.1 Conditions That Create Ringing 68
2.8 Transmission Line Termination Overview 70
2.9 RF Current Distribution 71
2.10 Analysis of RF Return Paths 73
2.11 Creating an Optimal RF Return Path 75
2.12 How RF Return (Image) Planes Work 76
2.12.1 Image Plane Implementation and Concept 77
2.13 Image or RF Return Path Violations 80
2.14 Layer Jumping–Use of Vias 82
2.14.1 Layer Jumping Concerns 83
2.15 Split Planes and Their Effect on RF Return Path Discontinuity 85
2.15.1 Digital-to-Analog Partitioning (Split Return Plane) 87
2.15.2 Using a Ferrite Bead versus Inductor in Split Plane Configurations 89
2.16 Flux Cancellation Concepts (Optimizing RF Current Return) 90

3 Inductance Made Simple 93
3.1 Types of Inductance 93
3.1.1 Self-Inductance 93
3.1.2 Mutual Inductance 95
3.1.3 Partial Inductance 98
3.1.4 Mutual Partial Inductance 98
3.2 Impedance and Transmission Line Behavior Related to RF Return Current.. 101
3.2.1 Typical Transmission Line Configuration 101
3.2.2 Path of Least Impedance 101
3.2.3 RF Return Current Travel in a Transmission Line 102
3.3 Inductance Concerns Related to PCB Layout and Trace Lengths 104
3.3.1 Loop Inductance 104
3.3.2 Loop Mutual Inductance 105
3.3.3 Decoupling Capacitor Mounting Related to Lead Inductance 105
3.3.4 Via Configuration and its Effect on Lead Inductance 107

4 Power Distribution Networks Made Simple 109
4.1 The Need for Optimal Power Distribution 109
4.2 Power Distribution Network as Transmission Lines 110
4.3 Primary Requirements for Enhanced Power Distribution 110
4.4 Defining Capacitor Usage on Printed Circuit Boards 111
4.4.1 Bulk Capacitor Description 112
4.4.2 Bypass Capacitor Description 113
4.4.3 Decoupling Capacitor Description 113
4.5 Review of Resonance (Basic Circuit Analysis) 114
4.5.1 Series Resonance 114
4.5.2 Parallel Resonance 115
4.5.3 Parallel *C*–Series *RL* Resonance (Anti-resonant Circuit) 115
4.6 Physical Characteristic of Capacitors 116
4.6.1 Capacitor Types 116

4.6.2 Commonly Used Dielectrics 117
4.6.3 Impedance Plots Based on Dielectric Composition 120
4.6.4 Effective Range of Decoupling Capacitor Families 121
4.6.5 Energy Storage Capabilities of Capacitors 122
4.6.6 Impedance (Actual Self-Resonant Frequency) 123
4.6.7 Resonance of a Capacitor When Installed on a Printed Circuit Board 126

4.7 Capacitors Placed in Parallel (Anti-Resonant Effect) 129

4.8 Power and Return Planes Providing Internal Decoupling Capacitance 132

4.8.1 Calculating Power and Return Plane Capacitance 133

4.9 Vias and Their Effects in Solid Planes 134

4.9.1 Combined Effects of Plane Capacitance with Discrete Capacitors 135

4.10 Effects of ESR and ESL in Decoupling Applications 137

4.10.1 Effects on Performance-Changes in *ESL* Values 137
4.10.2 Effects on Performance-Changes in *ESR* Values 138

4.11 Planes as RF Return Path for Transmission Lines 138

4.12 Multi-Pole Decoupling Methodology 139

4.13 Effects of Proper Decoupling Implementation 142

4.14 Simplified Description of the Capacitor Brigade 143

4.15 Radius of Operation-Effectiveness of Maintaining Voltage Levels 144

4.16 Equivalent Circuit Model of a Printed Circuit Board 145

4.16.1 Transmission line (trace) inductance 147

4.17 Conflicting Rules for Printed Circuit Board Decoupling 149

4.17.1 Where do we locate a decoupling capacitor? 150
4.17.2 Should we install decoupling capacitors where there are no components? .151
4.17.3 Should we use a capacitor with high ESR or low ESR? 153
4.17.4 Relationship between capacitance value and packaging dimensions? 154
4.17.5 Should we use one capacitor per power/return pair or share? 154

4.18 Inductance of Mounting Pads for Components and Capacitors 156

4.19 Bypass and Decoupling Value Calculation 159

4.20 Capacitive Effects on Signal Traces (Wave Shaping) 162

4.21 Bulk Capacitor Application 165

4.22 Buried Capacitance 167

4.23 Summary-Guidelines for Power Distribution Networks 171

5 Referencing Made Simple (a.k.a. Grounding) 175

5.1 An Overview on Referencing (a.k.a. Grounding) 175

5.2 Definitions 176

5.3 Defining Various Types of Grounding Systems 177

5.4 Common Grounding Symbols 178

5.5 Different Types of 0V-Referencing 179

5.6 Fundamental Grounding Concepts 179

5.6.1 Grounding Misconceptions 180

5.7 Primary Concerns Related to the Issue of Grounding and Referencing 180

5.7.1 Grounding for Product Safety 180
5.7.2 Signal Referencing for Components (AC Signals or RF Return Current) ..185

5.8 Grounding Methodologies 188

5.8.1 Single-Point Grounding Methodology 189
5.8.2 Multiple Connections to Single-Point Reference 194
5.8.3 Hybrid Grounding 197

5.9 Controlling Common Impedance Coupling Between Transmission Lines 199

5.9.1 Lowering the Common Impedance Path Inductane 200
5.9.2 Avoiding a Common Impedance Path in the First Place 201

5.9.3 Minimizing Ground Inductance 203
5.10 Controlling Common-Impedance Coupling in Power and Return Planes 204
5.11 Breaking Ground Loops 207
5.11.1 Transformer isolation 208
5.11.2 Optical isolation 209
5.11.3 Common-mode choke isolation 210
5.11.4 Balanced circuitry 210
5.12 Resonances When Using Multi-Point Grounding 213
5.13 Signal and Ground Loops (Not Eddy Current Loops) 216

6 Shielding, Gasketing and Filtering Made Simple 219
6.1 The Need to Shield 219
6.2 Basic Shielding Equations 220
6.3 Theory of Shielding Effectiveness – Made Simple 221
6.3.1 Technical Explanation-Shielding Theory 222
6.3.2 Shielding Effects 223
6.3.3 Near-Field Conditions 225
6.4 Losses Achieved with Shielding Material 225
6.4.1 Reflection Loss 225
6.4.2 Absorption Loss 227
6.4.3 Skin Effect and Skin Depth 229
6.4.4 Reflections Internal in Thin Shields 231
6.4.5 Composite Absorption and Reflection Loss 232
6.5 Apertures in Shield Barriers 234
6.5.1 Single Apertures 235
6.5.2 Multiple Apertures 239
6.5.3 Slot Antenna Polarization 241
6.5.4 Waveguide Below Cutoff 242
6.5.5 Waveguides or Transmission Lines Between Enclosures and Systems 243
6.6 Apertures in Shield Barriers 244
6.6.1 Proper and Improper Shield Penetrations 244
6.7 Cable Shield Grounding and Termination 247
6.7.1 Types and Applications of Cable Shields 248
6.7.2 Grounding the Cable Shields–One End or Both 251
6.7.3 Cable Shield Termination Overview–System Level 255
6.7.4 Implementing a Cable Shield into an Assembly 257
6.7.5 Terminating a Cable Shield Properly 258
6.7.6 Aspects to Consider When Specifying a Shielded Cable 259
6.8 Shielded Compartments 259
6.8.1 Board Level Component Shields 261
6.9 Gaskets-Application and Implementation 261
6.9.1 Material Composition and Performance 263
6.9.2 Common Gasket Material 264
6.9.3 Environmental Aspects of Gasket Use 268
6.9.4 Mechanical Problems When Using Gaskets 268
6.9.5 Installation Requirements 274
6.10 Conductive Coatings 276
6.10.1 Concerns When Using Coatings 277
6.11 Filters 280
6.11.1 What is an EMI Filter? 280
6.11.2 Insertion loss 280
6.11.3 Basic Passive Filter Elements 281

6.11.4 Parasitics related to filter components 284
6.11.5 Basic filter configurations..285
6.11.6 Common-mode and differential-mode filters 287
6.11.7 Signal line filter configurations 289
6.11.8 Criteria in selecting an EMI filter for AC mains applications.....................293
6.11.9 Using ferrite material for filtering 295
6.11.10 Use of ferrite material on cables..297
6.11.11 How to select a ferrite device for use on printed circuit board traces 298
6.11.12 Feedththrough capacitor filter..302
6.11.13 Three terminal capacitor filter 302
6.11.14 Installation Guidelines for Filters 303

Appendix A – Maxwell Made Simple...307
Appendix B – The Decibel...313
Appendix C - Fourier Analysis 317
Appendix D - Conversion Tables...321
Appendix E – Glossary 325
References..339
Index...343
About the Author 349

List of Figures

Figure 1.1 Voltaic pile. 3
Figure 1.2 Kirchhoff Current Law–Conservation of Electric Charge. 5
Figure 1.3 Kirchhoff Voltage Law–Conservation of a potential field. 5
Figure 1.4 RF transmission of an electromagnetic field. 14
Figure 1.5 Wave impedance versus distance from *E* and *H* dipole sources. 16
Figure 1.6 Wave propagation–electric and magnetic field elements. 17
Figure 1.7a Noise coupling model-electric field (dipole antenna). 17
Figure 1.7b Noise coupling method-magnetic field (loop antenna). 18
Figure 1.8 Right hand rule (Faraday's Law). 22
Figure 1.9 Closed loop circuit. 23
Figure 1.10 Concept of flux cancellation. 24
Figure 1.11 Differential-mode current model. 29
Figure 1.12 Differential-mode circuit configuration. 30
Figure 1.13 Common-mode current model. 33
Figure 1.14 Common-mode current injected into an I/O cable. 37
Figure 1.15 System equivalent circuit of differential and common-mode currents. 37
Figure 1.16 Differential-mode to common-mode conversion in an enclosure. 38
Figure 1.17 Various types of antenna configurations. 39
Figure 1.18 Component behavior at RF frequencies. 43
Figure 1.19 Real-word frequency response of a resistor. 45
Figure 1.20 Real-word frequency response of a capacitor. 47
Figure 1.21 Real-word frequency response of an inductor. 48
Figure 1.22 Operational characteristics of ferrite material. 48
Figure 2.1 Illustration-microstrip and stripline topology in a printed circuit board. 57
Figure 2.2 Balanced line in twisted pair configuration used with 2-wire circuits. 58
Figure 2.3 Balanced differential pair twisted uses with multi-wire circuits (Ethernet). 58
Figure 2.4 Star quad configuration of cable bundling for multi-conductor circuits. 58
Figure 2.5 Balanced twin lead for use with RF circuits, particularly antennas. 59
Figure 2.6 Basic description of a transmission line. 60
Figure 2.7 Variations on schematic representations for transmission lines. 61
Figure 2.8 Lossless transmission line equivalent circuit. 62
Figure 2.9 Lossy transmission line equivalent circuit. 63
Figure 2.10 Reflection effect in a transmission line causing overshoot and ringing. 65
Figure 2.11 Transmission line effects based on source and load impedance ratio. 68
Figure 2.12 Equivalent circuit for ringing and rounding of signal propagation. 69
Figure 2.13 Ringing effect in a transmission line. 70
Figure 2.14 Impedance matching requirements of a circuit. 71
Figure 2.15 Simplified network for drive current example. 71
Figure 2.16 Current density distribution from trace to reference plane 72
Figure 2.17 Typical PCB design without an optimal RF return current system. 74
Figure 2.18 Current return paths – DC and AC. 75
Figure 2.19 Loop area between components–single/double sided assemblies. 76
Figure 2.20 Schematic representation of a return path within a printed circuit board. 77
Figure 2.21 Image plane related to partial inductance and phantom current. 79
Figure 2.22 Image plane violation with traces. 80
Figure 2.23 Ground "loops" using through-hole components (slots in plane). 81
Figure 2.24 Passing RF return current across a split plane using a bypass capacitor. 82
Figure 2.25 Using ground stitch via during layer jumping. 84
Figure 2.26 Creating an RF return path on a four-layer printed circuit board. 85

Figure 2.27 Manual layer jumping to create an optimal RF return path next to a via.....85
Figure 2.28 Variations on split plane configurations......86
Figure 2.29 Localized ground plane–digital-to-analog partition......88
Figure 2.30 Ferrite material performance characteristics......89
Figure 3.1 Magnetic flux through a current loop with time variant current......94
Figure 3.2 Inductance and loop area physical dimensions......95
Figure 3.3 Concept of mutual inductance as a transformer model......96
Figure 3.4 Mutual coupling between two inductors or transmission lines......97
Figure 3.5 Loop area defining self and mutual partial inductance......99
Figure 3.6 Mutual partial inductance between two conductors......99
Figure 3.7 Simple example of a transmission line topology......101
Figure 3.8 RF current flow based on the path of least impedance......103
Figure 3.9 Method of moments results-current density at *1 KHz* (least resistance).....104
Figure 3.10 Method of moments results-current density at *1 MHz* (least impedance)..104
Figure 3.11 Loop inductance......104
Figure 3.12 Capacitor mounting dimensions......106
Figure 3.13 Typical via configuration and their relationship to lead inductance......107
Figure 3.14 Improvement on minimizing lead inductance for capacitor placement.....107
Figure 4.1 Simplified visualization of a power distribution system......110
Figure 4.2 Power distribution network represented as a transmission line......111
Figure 4.3 Series resonant circuit......114
Figure 4.4 Parallel resonant circuit......115
Figure 4.5 Parallel C–Series RL resonant circuit......115
Figure 4.6 Capacitor model with resistance, inductance and capacitance......116
Figure 4.7 Impedance curves due to different dielectric material......120
Figure 4.8 Effective range of typical decoupling capacitors......121
Figure 4.9 Theoretical impedance frequency response of ideal planar capacitors......124
Figure 4.10 Effects of lead length inductance within a capacitor......125
Figure 4.11 Self-resonant frequency of capacitors with radial or axial leads......128
Figure 4.12 Self-resonant frequency of SMT capacitors......129
Figure 4.13 Two different capacitor values in parallel showing anti-resonant effect....131
Figure 4.14 Bode plot of two capacitor in parallel......132
Figure 4.15 Physical parameters to calculate plane capacitance......133
Figure 4.16 Effect of vias in power/return planes causing change in capacitance......135
Figure 4.17 Decoupling effects of combined power/return planes with capacitors......136
Figure 4.19 Effect of *ESR* causing a ripple on the voltage rail with 100 nF capacitor..138
Figure 4.20 Multiple capacitor to achieve low impedance for large bandwidth......140
Figure 4.21 Multi-pole decoupling concept with various capacitor values......142
Figure 4.22 Power noise measured across power supply with capacitors removed.....142
Figure 4.23 Power noise measured across power supply with capacitors in place......143
Figure 4.24 Simplified illustration of the capacitor brigade......143
Figure 4.25 Power and return plane capacitive displacement current at 500 MHz......145
Figure 4.26 Equivalent representation of a PCB showing all parasitic elements......146
Figure 4.27 Power distribution model for loop area impedance calculation......147
Figure 4.29 Conflicting rules-of-thumb related to decoupling capacitor placement.....149
Figure 4.30 Loop area decoupling capacitor placement adjacent to power/return pin..150
Figure 4.31 Capacitor bodies with reverse aspect ratio......154
Figure 4.32 Sharing decoupling capacitors among pins and components......155
Figure 4.33 Simplified decoupling capacitor placement for large scale components...156
Figure 4.34 Comparison of connection inductance for SMT components......157
Figure 4.35 Comparison of connection inductance for various SMT components......157
Figure 4.36 Placement patterns for optimal performance–multilayer implementation. 158

Figure 4.37 Placement recommendation loop inductance when using a trace. 159
Figure 4.38 Capacitive effects on clock signals...162
Figure 4.39 Capacitor equations, charging and discharging...163
Figure 4.40 Backdrilled via implementation..168
Figure 4.41 Buried capacitance structure...168
Figure 4.42 Implementation of buried capacitance in a ten layer stackup.....................169
Figure 4.43 Mutual inductive coupling due to loop area between vias. 173
Figure 5.1 Typical ground symbols commonly found on PCB schematics.................179
Figure 5.2 Stray impedance in a circuit from a voltage source to chassis ground.......181
Figure 5.3 AC Line Filter Configuration. 182
Figure 5.4 International creepage and clearance distance requirements......................184
Figure 5.5 Referencing signal propagation for functional operation...........................185
Figure 5.6 Typical grounding between circuits. 186
Figure 5.7a Common-impedance coupling in a ground or return structure..................187
Figure 5.7b Conductive coupling of ground noise with external interconnect.............187
Figure 5.8 Single-point grounding methods. 191
Figure 5.9 Bad implementation of single-point (connection) referencing...................192
Figure 5.10 Poor implementation of single-point referencing......................................193
Figure 5.11 Routing single-point, star configuration – a poor design technique...........193
Figure 5.12 Multiple connections to a single point reference methodology.................195
Figure 5.13a Dipole antenna representation. 195
Figure 5.13b Dipole representation of planes in a PCB with multi-ground points. 196
Figure 5.14 Hybrid grounding methodology. 198
Figure 5.15 Typical cabinet configuration using hybrid grounding. 199
Figure 5.16 Current flow in a conductor – skin effect..200
Figure 5.17 Separation of grounds to avoid common impedance coupling..................202
Figure 5.18 Single-point grounding for lower frequencies/ multi-point for higher.......203
Figure 5.19 Mutual inductance between two parallel transmission lines. 205
Figure 5.20 Mutual capacitance between two parallel transmission lines. 205
Figure 5.21 Common impedance coupling in a power and return plane configuration.206
Figure 5.22 Ground loop between two circuits...208
Figure 5.23 Isolating a ground loop using a transformer..209
Figure 5.24 Isolating a ground loop using optical isolation..209
Figure 5.25 Isolating a ground loop using common-mode chokes...............................210
Figure 5.26 Isolating a ground loop using balanced circuit isolation...........................211
Figure 5.27 Circuit representing common-mode-rejection-ratio. 212
Figure 5.28 Resonance in a multi-point ground to chassis. 214
Figure 5.29 Problems grounding a printed circuit board by a screws to a standoff.......215
Figure 5.30 Mounting printed circuit board to an enclosure using screw and standoff.216
Figure 5.31 Ground loops within a printed circuit board assembly..............................217
Figure 5.32 Loop area between components. 218
Figure 6.1 Shielding effectiveness example. 221
Figure 6.2 Shielding theory illustrated as a transmission line. 222
Figure 6.3 Electric field effects on a boundary condition..223
Figure 6.4 Magnetic field effects on a boundary condition. 224
Figure 6.5 Magnetic field producing eddy currents within a boundary condition........224
Figure 6.6 Electromagnetic fields propagating around openings in a shield partition. 224
Figure 6.7 Reflection from a shield barrier...226
Figure 6.8 Reflection loss in copper shield relative to frequency................................227
Figure 6.10 Absorption loss for various types of metal and their thickness.................229
Figure 6.11 Multiple reflections internal to a shield barrier. 231
Figure 6.12 Re-reflection loss correction factor (*B*) for thin shields/magnetic fields....232

Figure 6.13 Shielding effectiveness of a 0.02-inch thick copper shield in the far field.233
Figure 6.14 Electric field, plane wave and magnetic field shielding effectiveness. 233
Figure 6.15 Slots in a shield barrier and effect of induced current flow........................234
Figure 6.16 Penetration of fields through apertures large compared to wavelength. 236
Figure 6.17 Penetration of fields through holes in thin barriers. 237
Figure 6.18 Huygen's principle - Diffraction of plane wave at opening at a barrier.....237
Figure 6.19 Shielding effectiveness vs. frequency and maximum slot length...............239
Figure 6.20 Penetration of fields through holes in a thin barrier..................................240
Figure 6.21 Slot antenna cross polarization effect on a radiated signal.........................241
Figure 6.22 Typical waveguide configurations and dimensions...................................242
Figure 6.23 Examples of honeycomb waveguides for air ventilation............................243
Figure 6.24 Possible shielding integrity violations from an enclosure or assembly......245
Figure 6.25 Proper and improper application of connectors penetrating shield barrier.246
Figure 6.26 Overview on how a shield appears on a cable assembly............................248
Figure 6.27 Various types of shields used for cable assembly. 250
Figure 6.28 Summary of shielded cable configurations and protection levels. 250
Figure 6.29 Basic antenna structures associated with cables and interconnects............251
Figure 6.30 Connection of cable shield at one end..254
Figure 6.31 Connection of cable shield at both ends. 254
Figure 6.32 Connection of cable shield - system level. 256
Figure 6.33 Terminating cable shields. 258
Figure 6.34 Barrier partitions with shielding protection within an enclosure. 260
Figure 6.35 Printed circuit board component shield samples. 262
Figure 6.36 Sample display of different gasket material. 267
Figure 6.37 Surface roughness of metal panels requiring a gasket for full continuity. .271
Figure 6.39 Example of galvanic corrosion. 274
Figure 6.40 Correct method to install gaskets in sheet metal enclosures. 275
Figure 6.41 Method of shielding a meter hole in a panel. 276
Figure 6.42 Transfer function of a typical filter..280
Figure 6.43 Capacitor behavioral characteristic and shunting filtering capabilities......284
Figure 6.44 Inductor behavioral characteristic and series filtering capabilities. 284
Figure 6.46 Common-mode filter topologies using multiple elements. 288
Figure 6.47 Differential-mode filter topologies using multiple elements......................289
Figure 6.48 Basic filter configurations and behavior curves. 290
Figure 6.49 Differential pair signaling with regard to CMMR definition. 291
Figure 6.50 Common-mode choke implementation-differential-mode configuration...293
Figure 6.51 AC mains line filter configurations. 294
Figure 6.52 Equivalent circuit for a ferrite device. 296
Figure 6.53 Typical impedance curves for a ferrite device..297
Figure 6.54 Performance characteristic ferrite material both resistive and inductive. ..301
Figure 6.55 Feedthrough capacitor construction and configurations.............................303
Figure 6.56 Three terminal capacitor configuration. 303
Figure 6.57 Installing a filter for AC mains protection...305

List of Tables

Table 1.1 Skin depth for copper substrate...25
Table 1.2 Physical characteristics of wire...26
Table 3.1 Mutual partial inductance between two parallel transmission lines...............101
Table 3.2 Mounting inductance for typical capacitor configurations. 106
Table 3.3 Mounting inductance of capacitors 106
Table 4.1 Summary of various capacitor types...117
Table 4.2 Typical usage of capacitor families and operating range..............................117
Table 4.3 Capacitor classification code (Class 2). 118
Table 4.4 Approximate self-resonant frequency of various packaging styles. 126
Table 4.5 Magnitude of impedance of a 15-nH inductor versus frequency..................127
Table 4.6 Various capacitor values and number to achieve 1 mΩ target impedance...141
Table 4.7 Inductance of a microstrip trace (per/pad, not including via inductance)......148
Table 4.8 Parallel plate capacitance - 10 in. (25.4 cm) square assembly, ε_r=4.1...........171
Table 5.1 Impedance of a 10x10 inch (25.4x25.4 cm) copper metal plane. 196
Table 5.2 Inductance of various conductors at 1 MHz. 207
Table 6.1 Table of skin depth for various metals...230
Table 6.2 Shielding effectiveness vs. frequency for a single aperture...........................238
Table 6.3 Properties-common types of RF gaskets and fingers....................................266
Table 6.4 Characteristics of common gasket materials..266
Table 6.5 Electrochemical groupings to prevent corrosion between various material. .273
Table 6.6 Characteristics of common surface coatings...278
Table 6.7 Comparison of metallizing techniques...278
Table 6.7 Basic filter configurations...286
Table 6.8 Permeability of ferrite material and frequency range supported....................297

Preface

- *EMC Made Simple* ®
- *Electromagnetic Compatibility Made Simple* ® and
- *Maxwell Made Simple* ®

These three phrases are a registered trademark of Montrose Compliance Services as well as my tagline and approach to designing products based on simplifying the most complex aspect of electrical engineering and design. Understanding electromagnetics is critical for designing products using electrical power to provide an intrinsic function of use. Although devices may use a DC power source, once it enters a product and arrives at circuitry, conversion to analog occurs. Analog typically is a motional wave function described with units of Hertz. The popular and non-specific definition of RF is anything greater than DC to daylight again, with units of Hertz. Once we convert to the analog world we enter the field of electromagnetic functions. Using digital technology, electrical interference from radiated emissions becomes a concern as well as immunity to perturbations and performance requirements.

There is no system I am aware of that does not convert DC energy or signals to the AC mode either through an onboard power supply providing both voltage and current to active components. We thus live in an analog world since it takes a finite amount of time for a digital signal to transition from low-to-high and high-to-low. For those who consider themselves to be a digital designer, remember the real [technical] definition of digital is "infinitely fast AC slew rate signal." With this definition, there is therefore no such thing as digital components. This is because between binary "1" and "0" is the realm of analogue radio frequency! Digital components in reality contain gain function devices, similar to op-amps driven to saturation for all inputs and outputs. This means they function using fast AC slew rates of operation, producing an AC spectral profile across a broad frequency range.

The focus of this book is to take a complex field of engineering and simplify it for those who never took courses in electromagnetics, or had difficulty in understanding calculus as it relates to applied engineering applications. Engineers and designers in this category must develop products to be functional and compliant to a multitude of standards that include regulatory compliance. In many companies, product design is performed not only by [experienced] engineers, but also by a wide variety of talent all with various areas of expertise. A person who spent years working in a manufacturing facility can be transferred to engineering working as a technician to help on time critical tasks. They may also be assigned to work in regulatory compliance that includes electromagnetic compatibility (EMC), product safety, telecommunication approvals, medical applications, documentation related to various directives (RoHS, WEEE, REACH and numerous other acronyms) plus other requirements. Technicians and recent graduates are generally not trained in product design yet they take on more and more responsibility that may include printed circuit board layout and integration to meet regulatory standards without being educated in the field of compliance: thus, they must learn everything in real-time often without having a mentor.

When employed at several major companies in Silicon Valley, California, technicians who worked with me did time consuming tasks with vigor and excellence, but usually not design, analysis or troubleshooting. My time was spent solving problems created by engineers that were easy to identify and fix, but which could have been performed by someone else only if this person had knowledge of *EMC Made Simple*®. When everyone in a team understands what needs to be done using common sense, along with a basic understanding of electromagnetics, our work would have been more efficient.

As a consultant, I find that most of my clients have limited expertise in solving complex problems or they delegate the job to someone else to get a product "certified." Even with junior engineers and technicians, the senior compliance engineer may be overworked trying to optimize a system for both functionality as well as compliance. In addition, they must spend time with those in product design by helping them understand where problem areas are and explaining electromagnetics using simple definitions and examples. If they had an easy to read book that could be given to team members, educating them would be an easier task.

Most of the time with these observations, EMC engineers must fix a poor design after printed circuit boards (printed circuit board) are assembled and ready for customer shipment, hoping they pass EMC tests. If designers and engineers understood electrical engineering in a "simplified manner," and consider all aspects of system development and not just a specific area such as printed circuit board or mechanical design, things could turn out differently. Engineers must now know about multiple aspects of engineering that includes how to design a printed circuit board using schematic capture and layout tools, material science, mechanical design, thermal issues, low-cost manufacturing requirements, along with testing, troubleshooting and support.

In years of working as an EMC consultant with hundreds of small to medium size companies, I find that many do not have resources, both in terms of senior or experienced engineers or money to purchase simulation programs to perform computational analysis. I have observed on a consistent basis that designers constantly experiment with trial and error fixes without understanding the problem, or means of finding a simple solution using Ohms law instead of Maxwell Equations. The incentive for writing this book comes from numerous students whom I teach professionally in the United States, Europe and Asia as well as clients that call me in under emergency situations, usually after a product fails EMC testing. Everyone wants to know "*How and why does EMI get developed within a printed circuit board*" in a simplified manner without use of math or computational analysis. Recognizing a need to fill a gap that currently does not exist within the published literature (at time of writing), I want to enlighten readers to a field of engineering that is considered to be a science of *Black Magic*.

My goal is to help companies and engineers worldwide that cannot afford to hire a consultant or experienced senior compliance engineers with design knowledge that includes circuit analysis at the schematic level, using *Maxwell Made Simple®*.

Given the observations above, we can simplify design engineering to achieve success quickly and at low cost. ***The target audience*** of this book is:

1. Those who never studied applied EMC engineering but now work in the field as a technician or engineer due to a shift in job assignment, or the need for a compliance engineer.
2. Managers of regulatory compliance staff who provide resources such as technical expertise or mentoring.
3. Those who have "never" performed simulation or computational analysis, nor probably ever will. [I estimate that close to 95% of my clients have never simulated anything for numerous reasons that include cost, time to market, lack of expertise to use any type software package in the time or frequency domain, etc.]. These are all generally smaller size companies.
4. Printed circuit board designers who must create a product from a schematic and bill-of-material and told to make it will pass EMC without the aid of a senior or experienced EMC engineer reviewing their layout. This includes third-party resources or staff that has minimal interaction with designers and EMC experts.
5. Application and technical support engineers that must solve customer problems in the field by explaining why they cannot get their design to work.

6. University professors who want to learn new techniques on how to simplify the process of teaching applied electromagnetics and design engineering in a manner that helps students visualize this complex field, while providing a solid foundation based on theory and applications.

This book is ***unsuited*** for:
1. Those who use computational analysis to analyze designs, both pre- and post-layout.
2. Senior engineers that already are experienced and understand the field of EMC at a high level.
3. Those who work in a company dealing with high-technology design having access to sophisticated instrumentation, software tools, and staff with considerable experience in design engineering.
4. Those who would rather hire a consultant or third party resource to do their regulatory compliance and design because the field of EMC is too difficult to understand.

About the book structure

During my career as a consultant, I estimate that somewhere between 80-90% of problems encountered deal with having a poor power distribution network on a printed circuit board or improper implementation and use of decoupling capacitors in addition to having poor RF return paths. Referencing and I/O interconnects take up the remainder of product design failures. Therefore, these three primary topics are covered extensively in this book.

For one wanting to learn about other aspects of printed circuit board design such as layer stackup, impedance control, signal integrity, terminations, I/O and interconnects, ESD protection, backplane and daughter card interfacing, plus other unique aspects of printed circuit board development, please refer to the Reference section for a list of outstanding sources of books and technical publications. This book "*is not written to be a complete publication on all aspects of printed circuit board design, but a simplified handbook to get a product to near perfect completion using solid theory and known design and layout techniques that work based on years of experience.*"

Having authored previous printed circuit board design related books for signal integrity and EMC, a different approach is taken with this book. Why compete with other publications that attempt to be all-inclusive covering every possible aspect of engineering design, or to those that just want to get the job done and do not care about electromagnetic theory and related mathematics. Therefore, this book covers critical aspects one needs to know for success. It is also important to emphasize that using this book will not guarantee success, it should however help make success easier to achieve.

Regardless of advances in technology, the topics identified above have not conceptually changed over the years except for unique aspects of device logic applications and implementations. Where technological advancements have occurred in semiconductor manufacturing, these usually are associated with having a stable power distribution network or ensuring sufficient power is available to minimize plane bounce. Losses in board material also contribute to poor signal integrity and development of common-mode RF energy, examined in Chapter 2, "*Transmission Line Theory Made Simple.*"

Regardless of specialty that an engineer or designer works within such as design, testing and manufacturing, a design team must produce a product that can be built quickly and at low cost. Frequently, more emphasis is placed on system level functionality to meet a marketing specification rather than the need to meet legally mandated EMC and product safety requirements. Redesigning anything significantly increases costs that includes, but is not limited to engineering manpower along with administrative overhead.

My focus as a consultant is to assist and advise in the design of high technology products at minimal cost by helping educate clients in *EMC Made Simple*®. Implementing suppression techniques into the printed circuit board saves money, enhances performance, increases reliability, achieves first time compliance with emissions and immunity requirements, in addition to having the product function as desired.

EMC engineers know various tricks of the trade on how to apply rework or a quick fix on a printed circuit board to pass a particular test. When a problem develops, there is little one can do except to re-spin the board. By understanding *EMC Made Simple*®, our work as design engineers will become fun again along with accolades from management for getting the job done correctly on the first prototype of the system.

Mark I. Montrose
Santa Clara, California
USA

Acknowledgment

It has been nearly eight years since I published my last book on EMC and expected it to be my last. This book is a comprehensive presentation that includes system level design in addition to printed circuit board layout to achieve EMC, unique in style and content. I want to acknowledge the following that played a part in the development of this book.

Ed Nakauchi, my co-author of our book *Testing for EMC Compliance-Approaches and Techniques*, who assisted in editing and clarifying content.

Elya Joffe, a long-time friend and colleague, who like all of my previous books, provided outstanding scrutiny during the review process. Comments were associated with very high-speed design issues in addition to the section of correlating Maxwell's Equations to Ohms Law.

To W. Michael King, PhD, who gave tremendous guidance and feedback. Michael is not only my mentor but life-long friend. Without Michael, I would not have been able to achieve the knowledge that has allowed me to become not only a consultant and educator, but also an engineer with the ability to view things differently than others in this rapidly changing field. A new viewpoint within electrical engineering is now required with today's technology and his ability to teach me how to think in three dimensions and not conform to standard means of presenting EMC design is greatly appreciated.

My most special acknowledgment is to Monica Maxwell who encouraged me to write this book. We originally discussed the content to be a brief summary of important areas of EMC engineering, a collection of Cliff Notes based on university seminars that I teach internationally. What started as a short manuscript culminated in a comprehensive book on multiple aspects related to EMC. This book is a well thought-out perspective on design engineering never before published appealing to engineers globally.

Mark I. Montrose
Santa Clara, California
USA

1 EMC or Maxwell Made Simple

EMC or Maxwell Made Simple®...To most engineers, and definitely university students, these three words do not exist in this sequential order or they are an oxymoron. How can Maxwell's equations be simple? After all, in engineering school a primary focus on learning theory is often math based. Much time is spent working with differential and integral equations that either takes time to solve showing us how electromagnetic fields exist in space, but in a non-visualized manner. There is little emphasis on application of use or how equations relate to real-world design engineering.

The key element to realize is that theory, supported by equations, helps us understand the field of electrical engineering. Trying to design a product based on Maxwell's equations alone cannot be done: real-world hardware or parasitics are omitted from consideration. Those who love doing the math may be better suited to research or teaching. For engineers that like to build things, a design concept and tools required to be creative are all that is needed; forget the need to actually solve mathematical equations, understanding what they tell us is the element of concern!

Superficially, it may appear that this book discounts Maxwell's equations as not being relevant in product design. On the contrary, understanding Maxwell's mathematical vocabulary is having keys to success. If for some reason the equations need to be solved to analyze a complex problem, we have sophisticated software and simulation tools to do the work for us, instead of slide rules and calculators that engineers in the past had to use.

The field of engineering now has more focus on extremely advanced levels of high-speed technology. These are attributable and based around advances in semiconductor design and manufacturing. In the future, processors will consist of even smaller (nanometer) technology projected to operate in the femtosecond range. The primary reason for advancing technology to higher levels of functionality is due to needs for faster applications and content delivery. Because the complexity of products is increasing, having a single function device will become a thing of the past. Products will take on multiple applications of use, essentially becoming a system-of-system (SoS) on a single chip.

An example of a system-of-system is wireless communication products. Our hand/cell phones do not only make phone calls, they include text messaging, streaming audio and video, multiple high-resolution camera with a flash unit, WiFi, Bluetooth, global position system (GPS), multi-mode data communication protocols (i.e., CDMA, GSM, etc.), compass, accelerators, high quality graphic screen, an FM radio, and more features not identified in this list. Everything used for commercial acceptance must have multiple functions, which means the device can technically be classified as a system-of-systems, all in one small lightweight package.

With a high level of complexity in product design, it is impractical for one person to perform all tasks related to system development. Consequently, engineers will increasingly classify themselves as either a digital designer or analog designer. Few cross boundaries and work across both domains. As detailed in this chapter, at least in terms of Fourier transforms, there is no such thing as digital engineers or analog engineers. In reality, we are electrical engineers that must understand and be proficient in both domains using Maxwell's equations and Ohm's law simultaneously. It all depends on how one views tasks, and what one feels most comfortable working with.

The secret to *EMC* and *Maxwell Made Simple*® is to understand electromagnetic theory and translate it to Ohms law, which is easier to solve and work with than Maxwell's equations. Digital engineers ensure circuits work at proper voltage levels as well as providing optimal drive current along with numerous other aspects related to signal integrity. We examine later in the book that voltage and current are only metric units of measurement and that in reality it is wave propagation through transmission lines that contain both electric and magnetic fields within a dielectric. With *Maxwell Made Simple*®, converting from frequency domain to the time domain is easy if one can visualize how field propagate in three dimensions.

In the sections below, we examine various aspects of electromagnetics and the relationships to ensuring products are compatible for use within their intended operating environment.

1.1 Time Domain vs. Frequency Domain

Engineers tend to think in either the time domain or frequency domain. The reason why one generally thinks in a single domain is based on how they were educated, or the manner in which they entered the work force. If an engineer understands digital design and circuits, the focus will be on digital components and technology. Processors and Field Programmable Logic Arrays (FPLAs) are fun, and are generally the heart of all products requiring use of these devices. Even if the product does not have high-speed components, there is still digital processing that occurs somewhere in the system.

Analog designers generally work in the development of power supplies and communication networks along with instrumentation and sensors. Antenna and microwave engineers fall into this category. Although there is crossover between job functions, few engineers are well versed in both analog and digital circuit design.

With advances in technology, it is becoming more difficult to become extremely knowledgeable with design techniques and requirements that are now different from what one has experienced. It takes time to learn about advances in printed circuit board material as well as transmission line theory. However, if one understands what Maxwell's approach and electromagnetic theory tells us, we can visualize what occurs during signal propagation. If a problem is observed through visualization, we can quickly identify the source without spending time doing trial and error troubleshooting.

Working in the time domain is generally easier than the frequency domain. It may be difficult for many to visualize an electromagnetic field propagating down a transmission line with units of volts/meter or amps/meter. What is however easy to comprehend is working in units of volts and amps in the time domain that are easily manipulated using Ohm's law instead of the calculus from Maxwell's equations.

1.2 History of Electromagnetics (Made Simple)

In order to understand electrical engineering it is useful to learn history about those that contributed to the field along with the importance of their work. When one understands who contributed to developing the field of electrical engineering and their relationship to the design of products, understanding electromagnetic theory becomes easier. This section is a refresher course in introductory electrical engineering. Many tend to forget history.

Assuming a circuit will work based solely on Ohms Law, along with Kirchhoff and Ampere's Law, applies to signal propagation in both the time and frequency domain. In other words there are two representations of propagation, or different perspectives when analyzing a signal.

What may work when designing a printed circuit board having components sending a signal between a source driver and load (time domain analysis) may create an electromagnetic interference (EMI) event that causes functional disruption (i.e., crosstalk at either the macroscopic or microscopic level), or unintentional radiated emissions by creating undesired common-mode current (frequency domain analysis). Still, it is the same signal that must be analyzed in two domains simultaneously. If there is failure in one domain it will most likely fail in the other.

We begin by presenting a brief biography of pioneers to help understand how and why circuits work the way they do. Knowledge of history goes a long way toward making *EMC Made Simple®*, since Maxwell's equations are based on the research of predecessors studying electrostatic and electromagnetic fields that lived before him.

Alessandro Volta (1745-1827). In 1800 Volta developed the voltaic pile, a forerunner of the electric battery that produced a steady stream of electricity. In honor of his work in the field of electricity, the electrical unit known as the volt is named in his honor.

The battery made by Volta is known as the first electrochemical cell that consists of two electrodes: zinc and copper. The electrolyte used was sulfuric acid or a brine mixture of salt and water. Zinc has a higher potential than copper in the electrochemical series. The positively charged hydrogen ions (protons) from the zinc captures electrons from the copper, forming bubbles of hydrogen gas making the zinc rod the negative electrode and the copper rod the positive electrode.

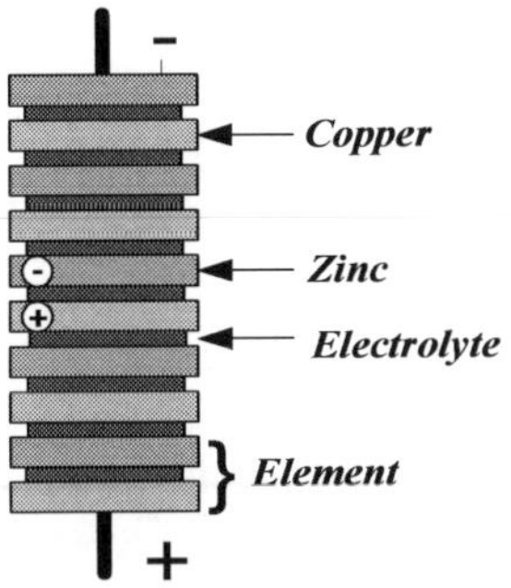

Figure 1.1 Voltaic pile.

André-Marie Ampère (1775–1836). A French mathematician and physicist considered the father of electrodynamics, now commonly referred to as electromagnetics, established the relationship between electricity and magnetism. Ampère became one of the first people to measure, rather than simply detect, electric currents by using a device of his own invention, quantitatively described the relation between a magnetic field and the electric current that produces it in a formulation that has come to be widely referred to as Ampère's Law. Ampère demonstrated that parallel wires carrying current attracts or repels each other depending on whether currents are traveling in the same direction (attraction) or opposite (repulsion). One Ampere is a measure of the amount of electric charge passing a point in an electric circuit per unit time at a rate of 6.241×10^{18} electrons (Coulombs). One Coulomb per second constitutes one Ampere, or one Amp.

Georg Ohm (1789–1854). Georg Ohm published a treatise in 1827 describing the measurement of applied voltage and current through simple electrical circuits containing various lengths of wire. He discovered that current within a conductor between two points is directly proportional to the potential difference across those two points. Introducing the constant of proportionality, he describes this relationship in (1.1), now called Ohm's law.

$$I=V/R \text{ or } V=IR \tag{1.1}$$

where:
V = potential difference measured across the conductor in units of Volts
I = current through the conductor in units of Ampere, and
R = resistance of the conductor in units of ohms.

More specifically, Ohm's law states that resistance in this equation is a constant value, independent of current that is present. Georg Ohm also presented a slightly more complex equation than (1.1) to explain his experimental results. Equation (1.1) is the modern form of Ohm's law and is associated with physical material such as a wire or transmission line providing fixed resistance.

In physics, the term *Ohm's law* refers to various generalizations of the law originally formulated by Ohm. The simplest example of this equation is shown in (1.2). This equation can be applied to electromagnetic field propagation when analyzed in the time domain. The reformulation of Ohm's law is due to Gustav Kirchhoff.

$$\boldsymbol{J} = \sigma \boldsymbol{E} \tag{1.2}$$

where:
$\boldsymbol{J}$ = current density at a given location in a resistive material
σ = material dependent parameter called conductivity, and
$\boldsymbol{E}$ = electric field at that location.

Gustav Kirchhoff (1824–1887). A German physicist who contributed to the fundamental understanding of electrical circuits, spectroscopy and emissions of black-body radiation created by heated objects. He coined the term "black-body" radiation in 1862 and defined two sets of independent concepts in both circuit theory and thermal emissions. These concepts are named "Kirchhoff's laws" as well as a law of thermochemistry. In 1857, Kirchhoff calculated that an electric signal in a resistance-less wire travels along the wire at the speed of light. Kirchhoff's circuit laws deal with two equalities that deal with the conservation of charge and energy in electrical circuits.

Kirchhoff's Circuit Law (KCL) - This principle of conservation of electric charge implies that at any node (junction) in an electrical circuit, the sum of currents flowing into that node is equal to the sum of currents flowing out of that node, or, the algebraic sum of currents in a network of conductors meeting at a point is zero (Fig. 1.2).

Per Kirchhoff's Current Law, there is always zero electromagnetic power at any *point* in space. The electromagnetic power entering a point must be exactly balanced out by the same electromagnetic power leaving it. This law is also described as "the sum of the currents at any point equals zero," and is equivalent to Ampere's Law.

Ampere's Law states that all current flows in closed loops. If some portion of current flow is able to find an alternate path on its own, never to return to the source, then at the location where it leaves the primary current path there would be an imbalance in current flow. The Law of Conservation of Energy does not permit this situation to exist.

With regard to the Conservation of Energy, we restate Kirchhoff's Current Law as "the sum of the electromagnetic power at any point must equals zero." This means that any node that sends current (power, signals, etc.) also simultaneously propagates an anti-phase current which is called the "return current."

Both source and return currents propagate through the impedance of various media (air, conductors, etc.). This defines a closed loop circuit. At any instant in time, the current in the both the source and return paths must balance, or cancel each other out.

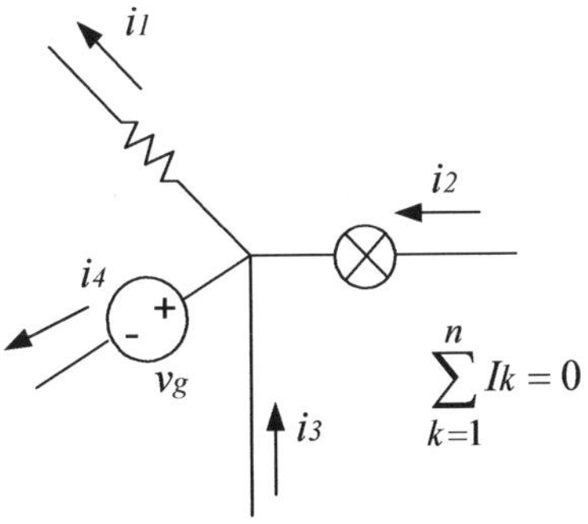

Current entering any junction is equal to the current leaving that junction. $i_1+i_4=i_2+i_3$.
Figure 1.2 Kirchhoff Current Law–Conservation of Electric Charge.

Kirchhoff's Voltage Law (KVL) - This principle of conservation of energy implies that the directed sum of the electrical potential differences (voltage) around any closed circuit is zero, or: more simply, the sum of the electromotive force (EMF) in any closed loop is equivalent to the sum of the potential drops in that loop. It other words, the algebraic sum of the products of the resistances of the conductors, and the currents in them within a closed loop, is equal to the total EMF available in that loop.

Given a voltage potential, a charge which has completed a closed loop doesn't gain or lose energy as it has gone back to initial potential level.

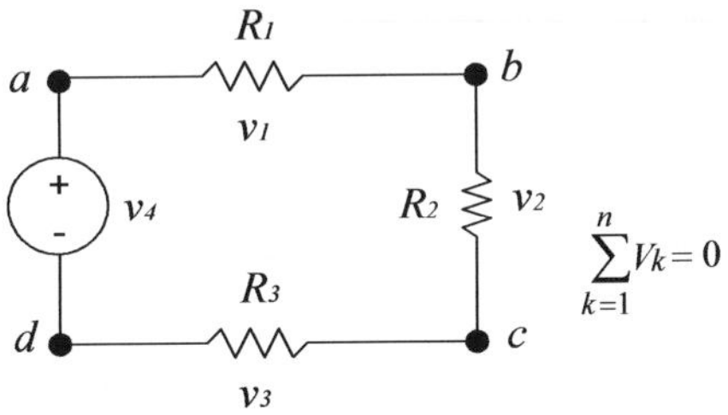

The sum of all the voltages around the loop is equal to zero. $v_1 + v_2 + v_3 - v_4 = 0$
Figure 1.3 Kirchhoff Voltage Law–Conservation of a potential field.

Heinrich Lenz (1804-1865). Heinrich Lenz was a Russian physicist most noted for formulating Lenz's law in electrodynamics in 1833. This law states that an induced electromotive force (EMF) always gives rise to a current whose magnetic field opposes the original change in magnetic flux (1.3). Lenz's law is shown with the minus sign in Faraday's law of induction, which illustrates that induced EMF (ε) and change in flux ($\Delta\Phi B$) have opposite signs.

$$\varepsilon = -N\frac{\Delta\phi B}{\Delta t} \tag{1.3}$$

where:

ε = induced electromotive force (EMF)
N = number of turns in the winding
$\Delta\Phi_B$ = change in magnetic flux, and
Δt = time rate of change of the flux in the circuit.

Lenz's law describes how transformers work. When an AC field is impressed on a primary winding, an equivalent AC field is induced in the secondary winding in the opposite direction whose amplitude is based on the ratio of the number of turns between primary and secondary. Understanding Lenz's law is a critical aspect of minimizing crosstalk between transmission lines and field coupling of circuits and systems.

Michael Faraday (1791–1867). Michael Faraday was an English chemist and physicist who contributed to the fields of electromagnetism and electrochemistry, and is considered one of the most influential scientists in history.

Faraday conducted research regarding the magnetic field surrounding a conductor carrying a DC electric current that established the basis for the concept of the electromagnetic field in physics. His discovery was improved upon by James Clerk Maxwell. He similarly discovered the principle of electromagnetic induction, diamagnetism, and the laws of electrolysis.

Faraday is best known for two discoveries: Law of Induction (or induced electromotive force) and the Faraday cage or Faraday shield.

Faraday's Law of Induction is applicable to a closed circuit made of thin wire and states that:

The induced electromotive force (EMF) in any closed circuit is equal to the time rate of change of the magnetic flux through the circuit.

or alternatively:

The EMF generated is proportional to the rate of change of magnetic flux within a transmission line.

EMF is defined as the energy available per unit charge that travels once around the wire loop (the unit of EMF is the volt). Equivalently, it is the voltage that would be measured by cutting the wire to create an open circuit and attaching a voltmeter to the leads.

Faraday's law of induction is closely related to the Maxwell–Faraday equation (1.4):

$$\nabla \times E = -\frac{\partial B}{\partial t} \tag{1.4}$$

where:

$\nabla \times E$ – curl of the electric field
$\boldsymbol{E}$ = electric field, and
$\boldsymbol{B}$ = magnetic flux density, or magnetic field.

To summarize in simple terms, *"Time variant current traveling in a transmission line creates time variant magnetic flux."* The Law of Induction does not apply to DC current or signal transmission, only AC transitions. AC current has units of Hertz, which is the same unit as RF energy which propagates as a sine wave somewhere in the frequency spectrum.

A Faraday cage or shield operates on the basis that an external static electric field will cause electrical charges within the cage's conducting material to redistribute so as to cancel the field's effects in the cage's interior. This item is an enclosure formed by conductive material or by a mesh of metal that covers the entire surface area. Such an enclosure blocks out external static and non-static electric fields.

Faraday shields cannot block static and slowly varying magnetic fields, such as Earth's magnetic field (a compass will still work inside). To a large degree though, they also shield the interior from external electromagnetic radiation if the conductor is thick enough and if any holes are significantly smaller than the radiation's wavelength. The reception of external radio signals, a form of electromagnetic radiation through an antenna within a cage, can be greatly attenuated or even completely blocked by the cage itself.

Carl Friedrich Gauss (1777-1855). Carl Gauss was a German mathematician and scientist who contributed significantly to many fields of science that includes electrostatics, number theory, statistics, analysis, differential geometry, geophysics, optics and astronomy. Gauss had a remarkable influence in many fields of mathematics and science and is ranked as one of history's most influential mathematicians.

Gauss's law, also known as Gauss's Flux Theorem, relates the distribution of electric charge to the resulting electric field or:

The electric flux through any closed surface is proportional to the enclosed electric charge.

The law was formulated in 1835 but not published until 1867. It is incorporated as one of the four Maxwell's equations which form the basis of classical electrodynamics. The other three laws contained within Maxwell's equations are "Gauss's Law for Magnetism," "Faraday's Law of Induction," and "Ampère's Law." Gauss's law can be used to derive Coulomb's Law and vice versa.

Gauss's Law of Induction is expressed by (1.5):

$$\Phi_E = \frac{Q}{\varepsilon_0} \qquad (1.5)$$

where:
Φ_E = electric flux on a closed surface
Q = total charge enclosed within the boundary area, and
ε_0 = electrical constant.

Electric flux Φ_E (1.6) is defined as the surface integral of the electric field.

$$\Phi_e = \oint_S E \cdot dA \qquad (1.6)$$

Because flux is defined as the *integral* of the electric field, this expression is called the *integral form*. Gauss's law can alternatively be written in the *differential form* (1.7):

$$\nabla \cdot E = \frac{\rho}{\varepsilon_0} \qquad (1.7)$$

where:
$\nabla \cdot E$ = divergence of the electric field
P = charge density, and
ε_0 = electrical constant.

Both integral and differential forms are related by the divergence theorem, also called Gauss's divergence theorem. Both forms can also be expressed two ways: 1) In terms of a relation between the electric field ***E*** and the total electric charge, or 2) in terms of the electric displacement field ***D*** and free electric charge.

Gauss's law can be used to demonstrate that all electric fields inside a Faraday cage have an electric charge. Gauss's law is an electrical analogue of Ampère's law, except it deals with magnetism.

James Clerk Maxwell (1831–1879). Maxwell is probably the world's most influential physicist and mathematician that ever lived, born in Scotland. His most prominent achievement was formulating classical electromagnetic theory that united previously unrelated observations, experiments and equations related to electricity, magnetism and optics into a consistent theory. Maxwell demonstrated that electricity, magnetism and light are basically the same phenomenon, namely an electromagnetic field. Subsequently, all other classic laws or equations became simplified cases of Maxwell's equations which is called the "second great unification in physics" after the discoveries by Isaac Newton.

Maxwell, while studying field theory, demonstrated that electric and magnetic fields travel through space in the form of waves at the speed of light. In 1865, Maxwell published *A Dynamical Theory of the Electromagnetic Field.* This famous treatise was where he first proposed that light was in fact, undulations in the same medium that is also the cause of the electric and magnetic phenomena. His work in developing a unified model of electromagnetism is one of the greatest advances in physics.

Maxwell is also known for creating the first durable color photograph in 1861 using prisms, the hologram, the color wheel, and numerous other advances in optics and is considered the to be the father of optometry. He also did foundational work on the rigidity of rod-and-joint frameworks used in bridge construction.

Maxwell is considered to be the 19th-century scientist with the greatest influence on 20th-century physics. His contributions to science are considered to be of the same magnitude as those of Isaac Newton and Albert Einstein.

A detailed analysis of Maxwell's Equations (*Maxwell Made Simple*®) is in the next section. The presentation in this section regarding Maxwell's life is a small fraction of his work as a scientist, scholar and mathematician. Numerous textbooks have been written about his life and contributions to the field of engineering, too numerous to list in a Reference section.

Oliver Heaviside (1850–1925). Oliver Heaviside was a self-taught English electrical engineer, mathematician, and physicist who adapted complex numbers to the study of electrical circuits. He invented mathematical techniques to solving differential equations (later found to be equivalent to Laplace transforms), reformulated Maxwell's field equations in terms of electric and magnetic forces and energy flux, and independently co-formulated vector analysis. Although at odds with the scientific establishment for most of his life, Heaviside changed the face of mathematics and science forever and is probably the most forgotten, or overlooked historical figure in all of his fields of expertise.

Oliver Heaviside developed transmission line theory (described by the "*telegrapher's equations*"). Heaviside showed mathematically that uniformly distributed inductance in a telegraph line would diminish both attenuation and distortion, and that, if inductance was high enough and insulation resistance not too high, the circuit would contain no distortion while currents at all frequencies would have equal speeds of propagation. Heaviside's equations helped further the implementation of the telegraph system.

Heaviside did much to develop vector methods and vector calculus. Maxwell's original examination and description of electromagnetism consisted of 20 equations with 20 variables. Heaviside employed the curl and divergence operators of the vector calculus to reformulate 12 of these 20 equations into four equations using four variables ($\boldsymbol{B}$, $\boldsymbol{E}$, $\boldsymbol{J}$, and ρ), the form by which they have been known ever since. Oliver Heaviside also created the step function and employed it to model the current in an electric circuit. Since Heaviside reformulated Maxwell's equations into the form they are today, he wanted no credit since his work was based upon others and felt Maxwell deserved full credit.

Oliver Heaviside coined the following terms related to electromagnetic theory:

- Admittance (December 1887)
- Conductance (September 1885)
- Electret for the electric analogue of a permanent magnet, or in other words, any substance that exhibits a quasi-permanent electric polarization (e.g. ferroelectric)
- Impedance (July 1886)
- Inductance (February 1886)
- Permeability (September 1885)
- Permittance (later susceptance; June 1887)
- Reluctance (May 1888).

Other major contributions to science and mathematics from Heaviside include independently discovering the Poynting vector. He was also the inventor of the coax cable.

In the next section, we examine Maxwell's equations and describe what they tell us, in the time domain, which is easier to understand for many engineers than the frequency domain.

1.3 Theory of Electromagnetics (Maxwell Made Simple)

Maxwell's four equations describe the relationship of electric and magnetic fields and are derived from Ampere's law, Faraday's law and two from Gauss. These equations describe field strength and current density within a closed-loop environment and require extensive understanding of calculus and higher order mathematics. Since Maxwell's equations are complex and described in both differential and integral form, we present an overview on simplified electromagnetic theory for those who never took a university course in electromagnetics. Understanding what Maxwell tells us provides insight toward further discussion. For a rigorous presentation and understanding of Maxwell's equations at an academic level, any textbook on the subject matter may be used in addition to those few publications provided in the Reference section as a starting point.

The target audience of this book is those that want to know what the *equations tell us, not how to solve them*. In addition, applying fundamental concepts on how electromagnetics works in simplified form based on these four equations is the focus of this book.

We begin with a very brief overview of what the four equations represent, with greater discussion within this Chapter. The manner of presenting *Maxwell Made Simple*® will be extremely unique, especially to those already well versed in the field of electromagnetics. This presentation is also very different from every college text or other engineering reference since it is not written to be at an academic level.

Maxwell specialized in studying wave theory, both photonic and electromagnetic among other interests of science and math, which is the core of electrical engineering. Maxwell's equations are one of three fundamental concepts that define our existence-electromagnetics. The other two fundamental concepts are Newton's Laws and the Laws of Thermodynamics. Understanding electromagnetics is critical to understanding how and why anything electrical works. Maxwell's equations are based on extensive analysis, verification, discoveries and publications from other brilliant minds that preceded him. He basically tied everything together.

The field of electromagnetics, along with the specialty sub-field of compatibility, is best described technically using complex mathematical concepts. It is impossible to design products by only solving mathematical equations. If for some reason equations must be solved to analyze a certain design requirement, calculations are easily performed using computational analysis and not manually.

Maxwell's equations are shown in (1.8) for completeness as a prelude to converting complex math into simplified algebra, discussed in detail later in this Chapter as well as Appendix A. Although detailed knowledge on solving Maxwell's equations is not a prerequisite for printed circuit board layout or system design, understanding how electromagnetic energy propagates between a source and load, along with its relationship to its environment which includes metal enclosures must be understood to design a product quickly and at low cost.

$$
\begin{aligned}
&\textit{First Law: Electric Flux (from Gauss)}\\
&\nabla \bullet D = \rho \qquad \varphi_e = \oint_s D \bullet ds = \int_v \rho \, dv = 0\\
&\textit{Second Law: Magnetic Flux (from Gauss)}\\
&\nabla \bullet B = 0 \qquad \varphi_m = \oint_s B \bullet ds = 0\\
&\textit{Third Law: Electric Potential (from Faraday)}\\
&\nabla \times E = -\frac{\partial B}{\partial t} \qquad \oint E \bullet dl = -\int_s \frac{\partial B}{\partial t} \bullet ds\\
&\textit{Fourth Law: Electric Current (from Ampere)}\\
&\nabla \times H = J + \frac{\partial D}{\partial t} \qquad \oint H \bullet dl = \int_s \left(J + \frac{\partial D}{\partial t} \right) \bullet ds = I_{total}
\end{aligned}
\tag{1.8}
$$

To describe Maxwell's equations in *simple* terms, a few fundamental principles are now examined. Letters ***J, E, B*** and ***H*** refer to vector quantities based on scalar analysis.

- Maxwell's equations describe the interaction of electric charges, currents, magnetic and electric fields.
- The Lorentz force relation describes the physical forces imposed by both electric and magnetic fields on charged particles.

All materials have a constitutive relationship to other materials. These include

1. Conductivity - relates current flow to electric field (essentially Ohm's law): $\boldsymbol{J} = \sigma \boldsymbol{E}$
2. Permeability - relates magnetic flux to a magnetic field: $\boldsymbol{B} = \mu \boldsymbol{H}$
3. Permittivity (dielectric constant) - relates charge storage to an electric field: $\boldsymbol{D} = \varepsilon \boldsymbol{E}$

where:
$\boldsymbol{J}$ = conduction-current density, A/m^2
σ = conductivity of the material
$\boldsymbol{E}$ = electric field intensity, V/m
$\boldsymbol{B}$ = magnetic flux density, Weber/m^2 or Tesla
μ = permeability of the medium, H/m
$\boldsymbol{H}$ = magnetic field, A/m
$\boldsymbol{D}$ = electric flux density, Coulombs/m^2
ε = permittivity of vacuum, 8.85 pF/m.

To "*overly simply*" what these equations represent, the discussion that follows helps illustrate from a unique perspective what Maxwell's equations tells us. This discussion is not meant to be technically thorough; it is understanding what the equations *tell us* that is important, not how the equations apply to field theory in *free space*.

Maxwell's first equation is known as the divergence theorem and is based on Gauss's law. This equation applies to the accumulation of an electric charge that creates an electrostatic field. This is best observed between two boundaries; conductive and/or nonconductive. The boundary condition behavior referenced in Gauss's law causes a conductive enclosure (also called a Faraday cage) to act as an electrostatic shield. At the boundary of the enclosure, electric charges are kept on one side of a barrier yet are at zero potential on the other side.

Maxwell's second equation illustrates that there are no magnetic charges, only electric charges (monopoles). Electric monopoles are either positively or negatively charged and propagate between two points in space. Magnetic monopoles thus cannot exist. Magnetic fields are produced through the action of electric currents and fields present within a transmission line propagating between two points. Magnetic fields are created by time variant current traveling in a closed loop environment per Maxwell's third and fourth equations.

Maxwell's third equation, also called Faraday's Law of Induction, describes time variant current traveling in a closed-loop circuit (or transmission line, and every transmission line contains inductance) generating a magnetic field. The third equation has a companion (fourth) equation, discussed in the next paragraph. This third equation describes the creation of an electric field from a *time variant changing* magnetic field in an orthogonal manner that in turn creates an electromagnetic field (***E*** x ***H***). Magnetic fields are commonly found in transformer windings such as electric motors, generators and the like. The interaction of the third and fourth equations is the primary focus for electromagnetic compatibility. Together, they describe how coupled electric and magnetic fields propagate (radiate) at the speed of light. This equation also describes the concept of "skin effect" which predicts the effectiveness of magnetic shielding. In addition, inductance is included but in an indirect manner. It is the inductance in the equation that describes how an antenna propagates an electromagnetic field.

Maxwell's fourth equation is Ampere's law and shows that magnetic fields arise from two sources. The first source is current flow in the form of a transported displacement current. The second source describes how changes in the electric field traveling in a closed-loop circuit create magnetic fields. These electric and magnetic sources describe the actions of inductors and electromagnetics. Of the two sources, the first is the description of how electric currents create magnetic fields.

To summarize, Maxwell's equations define the field of electromagnetics, namely that time-varying currents in a transmission line creates time-variant magnetic flux, which in turn creates an electric field, and eventually an electromagnetic field.

Static-charge distributions produce static electric fields, not magnetic fields. Constant, non-time variant currents do not produce electromagnetic fields. Only time-varying (AC) currents produce both electric and magnetic fields that propagate within a dielectric.

Static fields store energy. This is the basic function of a capacitor: accumulation of charge and retention. Constant current sources are a fundamental concept for the use of an inductor which stores and returns energy, providing they are an ideal inductor. Non-ideal inductors will lose some of the electromagnetic energy within its windings and construction, discussed in Chapter 2 and 3.

1.4 Antenna Definitions Related to Field Propagation from Source

The fields surrounding any antenna are divided into three principle regions:

- Reactive near-field
- Radiating near-field or Fresnel Region
- Far-field or Fraunhofer Region

With regard to electromagnetic compatibility, the far-field region is the primary area of concern since regulatory compliance specification limits are based on measurements taken in the far-field. Antennas communicate wirelessly by propagating an electromagnetic field between a transmitting source and a receiver. These three regions relate to a specific antenna distance from the source of RF energy development. Depending on the amount of undesired electromagnetic interference (EMI) present, one may prefer to do troubleshooting in the near-field to solve a far-field concern.

When performing EMC measurements, the distance of the antenna from the radiating source is critical to determine if the field measured is either magnetic or electric. EMC compliance is concerned with "electromagnetic" propagation, which is generally in the far-field and mainly electric in nature.

The near-field and far-field distance between antenna and receiver related to electromagnetic radiation are regions around the source of the energy. The boundary between the two regions is vaguely defined and depends on the dominant wavelength (λ) of the signal emitted by the source.

Both electric and magnetic fields observed in the far-field region are produced by a change in propagation of the other. The ratios of both field strengths are fixed and unvarying in the far-field. However, in the near field both electric and magnetic fields are nearly independent of each other. One cannot calculate a particular field strength from knowing the strength of the other. Both fields must be independently measured in the near-field which is extremely difficult to perform! Depending on the type of energy source, the *near-field* will be dominated by either a magnetic or electric field component, with magnetic the primary field of concern.

1.4.1 Reactive Near Field Region

In the immediate vicinity of any antenna, there exists the reactive near-field. In this region, both electric and magnetic fields are predominately reactively and independent which means the ***E***- and ***H***- fields are out of phase by 90 degrees to each other. Recall that for propagation of radiating energy, fields are orthogonal (perpendicular) and in phase. The magnetic field dominates due to the lower impedance of the region physically near the noise generating source.

Near-fields are dominated by a dipole-type electric, or a magnetic field propagating in a loop around the transmission line. The magnetic near-field, due to changing currents, must be in the form of a loop as magnetic "charges" (magnetic monopoles do not exist; only electric fields have monopoles). Although electric charges do exist and may create static electric fields, the oscillating electric part of an electromagnetic near-field, created by an electric potential in the radiator always shows a dipole configuration. This is because the source of the electric part of the electromagnetic near-field is created from an electrical neutral conductor only in a way that temporarily creates a dipole or multipole. Both positive and negative charges in a radiating source have no way to propagate. The charges are separated from each other by an excitation "signal" (a transmitting source). A classic example of this behavior is a radio antenna, which on average over time is electrically neutral and differs from this state only by temporarily becoming an electrical dipole (or multipole) under the influence of the signal from the transmitter, which separates charges within it for brief periods only. If an antenna has a static charge, this antenna cannot contribute to the electric near-field in a manner that varies with respect in time. The same is true for any constant currents that may flow in an antenna.

1.4.2 Radiating Near-Field (Fresnel) Region

The radiating near-field or Fresnel region is the physical area between the near- and far-fields. In this region, the reactive fields do not dominate and a unified radiating field begins to emerge. However, unlike the far-field region, the shape of the radiation pattern may vary appreciably with distance from the transmitting source.

The energy in the near-field is radiant energy, or a mixture of magnetic and electric components that are very different from both the near- and far-field pattern. Further away from the transmitting source out into the radiative region, or near-field (one-half to 1-wavelength from the source), both the ***E*** and ***H*** field relationship is more predictable, but still complex. Since the radiative near-field is still part of the near-field zone, there is potential for unanticipated (or adverse) conditions in the propagating field pattern. In contrast to far-field propagation, the diffraction pattern in the near-field typically differs significantly from that observed at infinity and varies with distance from the energy source. Within the near-field, the relationship between ***B*** and ***H*** becomes very complex. Also, unlike far-field propagation where electromagnetic waves are usually characterized by a single mode or propagation wave (horizontal, vertical, circular, or elliptical), all polarization can be present in the Fresnel Region.

1.4.3 Far Field (Fraunhofer) Region

The far-field is the region at a distance from the source that is used mainly for communication purposes. In this region, the radiation pattern does not change shape with distance although the fields decrease with distance (r), or $1/r$ (power dies off as $1/r^2$). This region is dominated by both electric (***E***) and magnetic (***H***) fields, with both fields orthogonal to each other.

In the far-field, the shape or angular field distribution of the antenna field pattern is independent of distance between source and antenna. The far-field is frequently referred to as the *radiation zone* or *free space*. The radiation zone is important because far-fields in general fall off in amplitude by $1/r$ where "r" is distance spacing between source and receive antenna. This means that the total energy per unit area at a distance r is proportional to $1/r^2$. The area of the sphere of the propagating field is proportional to r^2, so that the total energy passing through the sphere is constant. This means that the far-field energy actually propagates to an infinite distance. In general, antennas are used to communicate wirelessly for long distances using the far-field as a propagating media, however certain antennas specialized for near-field communication do exist.

1.5 Relationship Between Electric and Magnetic Sources

Viewing the process whereby changing currents create time variant magnetic fields and static-charge distributions create electric fields and dipoles, we next examine the relationship between alternating, or time variant currents and their relationship to creating an electromagnetic field. In order to understand this relationship, we examine how magnetic and electric fields are related based upon how they propagate within a transmission line, which includes free space.

Time-varying currents exist in two configurations:

- Magnetic sources (which are developed due to a closed loop circuit)
- Electric sources (which are propagated from a dipole antenna structure)

We first examine magnetic fields.

Consider a circuit containing a clock source (oscillator) and load (Fig. 1.4). We have current flowing around a closed loop path (signal trace and RF current return). We can easily calculate the electromagnetic field generated by modeling this closed loop circuit using simulation software or manually using equations provided later. The electromagnetic field produced by this loop circuit is a function of four attributes.

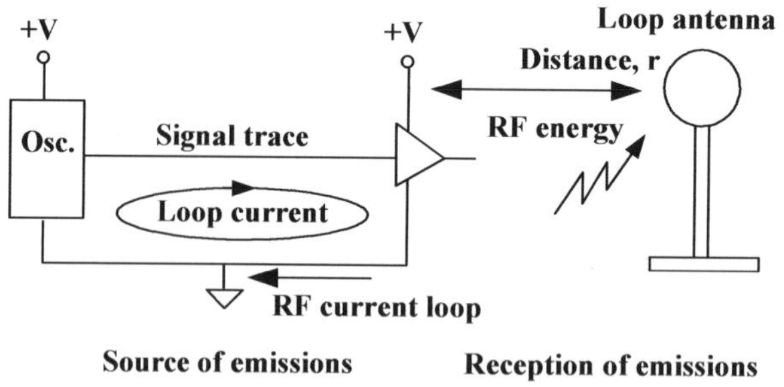

Figure 1.4 RF transmission of an electromagnetic field.

1. *Current amplitude.* The field strength of the propagating field is directly proportional to the total current flow within the loop circuit.
2. *Orientation of the loop circuit (which creates a magnetic field) relative to an observation point.* For a propagating signal to be measured, or observed, the polarization of the source loop current should match that of the measuring device if the measuring antenna is also a loop configuration. If the measuring antenna is a dipole, it must not be in the same polarization but rather cross-polarized. For example, if a loop antenna is horizontally polarized, it must be the same polarization; however, if the measuring antenna is a dipole, it must be vertically polarized.
3. *Size of the loop circuit (perimeter versus area).* If the perimeter of the loop is electrically small, much less than the wavelength of the generated signal or frequency of interest, the RF energy created by the magnetic field (Ampere's law) will be directly proportional to the area of the loop, not perimeter. The larger the physical size of loop the lower the frequency observed at the terminals of the antenna. For a particular physical dimension, any antenna will be resonant at a specific frequency.
4. *Field strength at a distance from the source of RF energy.* The rate at which the field strength decreases depends on the distance between the source of the RF energy and antenna placement. This distance can be either in the near-field or far-field. When the distance is electrically *"close"* to the source, the magnetic field falls off as the square of the distance while the electric field becomes stronger, but only for high impedance circuits. When the distance is electrically far away, we observe an electromagnetic field that decreases inversely with increasing distance ($1/r$). The point where magnetic and electric field vectors cross occurs at approximately one-sixth of a wavelength ($\lambda/2\pi$). The distance is also described as the speed of light (c) divided by frequency (discovered by Maxwell). The equation ($\lambda/2\pi$) can be simplified to $\lambda = 300/f$, where λ is in meters and f is in MHz.

For electric field propagation, in contrast to the magnetic field created by a closed loop circuit, the electric field is modeled by a time-variant dipole structure. This means that two separate, time-varying point charges of opposite polarity exist in close proximity. The ends of the dipole contain this change in electric charge. This change in electric charge occurs due to time variant current flowing throughout the dipole's length. Using the circuit described in Fig. 1.4, we can represent the electric source as the oscillator's output driving an un-terminated antenna with free space acting as the RF return path.

When examined in the context of low-frequency circuit theory, this circuit configuration is not valid. The finite propagation velocity of the time variant signal in the loop circuit (based on the dielectric constant of the nonmagnetic material) is not taken into account in addition to the RF currents that are created. This is because propagation velocity is *finite*, not *infinite*! The assumptions made above are that the wire or dipole antenna contains the same voltage potential and that the circuit is at equilibrium at all points instantaneously. The electric field created by this electric source is thus a function of four attributes.

1. *Current amplitude.* The electric field created in the circuit is proportional to the amount of time variant current flowing in the transmission line.
2. *Orientation of the dipole relative to the measuring device.* This is equivalent to the magnetic source variable described above with the need for the antenna to be cross polarized to the source.
3. *Size of the dipole antenna.* The electric field is proportional to the length of the current driven element located on one leg of the dipole antenna. This is true if the length of the trace is a small fraction of a wavelength. However the larger the physical size of the dipole antenna, the lower the frequency that is observed at the terminals with efficiency. The self-resonant frequency of a dipole is always at a particular physical dimension where maximum signal transfer will occur.
4. *Distance.* Electric and magnetic fields are related (Faraday's Law). Field strength decreases with distance from the point source. The amplitude of decrease is either: $1/r^3$, $1/r^2$ or $1/r$, depending whether distance between source and antenna is in the reactive near field, the radiated near field, or the far field (Fraunhofer) region. In the far field, the RF signal observed consists of almost exclusively the electric field, since the impedance of free space is low for the electric field and high for the magnetic field. When we move the antenna closer to the point source, both magnetic and electric fields have a greater dependence on the distance from the source although it is nearly impossible to distinguish if the field measured is predominately electric or magnetic.

The relationship between near-field (magnetic and electric components) and far-field (plane wave containing both magnetic and electric fields) is illustrated in Fig. 1.5. RF waves are a combination of both electric and magnetic field components. When both electric and magnetic field components are propagated together in the far field, this is call the Poynting vector (plane wave) and is only observed in the far-field or Franhoffer region.

The reason electromagnetic propagation of RF energy is termed a plane wave (or Poynting vector) is because that to an antenna several wavelengths from the source the wavefront looks nearly planar. An example on how a plane wave appears at an antenna in the far field is shown in Fig. 1.6, where the electric field propagates in the "*y*" axis, magnetic field in the "*z*" axis and the Poynting vector or plane wave is in the "*x*" axis. Fields propagate radially from the field point source at the velocity of light,

$$c = 1/\sqrt{\mu_o \varepsilon_o} = 3 \times 10^8 \text{ meters/second}$$

where:

c = speed of light ($3x10^8$ meters per second)

μ_o = $4\pi * 10^{-7}$ H/m (permeability of free space)

ε_o = $8.85 * 10^{-12}$ F/m (permittivity of free space).

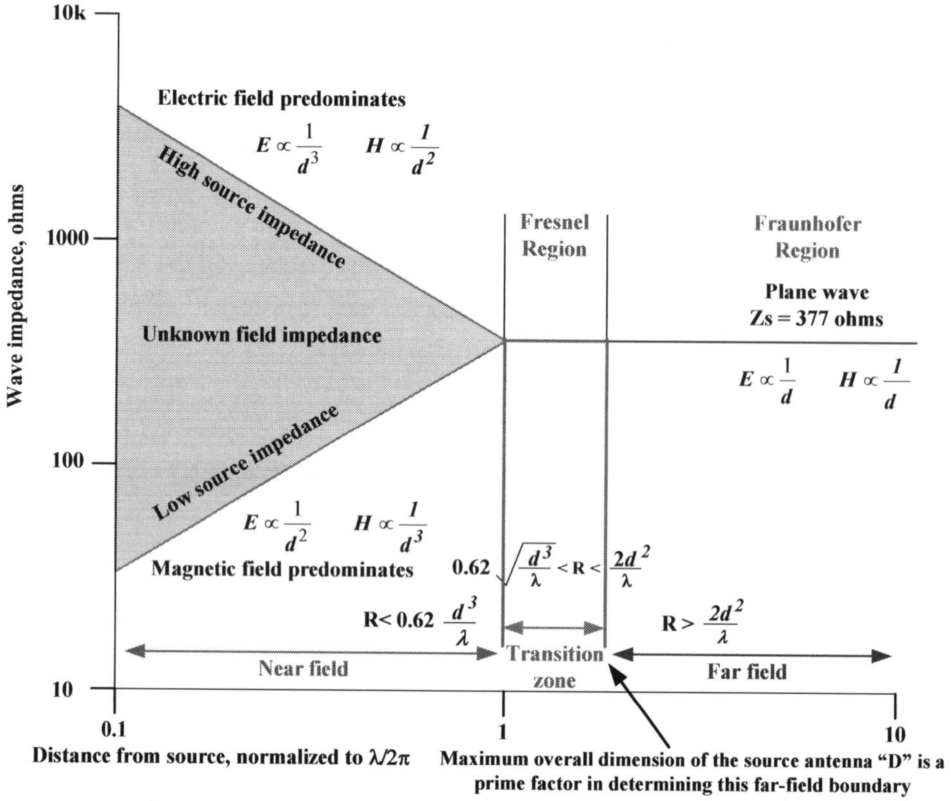

Figure 1.5 Wave impedance versus distance from *E* and *H* dipole sources. where "*d*" is distance between source and antenna

The electric field component is measured in *volts/meter* while the magnetic field is in *Ampere/meter*. The ratio of both electric (***E***) to magnetic field (***H***) is identified as the impedance of free space (1.8). The point to emphasize here is that the plane wave, or wave impedance Z_o, has a characteristic free space impedance that is independent of the distance from the source (reactive and near field) and does not depend upon the characteristics of the source once in the far field. The impedance of a plane wave in free space Z_o is described by (1.9):

$$Z_o = E/H = \sqrt{\mu_o/\varepsilon_o} = \sqrt{\frac{4\pi 10^{-7}\ H/m}{\frac{1}{36\pi}(10^{-9})\ F/m}} = 120\pi \text{ or } 377 \text{ ohms (exactly } 376.99 \text{ ohms)} \quad (1.9)$$

Power carried in the wave front is measured in watts/meter^2. Note, this is in frequency domain units of measurement, which will be converted to the time domain later in this Chapter. Regardless of time or frequency domain analysis, the numeric value will be identical.

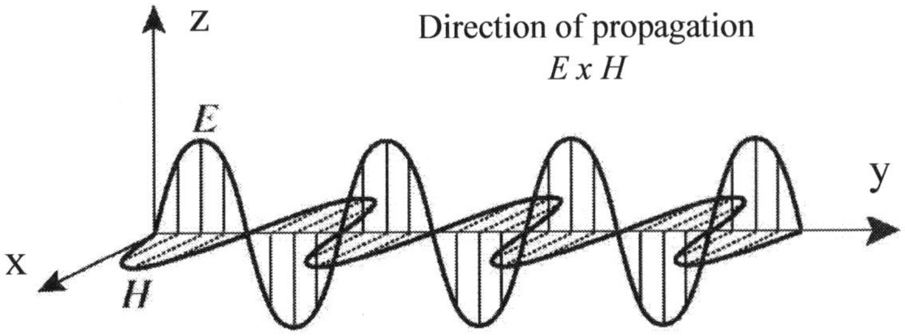

Figure 1.6 Wave propagation–electric and magnetic field elements creating a plane wave.

1.6 Electromagnetic Fields Represented as Antenna Elements

When applying field theory from Maxwell, both electric and magnetic fields propagate in the dielectric between source and load but only when a closed loop circuit exists, a requirement from Ampere. In order for RF energy to propagate we need a transmission line or noise coupling method that can be either a metallic interconnect (cable, wire, printed circuit board trace, power/return plane, etc.) or free space. The transmission line with the lowest impedance is the path taken by the electromagnetic field.

A transmitting antenna is merely a conductor that *intentionally* propagates electromagnetic voltages and currents. A receiving antenna is simply a conductor that *intentionally* picks up this propagating electromagnetic field.

The noise coupling mechanism, or antenna structure, is represented as equivalent component models. When a time-varying *electric* field is created per Faraday's Law, this electric field will somehow find a coupling mechanism to propagate from transmitter to receiver. This antenna is represented electromagnetically as a capacitor which by definition, are two metallic conductors separated by a dielectric (Fig. 1.7a).

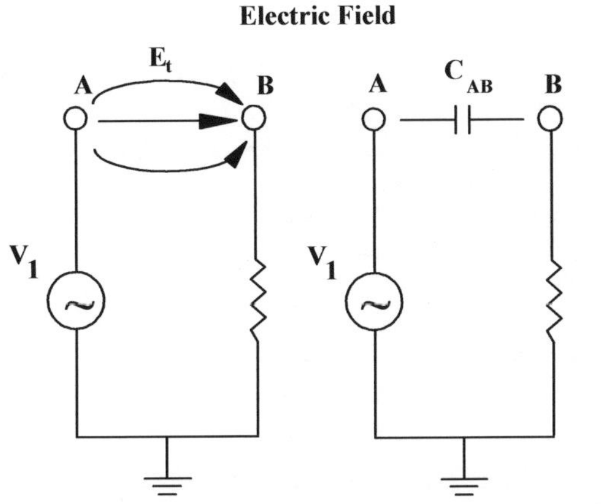

Figure 1.7a Noise coupling model-electric field (dipole antenna).

A time-varying *magnetic* field is represented as a loop antenna since a closed loop circuit must exist for signal flow to occur per Faraday's law (Figure 1.7b).

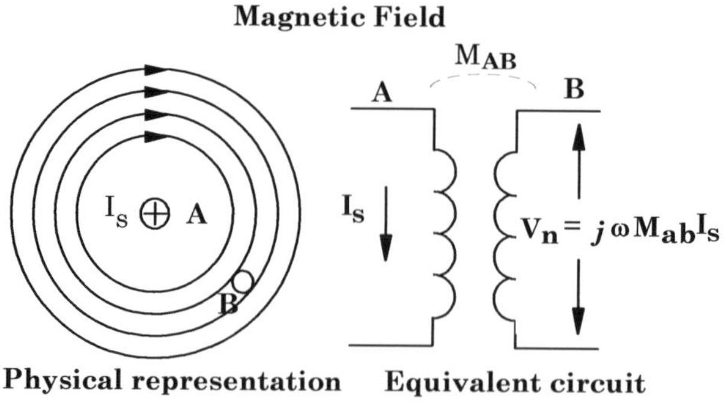

Figure 1.7b Noise coupling method-magnetic field (loop antenna).

When dealing with signal propagation, it is generally assumed that electromagnetic fields propagates internal to transmission lines such as a printed circuit board trace. In reality, the transmission line is only the path for electromagnetic field propagation. Electromagnetic fields physically exists in the *dielectric* that surrounds the transmission line. This dielectric is the physical properties material of the board (core and prepreg) or free space.

To further simplify Maxwell's equations related to antenna theory, it is easy for many digital design engineers to describe signal propagation on their schematics with units of *Volts* and *Amps*. These designers generally think and work in the time domain, making it easy to describe signal propagation with simple metric units of measurements. In reality, electric circuits must be analyzed in the frequency domain with units of *Volts/meter* and *Amps/meter*. By converting from the frequency domain into the time domain by dropping the units of "per meter," understanding how transmission lines function becomes easy. One must remember that working in one domain or another provides identical results.

For the noise coupling model to be valid, the physical dimensions of the circuit must be small compared to the wavelengths of the signals involved. When the model is not truly valid, we can use lumped component representations to explain field propagation for the following reasons.

Maxwell's equations cannot be efficiently applied directly for most real-world situations due to complicated boundary conditions. If we have minimal confidence in the validity of the approximation of the lumped models, the model is invalid.

Numerical modeling using computational analysis may not show how the RF energy is created, only that it exists. Modeling is dependent on system parameters including parasitics. Even if a modeling answer or result is computed, system-dependent parametric may not be clearly known or identified which makes results from simulation questionable.

When modeling electromagnetics for the majority of situations, configurations or applications, one need not attempt to achieve 100% accuracy since there will always be parasitic parameters that are not known or must be estimated. If we achieve a high percentage of accuracy, the probability exist that we have performed our job to a level required for customer shipment.

Why is this theory and discussion about Maxwell's equations important for printed circuit board and system design? The answer is simple. We need to know how electromagnetic fields are created and propagated in a simplified manner so that we can make RF field theory understandable, without higher order mathematics.

1.6.1 Conductive pathways

Electromagnetic waves propagate along defined pathways. The medium or dielectric that supports propagating waves have an important effect on the impedance of the transmission line in addition to the shape of the structures carrying time-variant current and the proximity of nearby conductors, especially the return path.

Electromagnetic wave propagation along conductors or a transmission line will experience impedances that are higher or lower than the desired impedance of the transmission path. Any impedance value in a transmission line different from the desired characteristic impedance means a change in propagating power. When signal loss occurs due to impedance differences, the time-variant electromagnetic field lost in the dielectric that surrounds the transmission line must find an alternative path, either through air and/or through another dielectric that is at a lower impedance to return to its source per Kirchhof's law. A circuit loop, especially one containing RF energy, makes it difficult for engineers to identify where propagating electromagnetic current may flow. *All* currents will divide somehow and flow in multiple paths, in proportions according to the admittances of each path. Admittance is the reciprocal of impedance.

Time-variant current within a transmission line, with finite inductance in addition to stray or parasitic capacitance, creates alternative propagation paths for electromagnetic energy to flow through. Conductors, components and devices all have both stray capacitive and inductive coupling creating undesired antenna structures. These unintentional antennas all have different admittances which make for perfect propagating structures. Our job as engineers is to reduce the propagation of unwanted electromagnetic waves that couple into nearby conductors via stray or parasitic capacitance and inductance.

When a conductor resonates in a way that contains high impedance, any capacitive path (dipole antenna) present may permit free space electric field propagation to occur, or if a loop antenna is present propagation of a magnetic field. Any electromagnetic field present causes most of propagating current to travel as displacement currents depending on frequency. Both source and return currents are generated simultaneously and must balance each other out throughout the closed loop circuit.

1.7 Maxwell Simplified-Further Still (Conversion from Frequency to Time Domain)

With a fundamental overview of Maxwell's approach presented above, it is possible to transform the mathematics of advanced calculus to easy algebra. To acquire full comprehension we must *"overly simplify Maxwell"* by thinking about electromagnetic theory in the time domain. Maxwell's equations are one aspect of frequency domain signal propagation. We can easily convert Maxwell's equations to Ohms law in a ***conceptual*** manner. Remember, ***the context of this material is to simplify a complex topic***! Technical accuracy is not important, at this time, but will be presented later in greater detail.

We begin by reviewing aspects of Ohm's law related to circuit analysis. There are two aspects of Ohm's law, low frequency and high frequency considerations. Both are valid at all times. The difference with regard to this discussion relates to aspects of signal propagation when dealing with either AC and DC circuits including both lump and distributed circuit models when analyzing circuit performance.

Ohm's Law

Time domain (DC currents)	Frequency domain (AC currents)	
$V = I * R$	$V_{rf} = I_{rf} * Z$	(1.10)

where:
V = voltage
I = current
R = resistance
Z = impedance ($R + jX$)
rf = radio frequency energy or time-variant AC voltage or current in the transmission line.

To convert *Maxwell Made Simple®* to *Ohm's Law* – if time variant current (AC) propagates in a transmission line which has a 'fixed impedance value' (Faraday's law)–time variant voltage will be created proportional to the current present. In the frequency domain, units of measurement is Volts/meter and Amps/meter. By removing the "per meter" as a unit of measurement, we ***conceptually*** convert transmission line theory associated with Maxwell's equation from the frequency domain into the time domain. This conversion allows us to understand complex electromagnetics using Ohm's Law.

Note that this conversion from Maxwell to Ohm occurs using lumped models during circuit analysis. At higher frequencies, losses within electromagnetic field propagation occurs due to other factors that Ohm's law does not address, since Ohm's law was originally developed for DC current and Maxwell for AC signal propagation or electromagnetic fields.

Notice that in the electromagnetics model (frequency domain), resistance I is replaced by impedance *(Z)*, a complex number that contains both resistance (the real component, R, represents DC current present) and reactance (the complex component, *jX*, represents AC current). In every transmission line there is both AC and DC current propagation. Depending on whether we are at a steady state value (DC) or time variant (AC), the value of *Z* will vary with respect to frequency.

For the impedance variable (*Z*), various forms exist depending on whether we are examining plane wave impedance, circuit impedance and the like. For a transmission line, total impedance (*Z*) is dependent on both inductance and capacitance (1.11).

$$Z = R + jXL + \frac{1}{jXC} = R + j\omega L + \frac{1}{j\omega C} \qquad (1.11)$$

where:
X_L = $2\pi fL$ (the component in the equation that relates only to transmission lines in free space)
X_c = $1/(2\pi fC)$ (present when the RF current has a return path nearby, capacitively coupled)
f = frequency of operation (Hertz)
ω = $2\pi f$.

When a physical component has known resistive and inductive elements, such as a resistor, capacitor, inductor, transformer, ferrite bead, semiconductor component with bond wires or any form of interconnect, parasitic elements will exist by virtue of their physical placement with regard to other conductive pathways or physical size of the circuit loop containing inductance. Equation (1.12) now becomes applicable, as the magnitude of impedance versus frequency must be considered since the reactive part of the impedance equation becomes a greater concern than resistance starting around 10-100 kHz.

$$|Z| = \sqrt{R^2 + jX^2} \tag{1.12}$$

For frequencies greater than a few kHz, the value of inductive reactance *(Z)* will eventually become much larger than *R*, except for very rare cases of operation. Current always takes the path of least impedance. Below a few kHz, the path of least impedance is always resistance. Above a few kHz, somewhere between 10 kHz and 100 kHz depending on transmission line characteristics, the path of least *reactance* becomes dominant swamping out the value of resistance, essentially making it an insignificant percentage of the total impedance of the transmission line. Because almost all digital components operate at frequencies above a few kHz, the belief that current takes the path of least resistance provides an incorrect concept of how current flow occurs within a transmission line at higher frequencies.

Since current always takes the path of least impedance, noted here for wires carrying currents above several kHz, the impedance of the transmission line becomes equivalent to the path of least reactance. If the input impedance of a load is much greater than the shunt capacitance of the transmission line path, the inductive reactance will becomes the dominant element in the impedance variable. This impedance variable is associated with Ohms law although a frequency domain element exist since we are dealing with a time variant signal that varies with respect to frequency.

Each and every transmission line has a finite value of impedance, mainly inductive reactance. Having a large value of transmission line inductance is only one reason how RF energy is created. Lead bond wires or other form of interconnect methods (Ball Grid Array, flip chip, etc.) that bonds a silicon die to an interposer or directly to a printed circuit board may have sufficiently high impedance (usually inductive) to cause an RF potential difference to be established across this impedance. This potential difference creates unintentional common-mode currents that will be established in proportion to that potential difference (caused by resistance or high value of inductive reactance). Discussion on creation of common-mode current follows in the next section.

Traces routed on a printed circuit board will exhibit various levels of inductance, especially when the propagation time is electrically long with respect to the round-trip delay of a transmitted signal, including the return path to the source before the next edge-triggered event occurs. An electrically long transmission line (trace) is one that exceeds approximately $\lambda/10$ of the frequency propagated within the line. Basically, when an RF voltage traverses through any impedance, undesired RF currents will be developed which is not desirable. It is this RF current, in the common-mode, that propagates and may cause functionality concerns (i.e., crosstalk) and non-compliance to emission requirements both radiated and conducted.

Per Maxwell's third equation, a moving (time variant) electrical charge in a transmission line generates an electric current that creates a magnetic field. *Magnetic fields*, created by this moving electrical charge are also called magnetic lines of flux. Magnetic lines of flux can easily be visualized using the Right Hand Rule, shown in Fig. 1.8.

To understand this simple Right Hand Rule, configure your right hand into a loose fist with your thumb pointing straight up. Current flow (electrons) is in the direction of the thumb (upwards or from bottom to top). The curved fingers encircling the transmission line illustrates the direction of the magnetic field that has been created, and are called lines of magnetic flux.

With any time-varying magnetic field (AC sine wave), an electric field is established orthogonally (90 degree in direction) which travels in the direction of the thumb. Total RF emissions are a combination of both magnetic and electric fields which will propagate by either radiated or conducted means. Note that the transmission line is only the physical path the electromagnetic field takes. The actual electromagnetic field exists in the dielectric that surrounds the transmission line which could be air, core or prepreg of a printed circuit board (the physical material) or the insulation on a strand of wire.

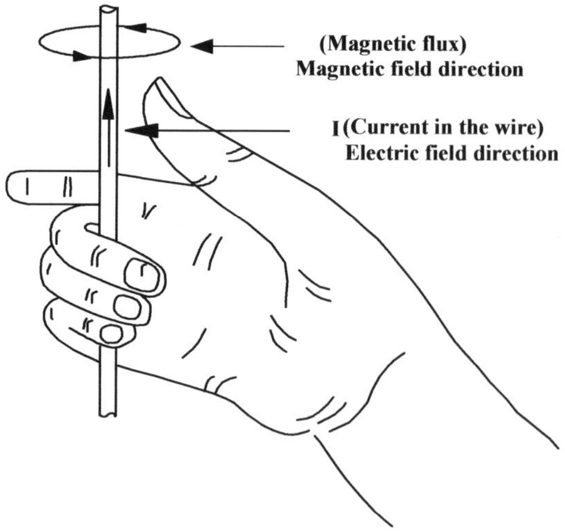

Figure 1.8 Right hand rule (Faraday's Law).

The magnetic field surrounds the physical transmission line, but can do so only if a closed-loop circuit is present, not shown in Fig. 1.8. Without time-variant current flow in a complete loop circuit there can be no magnetic field development.

In a printed circuit board, dynamic AC currents (at RF frequencies) are generated by a source driver and transferred to a load. RF current must return to their source (Ampere's Law) through a return path or second transmission line. As a result, this loop circuit permits the magnetic field surrounding the transmission line to propagate from this structure, which is in reality a loop antenna. A loop antenna does not have to be circular and may often be a convoluted shape. In this loop antenna, when a magnetic field is created due to time variant current, an electric field is also generated per Faraday's law. When we have both electric and magnetic fields present, we end up with an electromagnetic field discussed earlier in this Chapter.

In the near field, the magnetic field component dominates when the source of the field has low impedance and higher currents are found, such as the power and return planes in a printed circuit board. The same discussion applies to electric fields except they have high impedance at the source and are difficult to measure or observe until the far-field [Fraunhofer] region is reached. Regardless, both field structures exist simultaneously.

In the far field, the ratio of electric to magnetic field (defined as wave impedance) is approximately 120π ohms or 377 ohms, totally independent of the source impedance or any other impedance within the reactive and near field region of space. Magnetic fields are measured using an electrostatically-shielded loop antenna, providing the receiver has an adequate sensitivity level to observe the propagating field at low signal levels.

Another simplified explanation of how a time variant electromagnetic field exists within a printed circuit board is depicted in Fig. 1.9. We discuss this simple circuit in both the time and frequency domain to illustrate fundamental concepts of Maxwell's equations, or *Maxwell Made Simple®*.

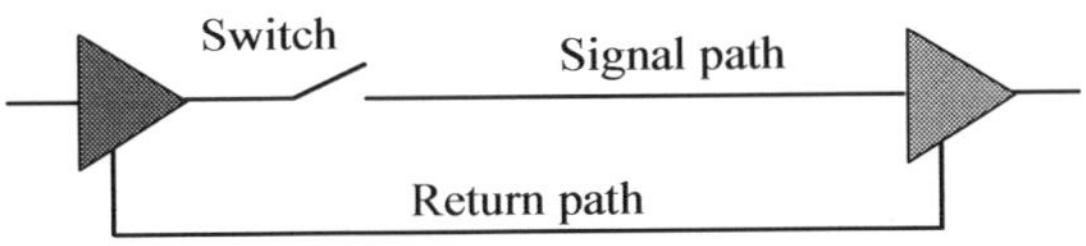

Figure 1.9 Closed loop circuit.

Ampere's law requires a closed-loop transmission line system if the circuit is to work. Kirchhoff's voltage law states that the algebraic sum of the voltage around any closed path in a circuit must be zero. Ampere's law describes magnetic induction at a physical point due to given currents in terms of their current elements and position relative to that point. In order for the circuit to work, the switch must be closed and energized through use of a power source (AC main, DC battery, or semiconductor output driver).

Without a closed-loop circuit, a signal would never travel through a transmission line system from source to load and return. When the switch is closed, the circuit is complete and AC and/or DC current flows.

There are two types of currents–AC and DC.

The only difference is the domain that we observe the propagation. If at steady state, DC is easy to understand and work with in the time domain using Ohms law. As soon as the current changes with respect to time we start to enter frequency domain analysis, again using Ohms law, which is essentially a subset of Maxwell's equations (two ways of viewing the same signal). Every transmission line system must conform to Maxwell's equations, Kirchhoff voltage/current laws, Ampere's and Ohm's laws regardless of the type of current present in either or both the time or frequency domain.

One simple way to *conceptually* convert from the frequency domain (Maxwell) to time domain (Ohm), is to take into consideration units of measurement, which is EMF potential (in volts), current (in amps) and resistance (as impedance). With regard to Ohms law, when asked what elements are contained in a transmission line-voltage, current or electrons, different answers will be given. In reality, voltage and current are metric units of measurement. Voltage is electromagnetic force. One Amp of current is equal to one coulomb/second, or roughly $6.241509324 \times 10^{18}$ electrons/second. Current traveling with respect to time in a transmission line creates a time variant magnetic field, which in turn creates a time variant electric field (Faraday's law). Thus the correct answer to the question is, "A transmission line propagates only electric and magnetic fields, not voltage, current or [non-time variant movement of] electrons. The correct units of measurement of energy transfer in the frequency domain is volts/meter and amps/meter. Since it is sometimes difficult to think in the frequency domain, removing the "per meter" unit of measurement makes the *concept* of "power" transfer easy to work with as we convert, conceptually, signal propagation into the time domain from the frequency domain.

Since power is now mentioned, recognize that in reality propagation of an electromagnetic field surrounding or within a transmission line, either a metallic interconnect or free space, is actually a *Power* level measurement, which is equal to *Voltage*Current* *(P=VI)*. In the frequency domain this is volts/meter and amps/meter which equals **watts/meter²**, but in the time domain we think easier in volts and amps which is equal to **watts**. When we remove of the words "per meter" from the units of measurement, then conceptually we are converting electromagnetic power propagation in the frequency domain to the time domain to make digital signal analysis easier to work with.

A spectrum analyzer's input, and almost all test instrumentation's input, measures only power with units of *dBm*. The value of 0 dBm is equal to 1 milliwatt. Since instrumentation's input is designed to only accept power as an electromagnetic field, the processor inside the instruments converts this input power level to either or both voltage and current elements for ease of measurement purposes. One can easily convert multiple units of measurement from the real value of power (*dBm*) relative to dB volts or any other metric value required. Appendix B and D provide details on decibel use as well as conversion tables.

1.8 Concept of Flux Cancellation (Flux Minimization)

We now examine how magnetic flux is developed (Faraday's law). Magnetic flux is created by time variant current flowing through a transmission line with finite impedance. Magnetic flux only exists in the dielectric surrounding the transmission and line and not physically internal to the wire such as the center core.

To prevent undesired magnetic flux from causing harmful disruption, *flux cancellation* or *flux minimization* is required. Although the term *cancellation* is used in this text, we may substitute the term *minimization (or containment)* since they represent the same thing, but in different context. Because magnetic lines of flux travel counterclockwise within a transmission line, if a return path is adjacent and in close proximity (another transmission line) to the source path, the lines of flux will appear to be in the opposite direction relative to each other. To restate this important concept, when a clockwise field is coupled with a counterclockwise field, a cancellation effect of the magnetic field occurs (Fig. 1.10). If magnetic flux between the source and return path is minimized, then undesired RF current cannot be developed except within the minuscule boundary of the trace, which for most applications is negligible relative to EMC compliance.

Details on implementing flux cancellation techniques are provided in Chapter 2.

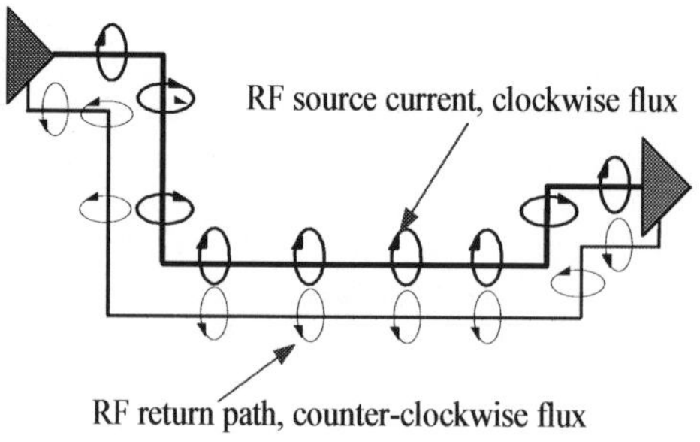

Figure 1.10 Concept of flux cancellation.

1.9 Skin Effect and Lead Inductance

A consequence of Maxwell's third and fourth equations is skin effect related to a voltage charge imposed on a homogeneous medium where current flows, such as a wire lead bond from a component or a printed circuit board trace. If the voltage level is maintained at constant DC amplitude, current flow will be uniform and at steady state, using the entire area of the conductor evenly distributed.

When the source voltage is *not* DC (steady state condition) but AC (time-variant), the density of current flow tends to concentrate in the outer portion of the conductor as we go higher up in frequency through the process called as skin effect.

Skin effect is the tendency of alternating electric current (AC) to become distributed within a transmission line such that current density is significantly greater near the surface of the transmission line than the center, decreasing in amplitude from the surface going inward per (1.13). The current present on the "skin" of the conductor between outer surface and internal level is called the skin depth. The small amount of thickness due to current flow causes the effective resistance of the conductor to increase at higher frequencies where skin depth is very small, thus reducing the effective cross-section of the transmission line as a carrier of current.

$$\delta = \sqrt{\frac{2}{\omega\mu_o\sigma}} = \sqrt{\frac{2}{2\pi f\mu_o\sigma}} = \frac{1}{\sqrt{\pi f\mu_o\sigma}} \qquad (1.13)$$

where:
ω = angular (radian) frequency ($2\pi f$)
μ_o = material permeability ($4\pi * 10^{-7}$ H/m)
σ = material conductivity ($5.82 * 10^{-7}$ mho/m for copper)
f = frequency (Hertz).

Skin effect is caused by opposing eddy currents induced by changing magnetic fields resulting from the alternating current waveform throughout the frequency spectrum. At 60 Hz in copper, skin depth is about 8.5 mm. At higher frequencies, skin depth becomes much smaller. Because the center portion of a large conductor carries little propagating current at higher frequencies, tubular conductors such as waveguides can be used to save weight and cost.

At very high frequencies, almost all RF energy is compressed on the outer skin of the transmission line with little current flow in the center. When current on the skin of the conductor becomes crowded, dielectric losses begin to affect signal propagation which minimizes the total power level of the electromagnetic field used to transfer electromagnetic intelligence between source and load. This loss of power through attenuation, measured as a ratio in dBs, can be severe and could cause a system to become non-functional.

Skin depth (δ) is defined as the distance to the point inside the conductor at which the electromagnetic field is reduced to 37% of the surface value, defined by (1.13)

Table 1.1 illustrates an abbreviated list of skin depth values at various frequencies for a 1-mil thick copper substrate (1 mil=0.001; 1 inch=2.54×10^{-5} m).

Table 1.1 Skin depth for copper substrate.

Frequency of operation	δ (copper)
60 Hz	0.0086 in. (8.6 mil, 2.2 mm)
100 Hz	0.0066 in. (6.6 mil, 1.7 mm)
1 kHz	0.0021 in. (2.1 mil, 0.53 mm)
10 kHz	0.00066 in. (0.66 mil, 0.17 mm)
100 kHz	0.00021 in. (0.21 mil, 0.053 mm)
1 MHz	0.000066 in. (0.066 mil, 0.017 mm)
10 MHz	0.000021 in. (0.021 mil, 0.0053 mm)
100 MHz	0.0000066 in. (0.0066 mil, 0.0017 mm)
1 GHz	0.0000021 in. (0.0021 mil, 0.00053 mm)

If any of the four variables in (1.13) increases, skin depth decreases which means more electrons will be present on the outer portion of the transmission line instead of the middle. The skin depth of conductors at higher frequencies is very small, typically 0.0066 mils or $6.6 * 10^{-6}$ inch (0.0017 mm) at 100 MHz. Current tends to be dominant in a strip near the surface of the conductor at a depth represented by the Greek symbol δ. When higher-frequency RF currents are present, current flow becomes concentrated into a narrow band near the conductor surface.

A wire's inductance equals its DC resistance value independent of the wire radius up to the frequency when the radius becomes on the order of a skin depth. Below this skin depth frequency, the wire's resistance *increases* as $\sqrt{f}$ or 10 dB/decade. Internal inductance of the wire is that portion of the magnetic field internal to the wire per-unit-length, where the transverse magnetic field contributes to the per-unit-length inductance of the line.

The portion of the magnetic flux external to the transmission line contributes to the total per-unit-length inductance of the line and is referred to as external inductance. Above this particular frequency the wire's internal inductance *decreases* as $\sqrt{f}$ or -10 dB/decade.

For a solid round copper wire, effective DC resistance is calculated by (1.14). Table 1.2 provides details on some of the variables used in (1.14). Signals may be further attenuated by the resistance of the copper in the conductor and by skin effect losses resulting from the finish or plating on the copper surface. In addition, the material that the copper is adhered to, either core or prepreg, has dielectric losses that affect electromagnetic propagation and increases skin depth loss. The resistance of the transmission line, in this case copper, may reduce steady-state voltage levels below functional requirements of components related to noise immunity requirements.

$$R_{dc} = \frac{L}{\sigma \pi r_{\omega}^{2}} \ \Omega \tag{1.14}$$

where:
R_{dc} = resistance of the wire (DC current)
L = length of the wire
σ = conductivity
r_{ω} = radius of the transmission line (Table 1.2).

Table 1.2 Physical characteristics of wire.

Wire Gage (AWG)	**Solid Wire Diameter (mils)**	**Stranded Wire Diameter (mils)**	**R_{dc} – solid wire (Ω/1000 ft) @ 25°C**
28	12.6	16.0 (19x40) 15.0 (7x36)	62.9
26	15.9	20.0 (19x38) 21.0 (10x36) 19.0 (7x34)	39.6
24	20.1	24.0 (19x36) 23.0 (10x34) 24.0 (7x32)	24.8
22	25.3	30.0 (26x36) 31.0 (19x34) 30.0 (7x30)	15.6
20	32.0	36.0 (26x34) 37.0 (19x32) 35.0 (10x30)	9.8
18	40.3	49.0 (19x30) 47.0 (16x30) 48.0 (7x26)	6.2
16	50.8	59.0 (26x30) 60.0 (7x24)	3.9

The units must be appropriate and consistent for the equation to work. As frequency increases, time variant current in the wire cross section will tend to crowd closer to the outer periphery of the conductor. Eventually, time variant current will be concentrated only on the wire's surface equal to the thickness of the skin depth described by (1.15) when the skin depth is less than the wire radius.

$$\delta = \frac{1}{\sqrt{\pi f \mu_0 \sigma}} \tag{1.15}$$

where at various frequencies:
δ = skin depth
f = frequency (Hertz)
μ_0 = permeability of copper ($4\pi * 10^{-7}$ H/meter)
σ = conductivity of copper (5.8×10^7 mho/meter).

A first approximation for inductance of a conductor at high frequency is described by (1.16):

$$L = 0.00511(2.38 \ln \frac{4l}{d} - 1) \tag{1.16}$$

where l is the conductor length and d is the diameter in the same units (inches or centimeters). Because of the logarithmic relationship of the ratio l/d, the reactive component of impedance for large diameter wires dominates over the resistive component above a few hundred Hertz. Thus, it is impractical to obtain a truly low-impedance connection between two points such as grounding a circuit using only a single wire. Such a connection would permit coupling of voltages between circuits due to current flow through an appreciable amount of common impedance.

1.10 What are Common-Mode and Differential-Mode Currents? (Made Simple)

In every transmission line, both common-mode (CM) and differential-mode (DM) currents exist. Both types of currents determine the total amount of RF energy present, some of which can cause harmful disruption to other electrical products or even itself in the form of crosstalk. There is significant difference between these two types of current modes. Generally speaking, differential-mode carries data or the signal of interest (information). Common-mode currents are usually a byproduct of unbalanced differential-mode signaling and are most troublesome for EMC compliance.

The definition of common-mode current in a transmission line is time variant current flowing in both source and return path traveling in the *same direction* at the *same time*. For the differential-mode definition, time variant current in both source and return path are *equal and opposite* and cancel, as discussed in Section 1.8 'Concept of Flux Cancellation (Flux Minimization)'. If there is any imbalance in differential mode signaling (a different scenario than differential pair signal routing such as Ethernet, USB, etc.), the magnitude of this imbalance creates undesired common-mode current based on Kirchhoff's law.

Common-mode currents is generated through either a loss in the transmission line media in the signal or return path due to impedance mismatches, absorption of the electromagnetic field in the physical material used to construct the board, crosstalk from one transmission line to another, or specific to a type of interface signaling protocol which generally occurs within a shielded cable assembly where clock and data are superimposed onto the two wire transmission line at the same time.

Common-mode currents, often magnitudes less than differential-mode currents in amplitude under the same operating conditions, produce unintentional electromagnetic fields that can cause harmful disruption to other electrical circuits and systems. The electromagnetic field propagating in the dielectric of a transmission line containing differential-mode currents will cancel each other, but not exactly since the two transmission paths are not 100% coincident. This is because a finite amount of physical separation between source and return path must exist. What is not exactly cancelled become common-mode propagation. Our concern as engineers is the magnitude of this undesired common-mode current and will it cause harmful disruption to circuits or EMI compliance. Not all common-mode current is bad, just that which exceeds a specific limit that permits harmful effects to occur such as EMI or circuit malfunction.

A number of factors create common-mode currents. Some of these factors include physical distance spacing between source and return path, physically different transmission line routed lengths associated with differential-mode (two-wire) signal propagation between circuits, dielectric losses in both the core and prepreg material, excessive inductance or loss of electromagnetic propagating power within the transmission line, impedance discontinuities, plus many other elements of concern. A very *small* amount of CM current will produce the same amount of undesired RF energy than a *significantly larger* amount of DM current, with numerical analysis provided later in this Section.

Differential-mode currents require a loop antenna structure to radiate from, which is generally an inefficient antenna on a printed circuit board due to typical size of circuit loops being very small, yet common-mode currents will drive a dipole or monopole antenna, which can be almost anything metallic without regard to a specific physical dimension.

When using simulation software or computational analysis to predict radiated emissions created on a printed circuit board, the differential-mode is usually analyzed based on parametric values describing the component's internal circuitry (IBIS or SPICE), or the assumption that there is an infinitely large ground plane present. An infinitely large return (ground plane) allows computational analysis to occur quickly and easily. It may not represent the actual configuration of the model during simulation, considering parasitics may exist and not be known. It is thus not practical to predict with 100% accuracy total radiated emissions based solely on differential-mode and power currents into devices which are typically in the common-mode and greater than the total value of all signal currents combined. Computational analysis is usually performed using a Fast Fourier Transform (FFT) to determine radiated emission levels. Simulation software using FFT analysis sometimes indicates that the electromagnetic field generated *internal* to a printed circuit board will fail emission limits, but generally does not. It cannot distinguish if the current is common-mode or differential-mode when the transmission line is routed stripline, or an internal routing layer of the printed circuit board.

If simulation is to occur on a printed circuit board to calculate total EMI generated, one must simulate each and every trace simultaneously and observe phasing effects in both radiated and conducted modes. Simulation analysis only provides insight into how a circuit may function and potential for EMI problems, but cannot be used for formal certification, verification or testing. In addition, computational analysis may indicate a problem locally that may not have an effect on the environment of use, such as a radiated field from a printed circuit board packaged in a metal enclosure with 60 dB of shielding effectiveness. They key item is understanding what is important and residual effects that are a “do not care” situation.

Calculated RF currents from simulation analysis can severely under predict the radiated emissions of printed circuit board traces and they often omit power currents sourced to the devices. There are many hidden parasitic parameters not generally known that creates CM currents from DM voltage sources. These hidden parameters usually cannot be anticipated and are generally unknown within a printed circuit board's structure, especially dynamically under formation of power surges in the planes during edge switching times. Details on hidden parasitics are provided later in this Chapter.

1.10.1 Differential-Mode Current Description

Differential-mode current (also called transverse or metallic mode) is the component of RF field propagation that is present on both signal and RF return paths that are equal and opposite each other as shown in Fig. 1.11. If a 180° phase shift is established precisely, the magnetic field present within the dielectric of the transmission line will cancel out leaving us with only a finite amount of residual magnetic flux, which in reality is so small that this can generally be ignored by designers. Common-mode effects may however be establish as a result of: any imbalance in the transmission line due to excessive inductance in either the source or return path, insufficient decoupling in the power distribution network, or dielectric losses in the physical material used to create a printed circuit board assembly.

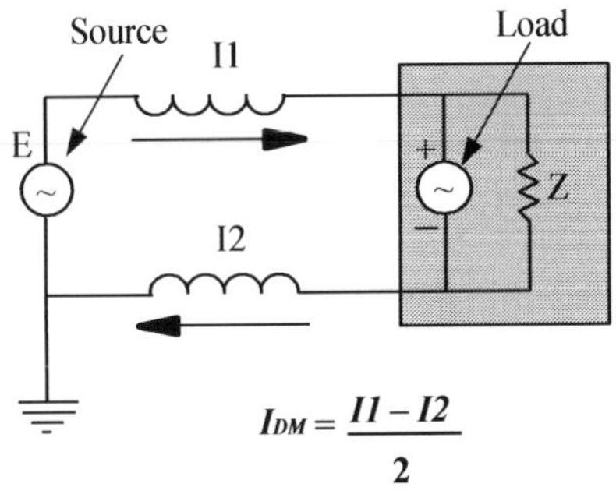

Figure 1.11 Differential-mode current model.

Differential-mode signals are thus:

1 The type of signaling used by digital components (a trace routed between source and load).
2 Conveys desired electromagnetic information by sending a [digital] signal from source to load with a return path that cancels magnetic flux between the two transmission lines, based on a closed loop circuit configuration.
3. Causes minimal EMI as the magnetic field generated on both transmission line paths oppose each other and cancel out, if properly designed into the printed circuit board.

With differential-mode signaling, a digital driver sends out current that is received by a load at a specific voltage level which is easily calculated using Ohms law (time domain). In reality, we should calculate this signal level using Maxwell's equation which gives the same results (Fig. 1.12) however, it is easier for digital designer to think in the time domain although both domains are identical with respect to signal propagation. An equal value of return current must be present to balance out the circuit without loss (Kirchhoff's law).

Currents traveling in opposite directions represent differential-mode operation. Because a printed circuit board can be made to emulate a perfect self-shielding environment (e.g., a coaxial confined structure when traces are routed internal or stripline to the assembly), complete ***E***-field capture and ***H***-field cancellation is possible. Any RF fields not coupled to each other or cancelled out are therefore the source of undesired common-mode EMI.

In the battle to control EMI and crosstalk in the common-mode, the key is to minimize excess electromagnetic field loss through proper source control and careful handling of the energy-coupling mechanisms. This ensures differential-mode signal integrity is present at all times. In Fig. 1.12, the return path is generally a plane internal to the printed circuit board (stripline configuration [signal layer between two solid conductive planes]). If a signal trace is routed stripline, radiated emissions cannot exist except from the board edge, a through-hole via and/or the component silicon die or package. The return path in Fig. 1.12 can also be an adjacent transmission line (trace) that captures nearby magnetic flux, including a wire within a bundled cable assembly. Essentially, any metallic transmission line path acts as an signal return path.

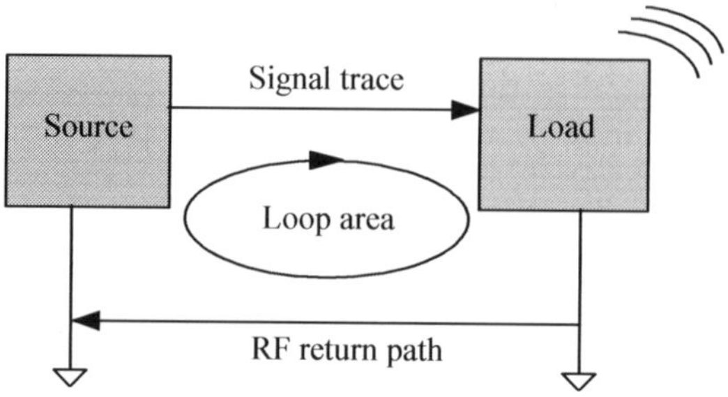

Figure 1.12 Differential-mode circuit configuration.

1.10.2 Differential-Mode Radiated Emission Equations

Differential-mode radiation is caused by the flow of RF current in a loop circuit. For a receiving antenna when operating in the far-field with a ground plane or return path between transmitter and antenna, RF energy can be numerically calculated by (1.17) [3, 4, 5]. Ground reflection can increase measured emissions by as much as 6 dB.

$$E = 2.63 * 10^{-14} \left(f^2 * A * I_s \right) \left(\frac{1}{r} \right) \textit{volts per meter} \tag{1.17}$$

where:
E = differential-mode radiated field (V/m)
A = loop area in cm^2
f = frequency (MHz)
I_s = source current in mA, and
r = distance from the radiating element to the receiving antenna.

In the majority of printed circuit board layouts, primary emission sources are created from currents flowing between components and in the power and return planes. When time variant return current is seeking a path home, it is observed that the return path does not have to be a [ground] plane; any metallic conductor works. Time variant current propagates differently than DC or steady state current. Thus, the voltage potential of a plane or RF return path is irrelevant. It could be at voltage or return (ground) potential. The fact a plane or large metallic structure exist at low impedance is all that the return current is looking for. A power plane acts as a perfect return path to time variant currents, the same as a ground plane. This is discussed in greater detail in Chapter 2.

Radiated emissions can be modeled as a small-loop antenna carrying time variant currents (refer back to Fig. 1.12). When the signal travels from source to load, a return current path must exist. A small signal loop is one whose dimensions are smaller than a quarter wavelength ($\lambda/4$) at a particular frequency of interest illuminated by current flowing within its structure. On printed circuit boards, loop areas are created by transmission line routing between components. These physically small dimensions are transparent for frequencies up to several hundred MHz (or even GHz when viewing the total loop area within circuit devices themselves). If one physically measures the perimeter ($C=\pi r$ or $r=C/\pi$) of a circuit loop path and calculates the area of that loop ($A=\pi r^2$ by extracting r from the perimeter equation), one will notice that the efficiency of the loop antenna will generally be in the upper GHz region. Therefore, the majority of loop areas between components are inefficient radiating antennas related to EMC compliance requirements.

The maximum loop area that will not exceed a specific specification level can be described by a simple equation (1.18), derived from (1.17).

$$A = \frac{380\, r\, E}{f^2 I_s} \tag{1.18}$$

or conversely, the maximum field strength created from a closed loop boundary area (1.19) is:

$$E = \frac{A f^2 I_s}{380 r} \tag{1.19}$$

where:
E = radiation limit (μV/meter)
r = distance between the loop and measuring antenna (meters)
f = frequency (MHz)
Is = current (mA), and
A = loop area (cm^2).

In free space, radiated energy decreases inversely proportional (distance wise) between source and antenna in the far field. The loop area formed by a specific current consuming component on the printed circuit board must be known in order to solve these equations, which can be difficult without knowing the total area of the specific circuit loop between source and return path. Equations (1.17) and (1.18) are for a single frequency. These equation must be solved for each and every loop (different loop-size area), and for each frequency of interest which can be time consuming if using computational analysis.

From (1.17), we can determine if a particular routing topology needs to have special attention as it relates to radiated emissions, generally with simulation software or a spreadsheet. This may include when designing a printed circuit board re-routing of transmission lines, changing routing topology, locating source and load components physically closer to each other, or providing external shielding of the assembly (containment). Detail on these layout techniques are provided in Chapter 2.

Example

Assume a convoluted shape exists between two components located on a printed circuit board without an RF current return path (dipole antenna model): $A=4\ cm^2$, $I_s=5\ mA$, $f=100$ MHz. Using (1.16), field strength I is 52.8 dBμV/m at 10-meter distance. Radiated emission limits for EN 55022[1], Class B, is 30 dBμV/m (quasi-peak limit). This loop area, which is a typical trace route length on a printed circuit board is 22.6 dBμV/m above the limit!

---

[1] IEC/EN 55022. "Limits and methods of measurement of radio disturbance characteristics of information technology equipment," an international emissions test specification.

1.10.3 Common-Mode Current Description

Common-mode current (also known as "longitudinal mode" or "antenna mode") is the component of time variant current present on both signal and return paths traveling in the same direction simultaneously, usually in common phase. This undesired electromagnetic field when evaluating a system to EMC compliance levels is due to the sum of currents that exist in all loop circuits including both signal and return, or time varying current not cancelled under differential-mode operation. The magnitude of any RF current not canceled out in the differential-mode is called "common-mode." Phase addition could be substantial under simple configurations although forms of phase-cancellation may randomly confuse the results.

Any amount of CM current in a transmission line system may be the major cause of undesired EMI. Common-mode current drives a dipole antenna. I/O cables with a ground wire or 0V reference is easily represented electromagnetically as a dipole antenna. Common-mode current, created by poor differential-mode cancellation of magnetic flux, is developed by imbalance or loss within the two signal paths. Common-mode loops are generally much larger than the differential-mode therefore, their ***E*** and ***H*** field patterns are much more widely spread.

Common-mode currents are thus:

1. The cause of both radiated and conducted EMI.
2. Drives a dipole antenna, which can be represented by a source transmission line relative to an RF return path (refer back to Fig. 1.7a).
3. Contains no useful information since signals between digital components are in differential-mode. Any imbalance in differential-mode signaling of a single-ended transmission lines is one element that creates CM currents.
4. Job security for those working in the field of EMC.

Common-mode development begins as the result of different currents mixing in a shared conductive structure such as the power and return (ground) planes. This happens because multiple currents are flowing through unintentional or unknown paths shown as the dotted line in Fig. 1.13 (parasitic return path). Common-mode energy is thus developed when return currents lose their pairing with their original signal path or transmission line (i.e., splits or breaks in planes where traces cross), or when several transmission lines (signal traces) share common areas of the return plane. Since planes have finite impedance, common-mode currents are developed which can easily be calculated by Ohms law (*Maxwell Made Simple®*). Any common-mode current developed will find a dipole or monopole antenna structure somewhere within the system to radiate from.

The most common cause of EMI is establishment of currents in conductors and shields of cables going to and from the printed circuit board or metallic enclosure. The key element for design engineers to remember is preventing common-mode development by controlling the path of all RF signal and return current flow. This ensures enhanced magnetic flux cancellation. A detailed presentation on RF return paths is found in Chapter 2.

Overly simplified explanation – how common-mode current is created
Maxwell Made Simple®

In Fig. 1.13, current source *I1* represents the flow of time variant current from source *E* to load *Z*. Current flow, *I2,* is the return current traveling in an adjacent trace, a plane (either power or return-it does not matter), a metallic chassis enclosure or by any other parasitic path available.

Common-mode current is caused by the summed contribution of both *I1* and *I2* traveling in the same direction in the return path. *How does the return current in I2 flip direction from differential-mode signal propagation*?

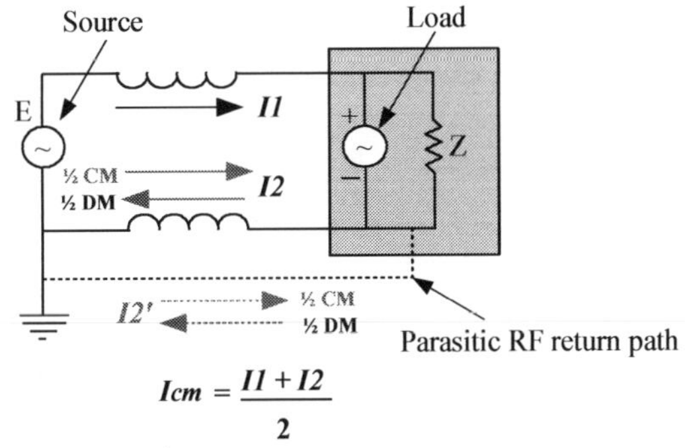

Figure 1.13 Common-mode current model.

The item of interest in Fig. 1.13 is what *I2* and *I2'* does at higher frequencies with time variant current. For this example, *I2* is the DC return path, usually through the power or 0V (ground) distribution network. DC current flows in the path of *least resistance.* We must be concerned about the path of *least impedance* related to EMC compliance. Digital components transition logic states between logic low (V_{IL}–voltage threshold level in the low condition) and logic high (V_{IH}–voltage threshold logic high). A stable 0V reference is mandatory for components to know when to transition states. For this discussion, we are only concerned about AC or time variant current propagation throughout the RF spectrum, or basically anything greater than DC to light.

Within every transmission line there are various types of losses presented in detail within Chapter 2. Losses include resistive, inductive, dielectric material construction, copper roughness and skin effect among many others items too numerous to list at this time. No matter what form of loss exist, differential-mode signal propagation in a transmission line is attenuated by a value that is usually not possible to calculate, measure or anticipate due to hidden parasitics or lack of vendor data related to loss parameters.

Power is sent from a source to load in path *I1*. For this simple example to help visualize a complex topic, we use units of "*watts*" (time domain analysis) and not "*watts/meter*^{2}" which is identical in value when performing frequency domain analysis. Power ($P=VI$) is the parametric value propagated in a transmission line described in terms of voltage and current; since both voltage and current exist we need to only consider total power propagated in the transmission line. It is easier for digital designers think in the time domain (watts) than frequency domain (watts/meter^2) if computational analysis is not being performed. The same results exists regardless of how one analyzes the circuit.

Assume 1-watt is injected into the source path and ½ watt dissipated or attenuated by some means due to inductive losses or parasitics; it does not matter for this simple example. Where does this ½ watt of lost or dissipated power go to? Per Newtons's Energy Law of Conservation, power (RF energy) is not lost but sent somewhere, only to return in a mysterious path to satisfy both Kirchhoff and Ampere's laws. This overly simplified *mysterious or parasitic path I2'* is now discussed, which is the source of undesired common-mode current development.

Kirchhoff's Law states that the directed sum of the electric potential differences (voltage) around any closed circuit is zero, and that they are equal to the sum of the potential drops within that loop. Conversely, Kirchhoff's current law states that at any node (junction) in a circuit, the sum of currents flowing into that node is equal to the sum of all current flowing out of that node, or zero.

Thus, this dissipated or lost ½ *watt/meter*^{2} of power (electromagnetic field) must somehow find its way back to the source to satisfy Ampere's law and Kirchhoff through a parasitic propagation path (the dotted line in Fig. 1.13) identified *I2'*. The propagation path containing this diverted ½ *watt/meter*^{2} of electromagnetic power may be through an adjacent transmission line, capacitive coupling to adjacent metallic structures or even free space as an example. Remember, we are dealing with electromagnetic field propagation (AC) and not steady-state current (DC) flow.

RF energy or electromagnetic field propagates through space using an efficient antenna structure if this is the diverted RF return path. For our simple example, assume this ½ watt/meter^2 returns to its source through space, shown as the dotted lines in the lower part of Fig. 1.13 (*I2'*). Ampere is satisfied since current is flowing in a loop.

In Fig. 1.13, we have ½ DM current traveling in the return path (*I2-right to left*) and ½ DM current flowing in *I2' (left to right)* to satisfy Kirchhoff, which means the total sum of the current throughout the two transmission lines (real and parasitic) is now equal zero. If the source and return path are physically located close together, a certain percentage of magnetic flux lines present will be cancelled out, which minimizes one form of electromagnetic field loss in the loop circuit.

In order to satisfy Kirchhoff's Law again for both return (*I2*) and parasitic paths (*I2'*), where the sum of all currents must equal zero, we now must have RF current flow in the opposite direction (reverse polarity) to make up the lost electromagnetic power in each particular transmission line path (visualize current traveling from *left to right* in the return path, while the desired path of current flow is *right to left*). Reverse current flow in the primary return path (*I2*), to satisfy Kirchhoff, is now in the "common-mode." This common-mode propagates in the same direction as the source path and adds in phase somewhere in the transmission line. Since we are dealing with AC or time variant currents, total RF energy does not care which direction it is flowing as the measurement point is taken at a particular location at a finite point of time. DC current in the return path will continue to travel only in one direction, from positive to negative, which is a different route than the AC path of least impedance which could be the parasitic path (includes free space).

With RF current now traveling in both directions (*I2, I2'*), the sum of the currents propagating in this path now equals zero, satisfying Kirchhoff. This is the same analogy as placing a current shunt or probe in a transmission line to measure total current flow. The shunt or probe does not know if current flow is from right to left or left to right. It only knows total current present regardless of direction. If using an ammeter that only measures DC current values, this simplified analysis does not apply since we are interested only in the AC current or the electromagnetic field present, not the DC bias when considering development of common-mode current.

Now that Kirchhoff is satisfied in path *I2*, the summation of AC current in return path *I2'* must also be equal to zero, with ½ being in the differential-mode and ½ in the common mode. Using the same analogy as the RF return path *I2*, there is common-mode current flowing in the same parasitic path (*I2'*). This is shown as ½ CM in Fig. 1.13. To visualize this discussion, think about wireless communication. We can both talk and listen at the same time which means the RF field is bi-directional within the media of propagation.

If there is any electromagnetic field loss in a transmission line, this loss must be made up somewhere, which will be with RF current flow in the opposite direction of the desired signal to satisfy Kirchhoff. The RF current traveling in the opposite direction may be in phase with the transmitted signal. Phase addition of both signals traveling in the same direction is in the common-mode, creating EMI problems.

To prevent common-mode current development, one must ensure the circuit has minimal loss by optimizing both source and return paths to operate in differential-mode, or be in perfect balance.

Summary

Any loss in the power level of a transmitted signal creates common-mode currents. The amount of attenuation, or loss, is directly proportional to the amount of common-mode current developed. This means the circuit is out of balance just like a scale with two unequal buckets on a fulcrum. If one side is heavier than the other, the scale tips and is out of balance (differential-mode=balanced scale). To minimize development of common-mode current, the fulcrum must be balanced. Although there may not be a loss in the transmission line network, common-mode current can still be generated from other sources or intentionally generated by unique circuit design or co-mingling of currents shared within a common path such as a plane.

As a side note, digital components have a common-mode rejection ratio (CMRR) value for all inputs, though this might not be clearly defined by the supplier. The CMRR circuitry internal to silicon die will attenuate or prevent common-mode currents from affecting differential-mode signal propagation. However, if common-mode current present on the input pins exceeds the total CMRR capabilities of the device, the component may not function properly. This is one signal integrity problem frequently encountered by digital design engineers.

The magnitude of many signal integrity problems, apart from other issues such as crosstalk and timing diagrams, is the magnitude of common-mode current developed when digital component operation is not in balance with optimal differential-mode signaling. Numerous loss factors develop when designing a printed circuit board that includes any change in impedance of a transmission line, signal traces jumping layers through vias, not having an optimal RF return path, a stable power distribution network and many other layout requirements.

With differential-mode currents in a closed loop configuration, the magnetic field component is the difference between $I1$ and $I2$, whereas common-mode current is the summation of the two. If $I1=I2$ exactly (no losses within the loop circuit) there will be no radiation or EMI from differential-mode currents that emanate from the circuit (assuming the distance from the point of observation is much larger than the separation between the two current-carrying conductors) except under certain conditions where common-mode energy is required for circuit operation. This occurs only if the distance separation between $I1$ and $I2$ is electrically small in physical dimensions.

Design and layout techniques for cancellation of EMI emanating from differential-mode currents are easily implemented in a printed circuit board and discussed in greater detail within Chapter 2. On the other hand, RF fields created by common-mode currents are more difficult to suppress and are the primary source of undesired EMI. Fields due to differential-mode currents are rarely observed as a significant radiated electromagnetic field.

1.10.4 Common-Mode Radiated Emission Equations

Common-mode (CM) radiation is caused by unintentional voltage drops, imbalances or losses within a circuit or transmission line relative to a reference potential that is generally called ground but in reality is a 0V reference. Cables connected to an interface circuit act as a dipole antenna and will radiate field components at CM potential. The far-field radiated signal is easily calculated by (1.20).

With time variant current and a fixed antenna length, the electric field at a prescribed distance is proportional to the frequency and physical length of the antenna. Unlike differential-mode radiation, common-mode is a more difficult problem to solve. In order to eliminate or reduce common-mode radiation that results from imbalances in differential-mode propagation, resulting common-mode fields must approach zero. This is achieved using a properly designed RF return path and referencing

$$E \approx 1.27 * 10^{-6} (f\ I_{cm}\ L) / R \quad \text{(V/m)} \tag{1.20}$$

where:
E = radiated field measured at a distance from the antenna (V/m)
f = frequency (MHz)
I_{cm} = common-mode current (A)
L = antenna length (m) (the driven element of the dipole antenna)
R = distance (m).

Emissions from the combination of both source and return path currents that do not cancel add in phase. It is easily calculated that for a 1-meter length of cable whose wires are separated by 50 mils (a typical ribbon cable configuration), differential-mode current of 20 mA (Eq. 1.16) or 8 μA common-mode current at 30 MHz (Eq. 1.19) will fail FCC Class B limits at 3 meters. These values illustrate a significant difference in magnitude between DM and CM current, with CM being much more severe by many orders of magnitude. Common-mode currents will radiate an electric field at 3 meters of 100 μV/m, which just meets the FCC Class B limit [3, 4, 5, 6], under these conditions.

There is a ratio of 2500, or 68 dB between DM and CM using this example. Therefore, a very small amount of common-mode current is capable of producing significant radiated emission levels whereas it requires a significant amount of DM current to produce the same results.

1.10.5 How Common-Mode Current Drives I/O Cables and Cause Radiated Emissions–Made Simple

Using the discussion above on how common-mode currents are developed in transmission lines within a closed loop circuit, we expand on Fig. 1.13 with Fig. 1.14. The difference between the two figures is that an I/O cable is now connected to the system using a connector represented by *Z*. Most connectors inject some loss in the transmitted signal causing a potential signal integrity problem if the interconnect is not designed properly and impedance controlled to the transmission line.

To help visually illustrate how a cable radiates EMI, consider a coaxial cable first introduced by Oliver Heaviside. A coaxial cable has a center conductor and "shield." The "shield" of a coax is actually a circumferential active image plane. We now change this coax configuration to a 2-wire system containing both signal and return path. Attach the cable to the I/O connector and separate the two wires into the shape of a dipole antenna (Fig. 1.14).

Like a coax or two wires in parallel, the electromagnetic field that propagates down the wires exists in the dielectric that separates the two transmission line paths, such as jacket insulation. A dipole antenna propagates an electromagnetic signal from the driven element to return through this dielectric. Every two-wire transmission line with one line at 0V potential (ground or return), and another at signal potential, stimulates this dipole antenna (capacitive antenna model-Fig. 1.7a) an efficient radiator of common-mode RF current.

Another illustration on how a printed circuit board drives an interconnect cable that allows common-mode current to be developed is shown in Fig. 1.15. In this figure, both differential- and common-mode currents are illustrated along with inductance caused by an internal voltage drop. Under this scenario, the system goes into an unbalanced state.

If any common-mode currents is present on the signal line, this undesired current will continue within this path as if the transmission line was physically extended in length, except for an impedance discontinuity at a finite time period in the picoseconds range at the junction of the connector interface. Without onboard filtering, the magnitude of the common-mode current on the source wire will be from the magnitude of common-mode current that drives the dipole antenna.

The return element of the dipole antenna is generally bonded to a reference that usually is at 0V reference or chassis ground. If the reference point is at DC levels, or not carrying RF (AC) current, then the magnitude of the common-mode current on the cable will be approximate the same as the driven element. If the reference is also carrying common-mode energy, there will be a summation of current from both paths which increases total radiated EMI.

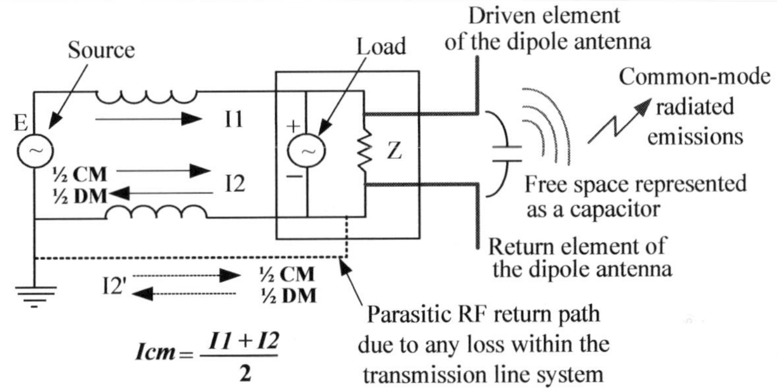

Figure 1.14 Common-mode current injected into an I/O cable.

There is also parasitic capacitance between the driven element and chassis ground which creates additional EMI, identified as I_{cm} in Fig. 1.15, caused by a voltage potential difference between the cable and chassis ground, identified as V_{cm}.

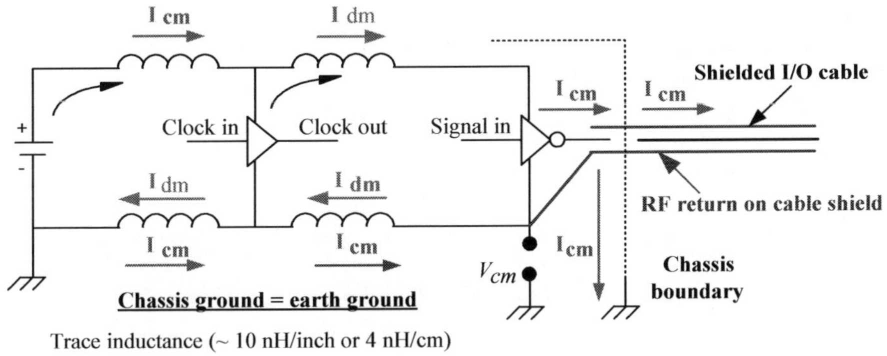

Figure 1.15 System equivalent circuit of differential and common-mode currents.

1.10.6 Conversion between Differential-Mode and Common-Mode Currents

Common-mode currents in a system may be unrelated to the intended signal source (i.e., they may be from other circuits and devices). Conversion between differential and common-mode may occur when two signal traces (a.k.a. transmission lines) both with different impedances are in close proximity and couple through both capacitive and inductive means relative to the physical placement of the transmission lines. For the majority of printed circuit board layouts, the designer has some control over minimizing capacitance and inductance within a network, thus minimizing differential- to common-mode conversion.

To illustrate this effect with a simple example on conversion from differential-mode signaling to common-mode RF development in a metallic chassis, refer to Fig. 1.16.

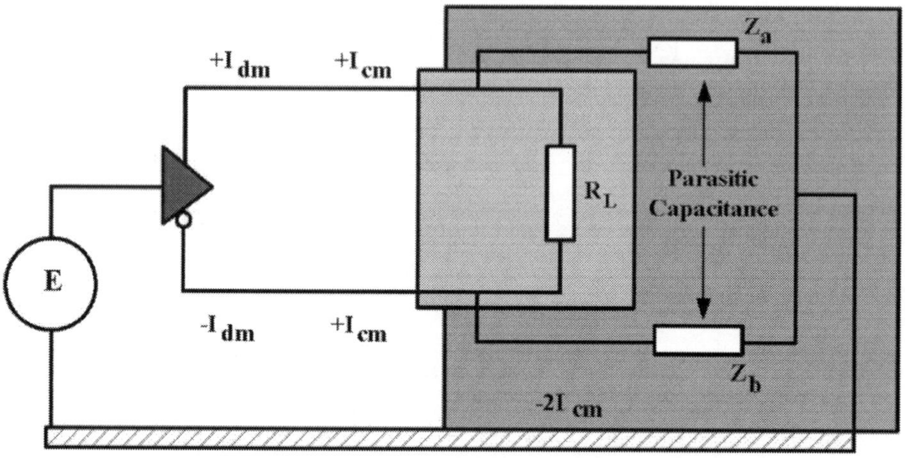

Figure 1.16 Differential-mode to common-mode conversion in an enclosure.

Differential-mode current is the desired signal of interest across R_L. Common-mode current, I_{cm}, will not flow through R_L directly if both source and return are in balance. However, common-mode current I_{cm} will flow through impedances Z_a, and Z_b which is *parasitic capacitance* in the system between the printed circuit board and metal chassis. Remember, a capacitor is simply two metallic conductors separated by a dielectric. If the printed circuit board is at one potential and the chassis at another, two potentials exist upon which RF energy can and will propagate across the parasitic capacitance with air as the dielectric. Once the field is in the dielectric it can radiate undesired EMI and cause harmful disruption in the near field to other electrical circuits (dipole model; capacitive structure).

Once the propagating RF field enters chassis ground, differential-mode signaling to common-mode RF current conversion will occur. For example, if an I/O cable shield is bonded to chassis ground through a pigtail and the signal line to a component through a connector interface that by itself presents an impedance discontinuity, the cable is now energized with chassis common-mode current on the ground wire, and which will now radiate as an efficient antenna at its' self-resonant frequency along with any other harmonic or wavelength that is efficient for the common-mode current to propagate from.

Impedances Z_a and Z_b are not physical components in Fig. 1.16, but illustrated to show stray parasitic capacitance or parasitic transfer impedance present. Parasitic capacitance could also come from a trace located on the outer layers of the printed circuit board reference to the metal chassis (second terminal of the parasitic capacitance configuration).

If $Z_a = Z_b$, no voltage drop is developed across R_L by I_{cm}. If there is any imbalance in the network $(Z_a \neq Z_b)$, a voltage difference will occur that is proportional to the difference in impedance according to Ohms law. Common-mode current is thus established due to V_{cm} and various transfer impedances Z_a *and* Z_b (1.21).

$$V_{cm} = I_{cm} * Z_a - I_{cm} * Z_b = I_{cm}(Z_a - Z_b) \tag{1.21}$$

Because of the need for balanced voltage, current and 0V [ground] references, circuits with higher-frequency signals that tend to corrupt other transmission line or radiate high-levels of RF energy (i.e., video, high-speed data, etc.), or other traces susceptible to external influences, must be balanced in such a way that stray or parasitic capacitances of each conductor are identical so as to cancel out undesired common-mode current.

1.11 Antenna Efficiency

Electromagnetic radiation requires a dynamic frequency source that establishes field propagation between two elements or conductors. For a dipole antenna these are the two legs or elements. For a half-wave dipole, one is the driven element with respect to ground. A loop antenna propagates primarily a magnetic (lower impedance) field and thus sets up time variant current flow due to the closed loop path present (Fig. 1.17).

For current to flow, a voltage difference must be present between two elements. If the source driver is placed at one extreme end of the antenna, minimal radiation would occur as the antenna is now an inefficient radiator. Enhanced performance is observed when the antenna is driven at its resonance frequency which is based on physical dimensions compared to the wavelength of the excitation source.

To minimize radiation from the antenna or to make it inefficient for propagating RF current:

1. Reduce the intensity of the RF source;
2. Reduce energy coupled to the antenna (use self-shielding trace routing over a reference plane to create a coaxial based transmission line);
3. Reduce effectiveness of the antenna (physically change antenna dimensions); or,
4. Keep all metal structures at the same RF voltage potential, including the antenna affixed to the unit.

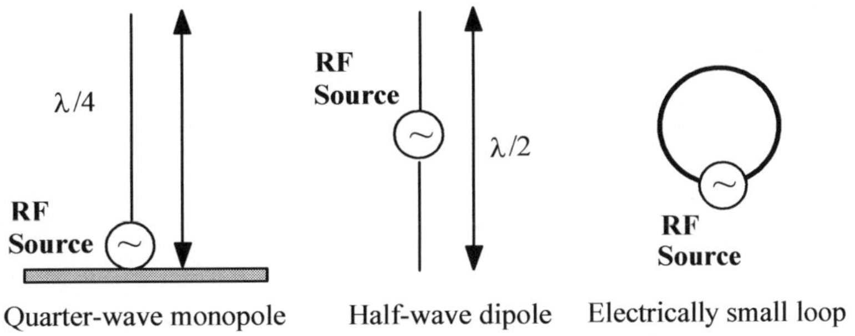

Figure 1.17 Various types of antenna configurations.

1.12 Fundamental Principals and Concepts for Suppressing RF Energy

1.12.1 Fundamental Principles of EMI Suppression

A fundamental principle related to development of radiated emissions involves common-mode noise developed within a printed circuit board. This fundamental principle is associated with RF energy transferred from source to load. Common-mode currents are developed by electrical circuitry and not necessarily just in the power distribution system which also creates a different form of common-mode current, discussed in Chapter 4.

Common-mode current by definition is RF energy common to any power, return, transmission and signal lines. For many products, a metal chassis is provided that may facilitate creation of common-mode current to propagate either into space or couple to other cables, circuits or interconnects located internal to the enclosure. Since RF current must return to its source per Ampere (direct or parasitic), common-mode current flow in a chassis is one method to close the loop for the RF return path. Since movement of electrical charge occurs through a transmission line (such as a trace, cable, wire, etc.), a voltage potential difference will be established across this impedance (per Ohms law). This voltage potential difference will then cause radiated emissions to occur if trace stubs, I/O cables, enclosure apertures and slots are present.

The following principles need to be understood. In later Chapters of this book these topics will be examined in greater detail.

1. For high-speed or higher frequency operating components, common-mode currents can be developed if there is an imbalance in differential signaling (this Chapter).
2. To minimize *distribution* of common-mode RF currents throughout a system, proper layout of printed circuit board traces, component placement, and provisions to allow RF currents to return to their source efficiently as an image current must be ensured to keep undesired RF energy from being created and propagated throughout the assembly. This is achieved by implementing proper transmission line theory (Chapter 2).
3. To minimize development and propagation of RF currents, proper termination of transmission lines must occur. At lower frequencies, RF currents are not a significant concern because the physical wavelength of the signal often does not couple efficiently to most antennas, especially those on a printed circuit board which are very short in length and only efficient at very high frequencies. At any frequency above DC, RF energy will be developed per Maxwell's equations, namely Faraday's law (Chapter 2).
4. Provide for an optimal 0V reference to achieve enhanced cancellation of undesired magnetic flux and to ensure digital components do not create common-mode currents due to common impedance coupling to other circuits and systems (Chapter 2).
5. To minimize development of *common-mode* RF currents, an optimal power distribution network (PDN) must be provided. If a high quality PDN is not implemented to minimize plane bounce, switching noise created by digital components both sourcing and sinking current at RF frequencies will also create common-mode current. All PDNs must be designed with forethought and not left to chance (Chapter 4).

1.12.2 Fundamental Concepts of EMI Suppression

One fundamental concept for suppressing RF energy within a printed circuit board deals with *flux cancellation or minimization.* As discussed earlier, time variant current that travels in a transmission line (or interconnect structure) causes magnetic flux to exist (Faraday's law). This *magnetic* field creates an *electric* field. Both field structures set up a propagating plane wave at a specific RF frequency depending on the switching edge rate of digital components (i.e. spectral distribution in Fourier terms-Appendix C), and which are observed at primary or harmonics of the switching frequency. If we cancel or minimize undesired magnetic flux, RF energy will not be present other than within the boundary between the trace and RF return path. Flux cancellation or minimization virtually guarantees compliance with regulatory requirements.

The following two concepts must be understood to minimize radiated emissions.

1. Minimize common-mode currents created as a result of a voltage drop or loss that occurs across any impedance within a transmission line.
2. Minimize distribution of common-mode currents throughout the network by cancelling unwanted flux.

Flux cancellation or minimization within a printed circuit board is necessary because of the following sequence of events.

1. An RF voltage drop is the product of RF currents and transients traveling through losses inherent within a transmission line (Ohm's law).
2. Common-mode RF current, created from the RF voltage drop established between two circuits due to losses in the transmission line, builds up on inadequate RF return paths between source and load per Ohms law (insufficient differential-mode cancellation of undesired RF currents).
3. EMI, both radiated and conducted, will propagate due to common-mode RF current that exists by any means possible as long as there is an efficient antenna within the loop circuit for the current to find.

1.13 Hidden Schematic or Parasitics of Passive Components

Traditionally, EMC has been considered a field of *Black Magic*. In reality, EMC is explained by mathematical laws. Even if computational analysis is performed most equations become too complex to solve for practical applications due to lack of information needed to build accurate models that contain hidden parasitics that may affect system-wide performance. In order to perform accurate computational analysis, simple models are generally utilized to describe how EMI is being created, or how EMC compliance can be achieved using data book values published for resistors, capacitor and inductors.

Many variables exist that cause EMI. This is because EMI is often the result of exceptions to the normal or expected rules of passive component behavior. The behavior of passive components at both high and low frequency of operation are illustrated in Fig. 1.18. Regardless of passive component used, a resonance will occur somewhere in the frequency spectrum as a zero or pole that may work either for or against achieving EMC.

When designing with passive components we must ask ourselves, "When is a passive element not a passive element?" We now examine resistors, capacitors and inductors, or passive components.

A resistor placed in a circuit operating at higher frequency will behave as a series combination of inductance along with resistance in parallel with a parasitic capacitor, detailed in Fig. 1.8 under high frequency behavior. In other words, the resistance of the device (a.k.a. impedance) changes somewhere in the frequency spectrum that may enhance or degrade component performance.

In reality, a resistor does not exist in the field of electrical engineering! A resistor, once installed on a printed circuit board, is technically a "*capacitor with very high Equivalent Series Resistance (ESR).*"

A capacitor at higher frequencies (above the self-resonant frequency of the device) is modeled as an inductor and resistor in series-parallel combination across the two capacitor terminals. A capacitor does not function as a capacitor at higher frequencies because it has changed its operational characteristics and now appears as an inductor to the circuit due to both lead and loop inductance within the transmission line structure. This resonance condition works perfectly for power distribution networks however, an anti-resonant frequency will exist that may be the source of common-mode current development and propagation.

Conversely, "When is an inductor not an inductor"? An inductor appears to function as a capacitor due to parasitic wire coupling at higher frequencies with parasitic capacitance in parallel across the two terminals and between each winding (series/parallel combination of capacitance).

To be a successful designer, one must recognize limitations of passive components and how they work in a real printed circuit board. Use of proper design techniques and knowledge to accommodate for these hidden features becomes mandatory, in addition to designing a product to meet a functional specifications.

Unknown behavioral characteristics and parasitics are referred to as the "*hidden schematic of components*" [7]. Digital engineers generally assume that passive components have a single-frequency response based on data book values. As a result, passive component selection tends to be based on functional performance in the time domain (DC steady state operation) without regard to how the passive device operates in the frequency domain (AC current). Many times, EMI exceptions as illustrated in Fig. 1.18[2] occur. In Fig. 1.18 for discussion purposes, the value of Z under "low frequency behavior" is assumed to be in the vicinity of 1-ohm, or low relative to its impedance at high frequency which can be in the ohms to thousands of ohms.

An alternate definition of Electromagnetic Compatibility is "*Working with everything that is not intentionally shown on a schematic or assembly drawing.*" This statement explains why the field of EMC is considered to be the Art of Black Magic due to unknown or unexpected parasitics.

Once hidden behavior of components is understood, it becomes a more defined process to design products that pass EMC requirements. Hidden behavior also takes into consideration the switching speed of active, not passive components along with their unique spectral characteristics, which also have hidden resistive, capacitive and inductive elements. We now examine each passive device separately.

---

2 Daryl Gerke & Bill Kimmel. "*The Designers Guide to Electromagnetic Compatibility.*" Reprinted from *EDN Magazine* (January 20, 1994). © Cahners Publishing Company, 1994. A Division of Reed Publishing USA.

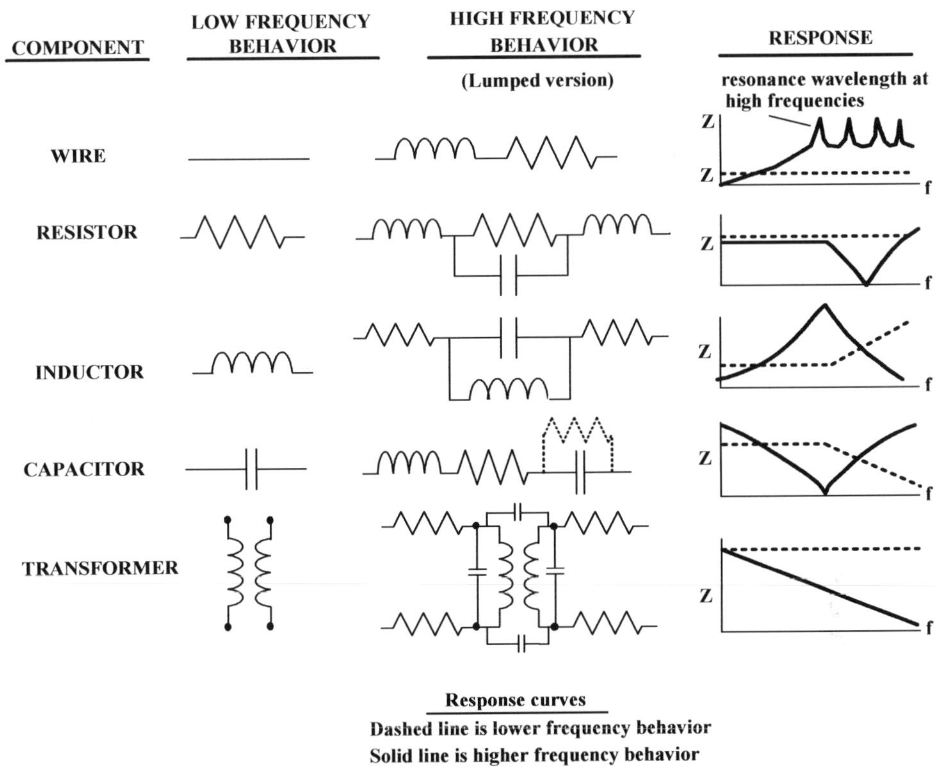

Figure 1.18 Component behavior at RF frequencies.

1.13.1 Wires, Printed Circuit Board Traces and Transmission Lines

One does not generally consider internal wiring, harnesses and printed circuit board traces efficient radiators of RF energy. Every transmission line has inductance and it is this inductance that establishes common-mode RF energy. A detailed presentation on transmission line theory is presented in Chapter 2 with a thorough discussion related to Inductance in Chapter 3. Inductance will always be present from the bond wires of a silicon wafer (die) to its package pins, vias that transition signals between layers of a printed circuit board, the terminals or lead wires of resistors, capacitors and inductors, in addition to any element, component or device that carries AC or time variant current. Every passive device regardless of application has *parasitic* capacitance and inductance, including wire. For simulation purposes, do we always know the exact value of these parasitics? If we assume parasitics do not exists or ignore them, will results of simulation be accurate enough or in other words, how accurate do we need to have simulated results be, or is close good enough to get the job done?

Transmission lines are described as *RLC* elements, or resistance, capacitance and inductance. The impedance of every transmission line, *Z*, contains both a real and complex elements (1.22). The resistive element ® describes DC or low frequency behavior whereas the complex element (*j*) is frequency dependent and which can get very large as we go higher up in frequency, generally several orders of magnitudes beginning around 100 kHz.

Parasitics (both capacitive and inductive) will affect wire impedance and are frequency dependent, shown in the complex portion of impedance equation (1.22). Depending on the value of both inductance and capacitance within the transmission line, a resonance will occur somewhere in the frequency domain. If there is common-mode RF energy at this resonant frequency, the transmission line now becomes an efficient radiating antenna provided the physical dimension of the antenna is also resonant at that frequency.

The impedance, *Z*, of a transmission line contains both resistance and inductive reactance (ignoring capacitive coupling) as:

$$Z = R + jX_L \approx j2\pi fL \quad (1.22)$$

where:
Z = total impedance containing both low and high frequency components (ohms)
R = lower frequency resistance (ohms)
L = inductance in the transmission line (ohms), and
f = frequency of operation.

At lower frequencies almost all wire is primarily resistive. The definition of lower frequency is generally anything less than 100 kHz, and more commonly 10 kHz. At higher frequencies generally above 100 kHz, every wire (transmission line) takes on characteristics similar to that of an inductor. The primary difference between a wire and printed circuit board trace is that wire is round and a trace is rectangular; both are transmission lines. A rectangular transmission line has significantly less impedance than a round wire with the same perimeter value. Wire inductance based on physical construction is presented in Chapter 3.

For the equation above, capacitive reactance $X_c=1/(2\pi fC)$ is not taken into account, at least for the higher frequency impedance response of the wire, at this time. For DC and lower frequency applications (generally below 100 kHz), the wire (or trace) is essentially resistive and inductance does not play a significant role in performance. At higher frequencies, the inductance of the transmission line becomes the critical element of the impedance equation. Above 100 kHz, inductive reactance ($j2\pi fL$) begins to exceed the resistance value ® and becomes the dominate portion of the equation since variable "*f*" is becoming large and is directly proportional to the value of "*Z*", increasing in value quickly as frequency increases.

For example, using (1.22) assume a 10-cm trace has a DC resistance of R=57 mΩ. Assume a typical value of a printed circuit board trace is 8nH/cm (or 80nH). Per (1.22), we have inductive reactance of 5 mΩ at 100 kHz. For traces containing frequencies above 100 kHz, our transmission line now exceed 57 mΩ by a significant amount or one order of magnitude higher. The DC resistance of the wire is now negligible and can usually be ignored for almost all application of use. A 10-cm trace is an efficient radiator above 150 MHz (λ/20 of 100 kHz). If the frequency changes to 1 GHz, using the same value of inductance for the 10-cm trace, total impedance becomes 503 ohms! This is significantly larger than 5 mΩ at DC.

If the impedance of free space is 377 ohms, and the transmission line is 503 ohms at 1 GHz, RF return current will occur through space, since space has significantly lower impedance than the trace at 1 GHz. Once this return current is in space we may end up with a radiated emissions event related to regulatory compliance or disruption to adjacent internal circuitry.

Per the example above, the transmission line is no longer a low-resistive connection but rather behaves similar to an inductor at higher frequencies. Hidden parasitics will set up resonances. As a general rule of thumb, any wire (or trace) operating above the audio frequency range increasingly with higher frequency behaves as an inductor not a resistor, and could be considered to be an efficient antenna to radiated RF energy.

Most antennas are designed to be an efficient radiator at ¼ or ½ wavelength (λ) at a particular frequency. Within the field of EMC, design recommendations are to not allow any transmission line to become an efficient radiator below λ/20. Inductive and capacitive elements can result in efficiencies through circuit resonance that mechanical dimensions do not describe.

1.13.2 Resistors

Resistors are one of the most commonly used passive component on a printed circuit board. Resistors have a limitation related to EMI. Depending on the type of material used for the resistor (carbon composition, film, mica, wire-wound, ceramic, etc.), this limitation exists related to frequency domain aspects of the component. A wire-wound resistor is not suitable for higher-frequency applications due to excessive inductance in the wire. Film resistors contain some inductance and are sometimes acceptable for higher-frequency applications due to internal lower lead inductance.

A commonly overlooked aspect of resistors deals with parasitic capacitance. Capacitance, again by definition, are two conductors separated by a dielectric. With metal terminals on both ends of a resistor and a dielectric that has a high resistive value, we essentially have a capacitive structure with equivalent series resistance (*ESR*) in the high ohms range, whereas the *ESR* for a typical [real] capacitor is usually several milliohms!

Parasitic capacitance can play havoc with extremely high-frequency designs, especially those in the GHz range. For most applications, parasitic capacitance between resistor leads is not a major concern compared to lead inductance present due to mounting of the resistor to the printed circuit board along with interconnect loop inductance.

An illustration on how parasitic capacitance can cause a resistor to change value is shown in Fig. 1.19.

One major concern for resistors lies in overvoltage stress conditions to which the device may be subjected to. If an ESD event is observed at the resistor's terminals, interesting results may occur. If the resistor is a surface-mount device chances are this component may arc-over or self-destruct at high voltage levels. For resistors with radial or axial leads, an ESD charge may see a high resistive and inductive path and may be kept from entering the circuit, protected by the resistor's hidden inductive and capacitive characteristics from a system point of operation.

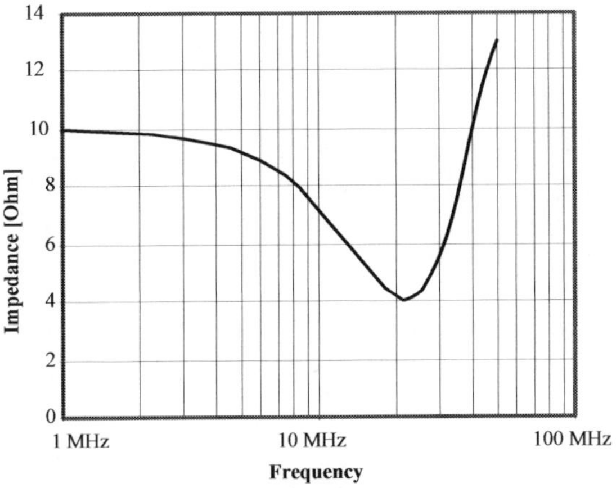

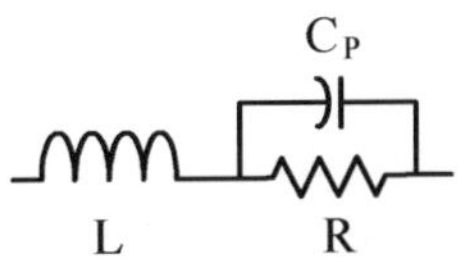

$$Z_R = \frac{1}{\left(j\omega C_P + \frac{1}{R}\right)} + j\omega L_S$$

Figure 1.19 Real-word frequency response of a resistor.
Frequency response: R=10Ω, L_S=50nH, C_P=1nF
(Plot courtesy: Elya Joffe)

When performing numerical analysis, if parasitic capacitance of a resistor is not considered, our simulation or computational analysis may be inaccurate. We may not know where within the frequency domain the resistor achieves resonance and assumes a low impedance value relative to the ohmic number printed on the device, or identified by the manufacturer which is always stated at a DC voltage level.

1.13.3 Capacitors

Chapter 4 presents detailed discussion on capacitors and their hidden characteristics. This section thus provides only a *brief overview* on parasitic aspects for completeness while discussing hidden parasitics. Capacitors are generally used for power bus decoupling, bypassing, bulk storage, DC isolation or charge applications. Capacitors are also used in filters and analog devices to establish a resonant condition for desired circuit operation.

Capacitors remains capacitive up to its self-resonant frequency. Above this frequency the capacitor exhibits inductive effects. This is described by two equations (1.23).

$$X_c = 1/(2\pi fC) \quad X_L = j2\pi fL \tag{1.23}$$

where:
Xc = capacitive reactance (Ohms)
X_L = inductive reactance (Ohms)
f = frequency (Hertz)
C = capacitance (Farads)
L = inductance (Henrys)

If $X_c = X_L$, a resonance exists and the impedance of the capacitor approaches zero ohms, or at least down to the value of internal *ESR* which is generally very small. This resonance condition is exactly what we want for power bus decoupling (low transfer impedance) depending on bandwidth of operation and the Q of the component. Chapter 4 discusses this topic in extreme detail; resonance, bandwidth of operation and Quality Factor.

Once the capacitor achieves self-resonance, the device now starts to behave as an inductor since frequency is increasing. If there is any inductance in the closed-loop or interconnect, development of common-mode current will be established, discussed extensively in Chapter 4.

To illustrate the impedance of a capacitor without taking into account lead or loop inductance, a 10 µf electrolytic capacitor has a capacitive reactance of 1.6 Ω at 10 kHz [$X_c = 1/(2\pi fC)$], which decreases to 160 µΩ at 100 MHz. At 100 MHz, a short-circuit condition would exist which is wonderful for power bus decoupling for the switching regulator. However, electrical parameters of electrolytic capacitors with high values of equivalent series inductance and resistance limit the effectiveness of this particular type of capacitor to operation below 1 MHz.

Real-world characteristic of a capacitor is provided in Fig. 1.20 and is similar in form to a resistor. Recall that a resistor is in reality *a capacitor with a very large value of ESR*.

Another aspect of capacitor use lies in lead inductance and body structure. This subject is discussed again in detail within Chapter 4. To summarize, inductance in the capacitor's wire bond leads causes the capacitor to function as an inductor above self-resonance and ceases to function as a capacitor for its intended function.

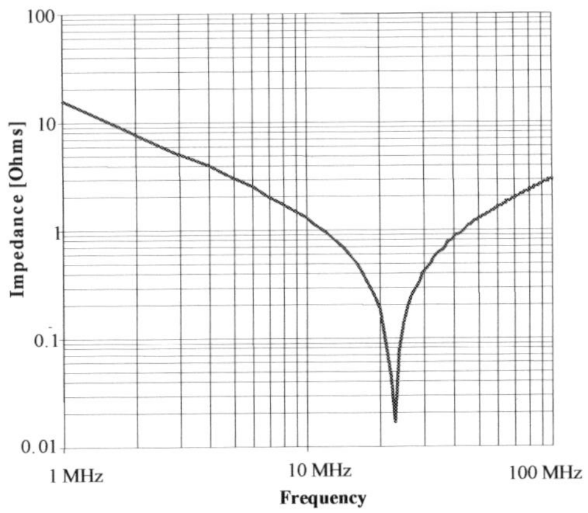

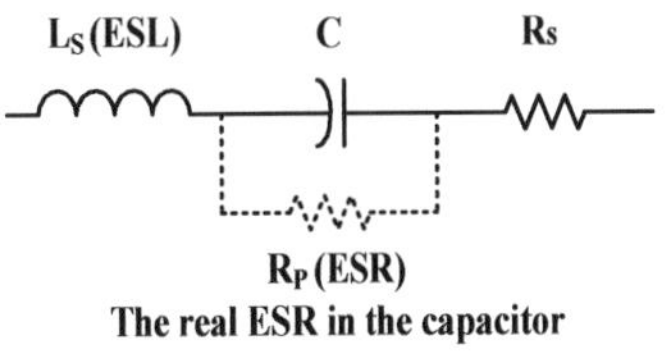

$$Z_C = \frac{1}{\left(j\omega C + \frac{1}{R_P}\right)} + j\omega L_S + R_S$$

Figure 1.20 Real-word frequency response of a capacitor.
Frequency response: C=10nF, L_S=5nH, R_S=2mΩ
(Plot courtesy: Elya Joffe)

1.13.4 Inductors

When using an inductor or inductive element, the characteristic behavior of this passive device is critical for successful application of use. One application is filtering or removing undesired RF magnetic field current in a transmission line. As for the characteristics of a basic inductor, inductive reactance increases linearly with increasing frequency, although this behavior when viewed as impedance is modified by distributed capacitance. This is described by $X_L=2\pi fL$, where X_L is inductive reactance (Ohms), f is frequency in Hertz, and L inductance (Henries).

For example, using $X_L=2\pi fL$, an "ideal" 10 mH inductor has inductive reactance of 628 ohms at 10 kHz. This inductive reactance increases to 6.2 MΩ at 100 MHz and now appears to be an open circuit. If we want to pass a digital signal at 100 MHz, great difficulty occurs related to signal quality.

Within every transmission line there is both trace and loop inductance. Across this inductance, a voltage potential difference will be established. With a voltage potential difference, development of common-mode current is created which is exactly what we do not want!

Like a capacitor, the electrical performance of any inductor (which has parasitic capacitance between each winding) limits this particular discrete device to less than 1 MHz. If an inductor has 1,000 turns or windings on a core base material, there are 999 capacitors in series which themselves are in parallel with free space capacitance between the two terminals, or one larger capacitor. It is this parasitic capacitance that limits the frequency range of a wire wound inductor to lower frequencies. Figure 1.21 shows the frequency response of an inductor taking into account parasitic capacitance.

If wire wound inductors are ineffective at higher frequencies to suppress RF current, which it does so by absorbing the magnetic field and converting it to heat, what type of inductive device can we use in a transmission line or printed circuit board trace without creating a loss, the magnitude which in turn develops undesired common-mode current? Ferrite beads are now the component of choice. Ferrite material consists of alloys of iron/magnesium, iron/nickel or other material with ferrous properties. Material with ferrous properties provide high impedance at a desired range of frequencies based on construction and material composition. Ferrite beads may exhibit minimal parasitic capacitance with careful layout so they are very useful at higher frequencies, whereas a regular inductor is perfect for low frequency power systems. Ferrite beads are further discussed in Chapter 6.

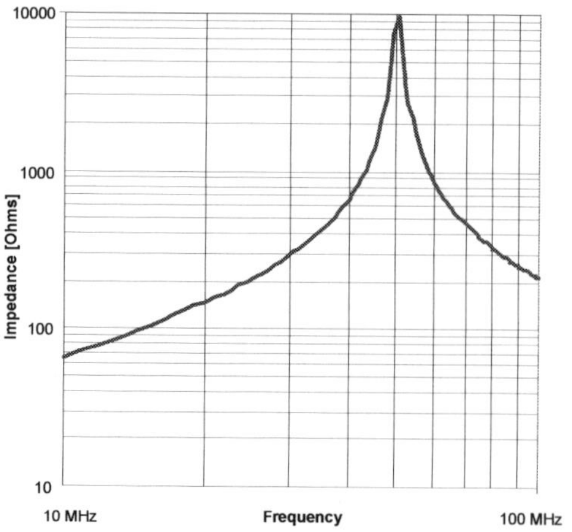

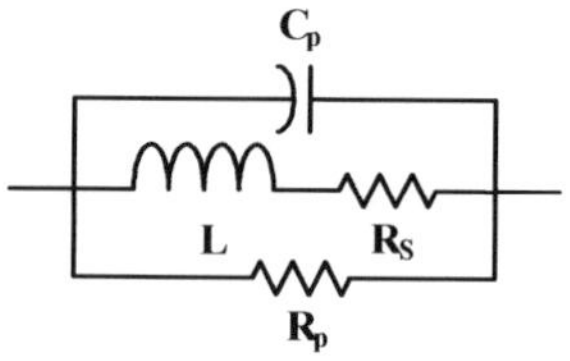

$$Z_L = \frac{1}{\frac{1}{R_P} + \frac{1}{j\omega L + R_S} + j\omega C_P}$$

Figure 1.21 Real-word frequency response of an inductor.
Frequency response: L=1μH, R=10mΩ, C_P=10pF
(Plot courtesy: Elya Joffe)

At higher frequencies, ferrite material becomes reactive and frequency dependent. This is graphically shown in Fig. 1.22. In other words, ferrite beads are high-frequency attenuators of RF energy.

Ferrites are better represented by a parallel combination of a resistor and inductor. At lower frequencies, the resistor is "shorted out" by the inductor, whereas at higher frequencies the inductive impedance is so high that it forces RF current flow through the resistive element which dissipates this energy as heat.

Ferrites function as "dissipative devices" which is explained by the resistive, not the inductive, elements of the material.

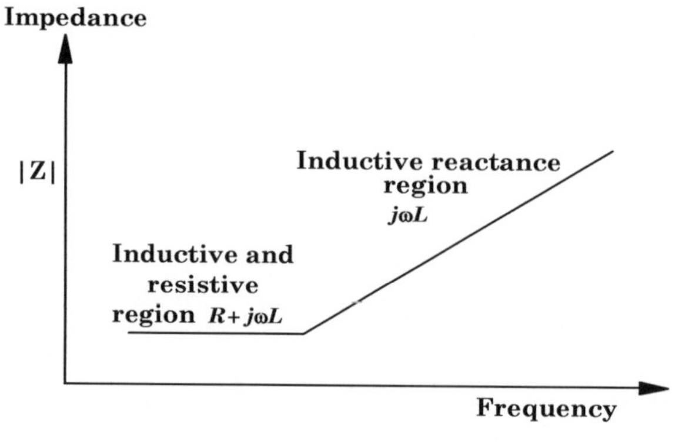

Figure 1.22 Operational characteristics of ferrite material.

1.13.5 Transformers

Transformers are generally used in power supply applications in addition to isolation for data signals, I/O interconnects, impedance matching (baluns) and power interfaces, among many other uses. Between the primary and secondary of the transformer parasitic capacitance is present that can permit undesired RF current to propagate between the two sets of windings.

Depending on type and application of use, an electromagnetic barrier or grounded Faraday shield between the primary and the secondary greatly reduces the coupling of common-mode noise due to this parasitic capacitance. This is accomplished by having another winding or a metal strip surrounding the windings connected to 0V reference, or ground. This Faraday shield barrier, connected to a chassis ground reference, is designed to prevent capacitive coupling between the two sets of windings.

Transformers are also widely used to provide common-mode (CM) isolation, particularly at lower frequencies where the influence of distributed capacitance is reduced. These devices depend on differential-mode transfer (DM) across their input to magnetically link the primary windings to the secondary windings in their attempt to transfer RF energy. This process of coupling occurs per Lentz's Law. As a result, CM voltage across the primary winding is rejected. One flaw that is inherent in the manufacturing of transformers is signal source and inter-winding capacitance. As the frequency of the circuit increases so does capacitive coupling; circuit isolation is now compromised. If enough parasitic capacitance is present, higher frequency RF energy (fast transients, ESD, lighting, etc.) may pass through the transformer and cause an upset in the circuits on the other side of the isolation gap that received this transient event.

Having examined the hidden behavior characteristics of components, we will explore why these hidden features create EMI within a printed circuit board in future Chapters of this book.

REFERENCES

1. Montrose, M. I. 1999. *EMC and the Printed Circuit Board-Design, Theory and Layout Made Simple*, Hoboken, NJ: Wiley/IEEE Press.
2. Montrose, M. I. 2000. 2nd ed. *Printed Circuit Board Design Techniques for EMC Compliance-A Handbook for Designers*, Hoboken, NJ: Wiley/IEEE Press.
3. Ott, H. 1988. *Noise Reduction Techniques in Electronic Systems*. 2nd ed. New York: John Wiley & Sons.
4. Ott, H. 2009. *Electromagnetic Compatibility Engineering*. New York: John Wiley & Sons.
5. Paul, C. R. 2006. *Introduction to Electromagnetic Compatibility*, 2nd ed. New York: John Wiley & Sons.
6. Paul, C. R. 1989. "A Comparison of the Contributions of Common-Mode and Differential-Mode Currents in Radiated Emissions." *IEEE Transactions on Electromagnetic Compatibility*. Vol. 31(2): pp. 189–193.
7. Gerke, D., and W. Kimmel. 1994. (January 20). "*The Designers Guide to Electromagnetic Compatibility*." EDN.
8. Kraus, J. 1984. *Electromagnetic*, 3rd ed. New York: McGraw Hill.

2 Transmission Line Theory Made Simple

2.1 The Definition of Signal Integrity

Before discussing signal integrity as it applies to the field of electromagnetic compatibility and transmission line theory, a definition needs to be provided. Signal integrity can be described as:

The ability of electrical signals to travel from a source to load without loss of signal amplitude or parametric values

The discipline of signal integrity is to ensure quality of an electrical signal to transfer intended functions between locations. In digital applications, data consists of a stream of binary values represented by a voltage (or current) value. However, digital signals between the "one's and zero's" state (rising and falling edge) presents the Fourier spectra of an analog waveform that creates an electromagnetic field propagating down a transmission line within the dielectric that surrounds the propagation path.

Why is the emphasis on transmission line theory and signal integrity presented in an EMC book? Signals are sent between locations. If there is a design deficiency or problem with routing such as not being able to maintain constant impedance throughout the transmission line path, undesired common-mode RF current may be developed, which then in turn may create an EMI event. Without transmission lines (signal or power transfer), there would be no field of electrical engineering. Thus, one of the most important aspects of electrical design is structuring a transmission line that does not develop undesired common-mode current and hence, undesired field structures. This line of work is known as optimizing signal integrity. Also, with poor or compromised signal integrity, systems may not function properly resulting in non-desired products and services.

Over short path distances and at low bit rates (lower frequency and longer rise and fall times), a transmission line propagates a signal without difficulty. At higher data rates (with faster rise and fall times) and over longer path distances or through different mediums, various effects can degrade the signal to the point where less-than-ideal conditions (errors and signal quality) occur, potentially causing the system or digital (signal) operations to malfunction.

Signal integrity engineering is the process of analyzing and mitigating effects that prevent optimal signal transfer. Signal integrity engineering involves all levels of electronic packaging and assembly, from internal connections of an integrated circuit (IC), the package, the printed circuit board (printed circuit board) design and layout, backplane design and inter-system connections. Signal integrity also includes other aspects such as thermal consideration and material science that supports electromagnetic field propagation. Signal integrity issues exists in cable assemblies as well as printed circuit board interconnects and circuit assemblies, essentially all transmission lines regardless of media used (metallic interconnects such as wire, a printed circuit board trace or air).

While there are common elements in a transmission line, there are also practical considerations during design and layout, in particular the interconnect flight time (signal propagation time) versus the bit period that cause substantial differences in the approach toward achieving optimal signal integrity.

Many items may cause a signal integrity problem to occur.

It only takes *"one"* item listed below to cause a signal integrity or EMI problem:

- Incorrect (inefficient) transmission line routing
- Improper terminations
- Power and/or return plane bounce
- Improper RF (imaging) for the return current path
- Mode conversion (differential-mode signals creating common-mode current)
- Rise and/or fall time degradation
- Transmission lines losses at higher frequencies due to board construction material chosen (core and prepreg)
- Inadequate power distribution network charge capability
- Hidden parasitics (*RLC*) [Discussed in Chapter 1]
- Propagation delay during differential signaling, capacitive loading or serpentine routing
- Skin depth and dielectric losses
- Non-monotonic edges
- Excessive inductance in the transmission line, both source and return
- Excessive ringing and reflection
- Overshoot and undershoot
- Impedance discontinuities from changing trace widths or propagation through a non-impedance controlled via
- Delta I noise
- Simultaneous switching noise induced into the power distribution network
- RC delay
- Crosstalk
- Reflective stubs (including vias) and their lengths
- Excessive capacitive loaded transmission lines
- IR drops
- Gaps in planes
- Dispersion
- Intra line skew
- Inappropriate printed circuit board material, considering applied signal bandwidths.

Not all of the items above are examined in this Chapter due to the complexity of the field of signal integrity and the target audience for this book. There are however many more elements of concern for high-technology design related toward achieving signal integrity and EMC.

The primary difference between signal integrity and EMC is simple, yet may appear to be difficult for many who consider themselves strictly an *analog* or *digital* designer, and not an *electrical* engineer. When we work essentially in the time domain using DC voltage levels such as logic high or low, or looking at circuit diagrams in a schematic, we generally concern ourselves with signal integrity, not EMC. Computational analysis using programs such as SPICE utilizes mainly RLGC parameters which are a time domain aspect of signal propagation. Units of measurement for those working in the field of signal integrity are volts and amps, or watts.

The field of EMC deals with electromagnetic field propagation with units of volts/meter and amperes/meter, or *watts/meter*^{2}. Maxwell equations, the heart of electrical engineering, discusses field propagation in the frequency domain which is what all signal propagation really is. In the frequency domain when analyzing a time variant (AC signal), and converting it to the time domain, we can thus redefine the digital profile or definition of a digital transition as an "*infinitely fast AC slew rate signal.*"

To learn details on solving signal integ*rity problems not presented herein including computational analysis, References [1, 2] is highly recommended. In this Chapter, we examine fundamentals of transmission line theory that if followed, should optimize signal integrity and address most of the items as a byproduct from proper design engineering. It all comes down to understanding the need to implement optimal signal transfer between a source and load within a transmission line structure.

2.2 Primary Concerns Related to High-Speed Signal Integrity Problems

There are four primary aspects of high-speed signal propagation that can cause a signal integrity concern along with development of unwanted common-mode current, otherwise known as EMI.

1 **Signal quality:** Reflections or distortions from impedance discontinuities in either the signal or return path can affect the quality of a propagated signal. A transmitted signal must travel through a constant impedance transmission line that includes vias and interconnect discontinuities. Should there be any discontinuity in the transmission line, or change in impedance, signal integrity may not be ensured and common-mode RF current will result. If the magnitude of common-mode current generation is significant then the potential for having an EMI event may occur.

2 **Crosstalk between nets**: Crosstalk is developed through both mutual capacitance and inductance between two transmission lines. The spacing of traces must be maintained at a specific distance separation value (broadside coupling) to minimize mutual inductance and capacitance while keeping the RF return (imaging) path impedance as low as possible (distance to an adjacent plane in the z-axis). A crosstalk event couples undesired energy from a source (transmission line) into a victim (transmission line) which when phase added/subtracted, can cause common-mode RF current to be created. This in turn creates a possible undesired mode of operation. Crosstalk is observed in two forms; near-end where the disrupted electromagnetic field returns to the source, and far-end (load) when the signal changes the voltage input level of a component at the physical end of the transmission line.

Crosstalk also occurs through conductive paths (common impedance coupling). Grounding and crosstalk effects are discussed in Chapter 5. We sometimes split analog ground planes from digital for this reason; to minimize common-impedance coupling.

3 **Rail collapse:** Rail collapse refers to a voltage or return (ground) drop within the power distribution network when digital components switch logic states, consuming (source or sinking) more current than can be uniformly delivered. One must minimize the impedance of the power and return path along with *Delta-I* (current) to ensure there is sufficient charge to keep all voltage margin levels intact. A bouncing power distribution network will develop EMI.

4 **EMI:** Development of undesired common-mode currents can be the cause of a signal integrity problem, including the performance of the power and return plane impedance when viewed as a "power distribution" transmission line. It is the designer's responsibility to consider appropriate bandwidth, ground impedance and common-mode coupling in addition to keeping all transmission lines balanced for both source and RF return paths.

In order to solve a signal integrity problem one must first understand transmission lines and how they function both in theory and reality. With today's high-technology products and faster logic devices, printed circuit board trace effects become a limiting factor for proper circuit operation. A trace routed adjacent to a reference plane forms a simple transmission line, detailed in the next Section of this Chapter.

Consider the case of a multilayer printed circuit board. When a trace or transmission line is routed on an outer layer (top or bottom), a microstrip topology is established. When a trace or transmission line is routed on an internal layer sandwiched between two planes, this configuration is called stripline topology (single or dual layers, symmetrical or asymmetrical).

A transmission line is a system of conductors such as wires, waveguides, coaxial cables or printed circuit board traces suitable for conducting electromagnetic power or signals efficiently between two locations.

To meet the challenges of high-speed digital processing, today's multilayer printed circuit board typically must:

- Reduce propagation delay between devices
- Manage transmission line reflections and minimize crosstalk (signal integrity)
- Reduce signal losses (decrease in amplitude of the electromagnetic field within the transmission line)
- Allow for higher density interconnections.

A transmission line allows a signal to propagate from one device to another at approximately the speed of light within a medium, as modified (slowed down) by the capacitance of the transmission line and by active devices in the circuit. The parametric value within printed circuit board construction material that causes a propagating field to slow down is called dielectric constant (DK or ε_r). This propagating electromagnetic field contains some form of energy. A question to consider is "Does this energy consist of electrons, line voltages and currents, or something else?"

In a transmission line, electrons do not travel in the conventional sense. An *electromagnetic field* is the component present within and around every transmission line, not voltage, current or electrons. This electromagnetic energy physically exists in the dielectric that surrounds the transmission line such as core or prepreg, the insulation of a wire or even free space.

We use simple units of measurement to describe how transmission lines sends intelligence between devices. The units of measurement commonly spoken are voltage and current, nothing else. In reality, *Voltage* is a unit of measurement whose spatial derivative describes the electrostatic force exerted on all electrons present. *Current* is a unit of measurement that describes how many electrons flow in a transmission line during a specific time period. Neither unit correctly describes the *electromagnetic field* or *electromagnetic wave* present. However, using metric units of measurement for voltage and current when associated with transmission line theory is easy to visualize by engineers when working in the time domain (simple use of Ohms law).

In reality, the real element of energy present in a transmission line is a complex frequency domain issue described by Maxwell's equations (Chapter 1). It is easily noticed in manufacturer's data sheets describing digital components, in their parametric table section, that they do not use volts/meter and amps/meter when stating output drive voltage or line current. They use volts and [milli]amps which are *incorrect units of measurement!* Electrical engineering is based on Maxwell's equations (frequency domain) yet we prefer to work with Ohms law (time domain). One must be well versed in both domains, yet most work in the time domain since it is easier than the frequency domain.

An electromagnetic field common to the environment consists of the following "partial list":

- AM/FM radio waves
- Television waves
- Light waves
- Cellular telephone/pager waves
- Microwave and radar transmissions
- EMI/RFI created as a byproduct (unwanted energy) of digital components

All waves travel at or near the speed of light in free space. EMI/RFI is included in this list to show that electromagnetic energy is a waveform spectra that may cause harmful interference to other electronic equipment as well as being susceptible to external electromagnetic field disruption.

If a transmission line is not properly terminated, circuit functionality and EMI concerns may develop. Any concern may severely compromise switching operations and signal integrity. Transmission line effects must be considered when the round-trip propagation delay exceeds the switching-current transition time. Faster logic devices and their corresponding increase in edge rates are becoming more common in the sub-nanosecond range and in the future, femo-second range. A very long trace in a printed circuit board can become an antenna for radiating RF currents, or cause functionality problems if proper circuit design techniques is not implemented during the design period. An overview on design techniques are provided throughout in this book as well as References [3, 4].

2.3 Defining Transmission Line Structures

Every transmission line propagates an electromagnetic field (alternating current waveform) throughout the frequency spectrum using electromagnetic energy (magnetic and electric field propagation). This propagating field is commonly called RF (radio frequency). Any RF current propagating throughout the frequency spectrum does so as an electromagnetic wave, discovered by Maxwell. The majority of his life work and publications was based on wave propagation (optics and electromagnetics).

Transmission lines are used for sending intelligent signal communication between a source (transmitter) and load (receiver) through a media or dielectric. Radio frequency currents tend to reflect back to their source from discontinuities in the transmission line such as connectors and vias. Radar operates by having a signal reflection return back to the source from a conductive (reflective to radio frequency) object at a distance away. If any signal is reflected back to the source, the total power intensity of the signal is diminished.

Within a digital system, the magnitude of signal loss due to reflections from a high-impedance load is the magnitude of common-mode RF current created, and which will find a parasitic means or antenna structure to propagate from.

Transmission lines use specialized construction that includes dimensions and spacing. Transmission lines also must have an appropriate characteristic impedance to carry electromagnetic signals with minimal reflections and power loss.

2.4 Types of Transmission Lines

A transmission line requires a pair of conductors. These two conductors are the path from source to load and then a return path from load to source. This configuration is known a *uniform* transmission line. However, any pair of conductors supporting transverse electromagnetic field (TEM) propagation is actually a transmission line although there may not be an obvious return path. By definition there is no "single wire" transmission line, unless a ground plane in a printed circuit board is involved, the return path is some distance away or parasitic in nature. A single conductor cannot support wave propagation per Ampere's law.

There are several types of transmission lines, some of which in this section appear to be a single conductor when in fact, there must be a second conductor somewhere such as earth ground for free space propagation. These include coaxial cable, dielectric slabs, microstrip, embedded microstrip, stripline, optical fiber and waveguides, among many other types not listed below.

2.4.1 Coaxial cable

A coaxial cable confines the substantial majority of a propagating signal (electromagnetic wave) to the area inside a concentric shield. In RF applications up to a few GHz, the electromagnetic wave propagates in the transverse electric and magnetic mode (TEM) only, which means that both electric and magnetic fields are perpendicular to each other in the direction of propagation. The electric field is radial and magnetic field circumferential. At frequencies at which the wavelength is significantly shorter than the circumference of the cable, transverse electric (TE) and transverse magnetic I waveguide modes can also propagate. When more than one mode exists, bends and other irregularities in the cable routing or geometry can cause RF power to be transformed from one mode to another.

2.4.2 Microstrip

A microstrip transmission line is a conductor parallel to a single reference or image return path. Microstrip transmission lines are generally the top and bottom layers of a printed circuit board with air, soldermask or conformal coating on the outer side only. The width of the trace, thickness of the metal conductor, distance spacing to a reference plane as well as the dielectric constant of the insulating layer determines characteristic impedance. Microstrip is an open transmission line whereas the coaxial cable is a closed structure due to having a shield that covers the entire length of the transmission line path, yet both function essentially identical with regard to signal propagation. One is concentric the other flat yet both function nearly identical with regard to performance.

2.4.3 Embedded Microstrip

Embedded microstrip is similar to regular microstrip except capped by a thick dielectric material placed above the surface of the line that is equal to or greater than the thickness of the dielectric between the trace and the image return (ground) plane below it. Embedded microstrip provide a signal integrity advantage since the electric field above the line is captured into a media that is generally identical to the media below the line, thereby containing electric fields and circumventing impedance mismatches that air or solder-mask would create.

2.4.4 Stripline (Single and Dual)

A stripline topology uses a conductor that is located between two reference (image) planes regardless of potential. The insulating material of the substrate forms a dielectric between the plane and transmission line. The width of the trace, thickness of the metal conductor, distance spacing to both reference planes as well as the dielectric constant of the insulating layer determines characteristic impedance. Stripline is essentially a closed transmission line system and functions approximately similar to a coaxial cable, under most conditions of operation.

A visualization of microstrip and stripline topologies within a printed circuit board is shown in Fig. 2.1.

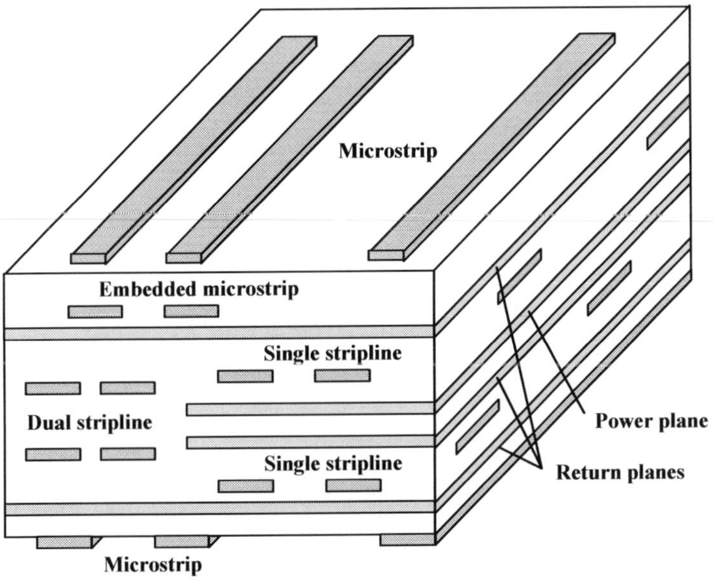

Figure 2.1 Illustration-microstrip and stripline topology in a printed circuit board.

2.4.5 Balanced lines

A balanced line is a transmission line that consists of two conductors of the same type with equal impedance to a reference source and other circuits within the network. There are many configurations of balanced (differential mode) lines. These include twisted pair, star quad and twin-lead, among many others.

2.4.5.1 Twisted pair

A twisted pair transmission line is commonly used for communications purposes when the need to send a signal is not based on a reference source or dedicated return. For most applications, multiple twisted wire pairs are grouped or bundled together in a single cable assembly. Twisted pair cable is commonly used for network communication (i.e., Ethernet, USB, etc.) and signals that travel between locations remote from each other such a digital display panels (enhanced signal integrity) or buildings hosting electrical products that are on a different AC mains distribution network (ground shift-Chapter 5) to prevent hazard of electric shock. This cable assembly may contain an overall shield (STP-shielded twisted pair) or unshielded (UTP-unshielded twisted pair).

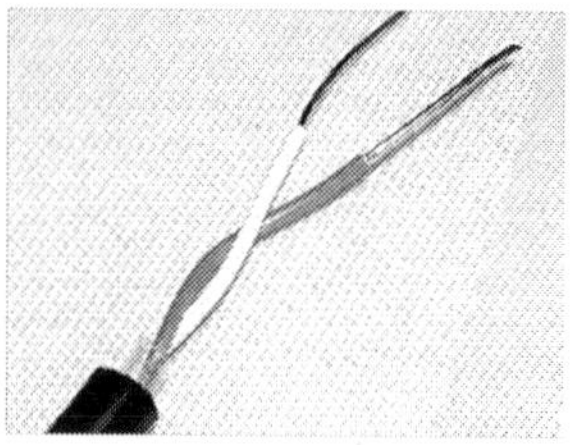

Figure 2.2 Balanced line in twisted pair configuration used with 2-wire circuits. Courtesy-Internet (public domain)

Figure 2.3 Balanced differential pair twisted uses with multi-wire circuits (Ethernet). Courtesy-Internet (public domain)

2.4.5.2 Star quad

Star quad is another balanced transmission line used for low frequency communication. Applications include 4-wire telephony and microphone circuits. Two pairs are provided by four conductors, which are twisted together around the cable's axis. Each pair uses non-adjacent conductors.

RF interference or noise that is observed by the cable arrives as a virtually perfect signal with no undesired common-mode signal. If common-mode energy is present, this is easily removed by coupling transformers or filtering. Because conductors are always the same distance from each other, crosstalk is minimized relative to cables that contain only two separate twisted pairs.

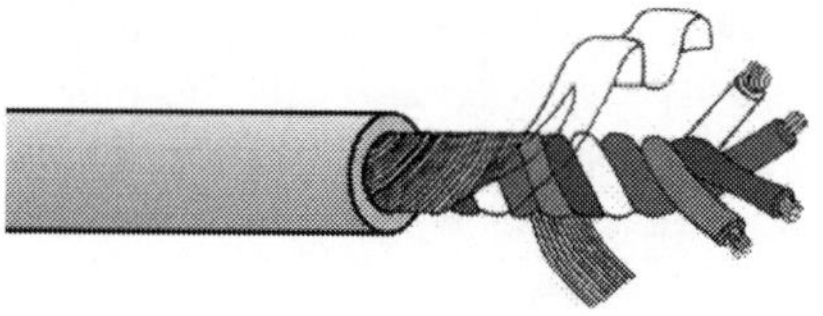

Figure 2.4 Star quad configuration of cable bundling for multi-conductor circuits. Courtesy-Internet (public domain)

2.4.5.3 Twin-lead

A twin-lead transmission line consists of a pair of conductors held apart by a continuous insulator, commonly found in 300 ohm lines used for roof top television antennas, obsolete in many locations within the world now due to cable and satellite connections from a service provider using digital technology instead of over-the-air analog transmissions.

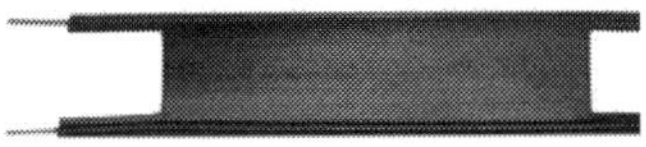

Figure 2.5 Balanced twin lead for use with RF circuits, particularly antennas. Courtesy-Internet (public domain)

2.4.6 Lecher lines

Lecher lines are a form of parallel conductor used at UHF (ultra-high frequencies) for creating resonant circuits and were used to measure the wavelength of radio waves, mainly at UHF and microwave frequencies in the early 1900's. Lecher lines fill the gap between lumped components used at High Frequency (HF)/Very High Frequency (VHF) and resonant cavities used at UHF/Super High Frequencies (SHF).

2.4.7 Single-wire

Single, unbalanced lines were formerly used for telegraph transmission many years ago with earth ground being the RF return path. Single-wire transmission lines are now used when many wires are bundled into the same cable assembly with only one return conductor provided, typical of many I/O interfaces such as RS-232 or equivalent configurations. There is no (or minimal) twisting involved to remove common-mode RF current. Single wire transmission lines work well for very low (i.e. audio) frequency circuits. All wires use a common RF path for the return current to travel through.

2.4.8 Waveguide

Waveguides are a rectangular or circular metallic tube in which an electromagnetic wave is propagated and is confined by the assembly using the process of skin effect travel. Waveguides are not capable of transmitting the TEM signal that is typically propagated in copper lines. They use various modes such as quasi-TEM where the signal propagating in the transmission line bounces between edges of the waveguide creating a mixed-mode means of signal communication. One thing to note about a unique characteristic of waveguides are that they are NOT a true transmission line, since there is no RF return path but are commonly considered to have one!

Waves in open space propagate in all direction in a spherical manner. When waves propagate like this they lose their power proportionally to the square of the distance (inverse square law); that is, at a distance R from the source. The power measured at the load is source power divided by R^2. Waveguides confines the wave to propagate in one dimension so that (under ideal conditions) the wave loses no power. Conductors used in waveguides have small skin depth and hence large surface resistance.

Waves are confined inside the tube due to total reflection from the waveguide wall. Thus the propagation inside the waveguide can be described approximately as a "zig-zag" between the walls. This description is exact for electromagnetic waves propagating in a hollow metal tube with a rectangular or circular cross-section.

Waveguides can be constructed to carry electromagnetic fields over a wide portion of the frequency spectrum, and are especially useful in the microwave and optical range. Depending on frequency of operation, waveguides can be constructed from either conductive or dielectric materials and are used for transferring both power and communication signals.

The most common waveguide is one that has a rectangular cross-section, not square. It is common for the long side of this cross-section to be twice as long as its short side.

These waveguides are useful for carrying electromagnetic waves that have either a horizontal or vertical polarization.

2.4.9 Optical fiber

Optical fiber is a solid transparent fiber of glass or polymer that carries a signal in the optics region of the frequency spectrum. Optical fiber is a form of a waveguide and is commonly used as the backbone of modern terrestrial communications networks due to their reasonable cost, low loss and high signal bandwidth (high data rate).

2.5 Description of a Typical Transmission Line System

Transmission lines are used when a propagating signal frequency is high enough that the wavelength of the RF signal begins to approach the physical length of the transmission path used to carry the signal. To propagate electromagnetic energy at frequencies above the standard communication range (up to 26 GHz), millimeter waves, infrared and light, the wavelength of these signals become much smaller than the dimensions of the structures used to guide them. Under this situation, standard transmission line routing techniques become inefficient. Fiber optic cable must now be used.

To describe a transmission line, this is modeled as a two-port network (also called a quadrupole network), shown in Fig. 2.6.

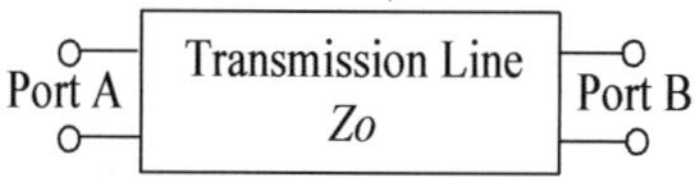

Figure 2.6 Basic description of a transmission line.

In the simplest configuration the network is assumed to be linear. This means the complex voltage across either port is proportional to the complex current flowing, especially if there are no reflected waves. The two ports are assumed to be interchangeable. If the transmission line is uniform, its behavior is described by a parametric value known as characteristic impedance (Z_0). The characteristic impedance of a transmission line is the ratio of the complex voltage of a given wave to the complex current of the same wave at any point in the line. Typical values of Z_0 are 50 or 75 ohms for coaxial cable. We commonly use 100 ohms for a twisted pair of wires (network communication networks) and 300 ohms for a type of untwisted pair used in radio transmission or outdoor television antennas (twin-lead).

When sending an electromagnetic field though a transmission line, it is required that full power be consumed by the load with as little energy as possible reflected back to the source. This is facilitated by having the load impedance equal to the transmission line impedance, Z_0. Under this condition, the transmission line is said to be matched or balanced.

It is also desired that all power provided to a transmission line is not lost in the medium of travel. One cause of loss in a transmission line is resistance, commonly called *ohmic* or *resistive loss*. At very high frequencies, typically >1 GHz, a phenomenon called "skin effect" will increase the DC value of the resistance by a multiple factor. At higher frequencies *dielectric loss* becomes significant, adding to the losses caused by the existing resistance. Dielectric loss occurs when the insulating material between the source and return path of the transmission line absorbs electromagnetic energy from the alternating electric field, converting it to heat.

Transmission lines are modeled with basic electrical elements, namely resistance I and inductance (*L*) in series with capacitance I and conductance (*G*) in parallel. Both resistance and conductance contribute to loss in a transmission line.

The total loss of power in a transmission line is often specified in decibels per meter (dB/m), and usually depends on the frequency of the signal. Manufacturers often supply a chart showing the loss in dB/m throughout the frequency spectrum that the cable is designed to operate within. A loss of 3 dB corresponds to 50% energy (power) loss.

There are several basic configurations that define transmission lines. These are shown in Fig. 2.7. The top figure is a two-wire system, the middle is balanced pair and the bottom is coaxial based.

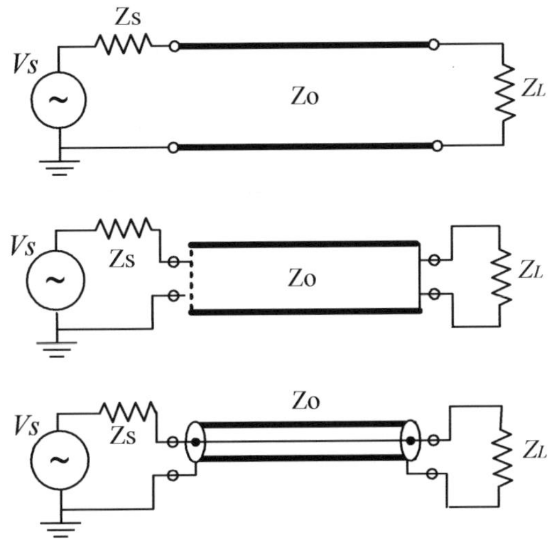

Figure 2.7 Variations on schematic representations for transmission lines.

2.6 Transmission Line Structures in a Printed Circuit Board (Lossy and Lossless)

This section presents different types of transmission lines found within a printed circuit board and how they relate to signal integrity (ensuring circuits work as desired), along with the creation of undesired common-mode EMI.

2.6.1 Lossless Transmission Line

Figure 2.8 shows conceptually how a typical *lossless* transmission line appears within a printed circuit board. Resistance, *R*, is omitted for simplicity. One will find this configuration in textbooks describing basic transmission line theory at lower frequencies, generally below 100 kHz. This transmission line configuration does not exist in printed circuit boards with frequencies above 100 kHz, detailed in the next section.

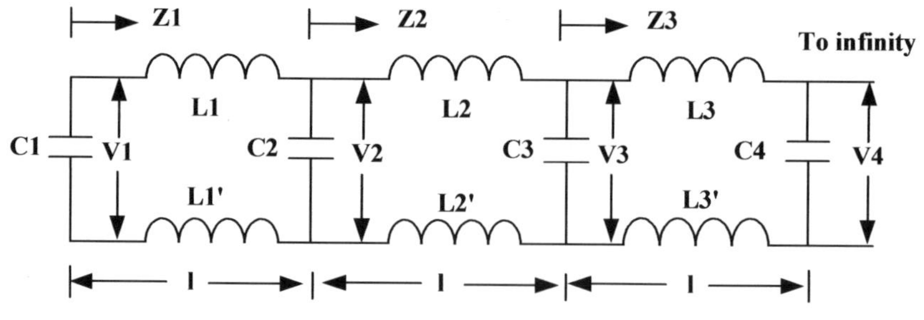

Figure 2.8 Lossless transmission line equivalent circuit.

Examining Fig. 2.8, we note that this configuration describes the basic concept of transmission line theory as it applies to the field of electrical engineering. Without this configuration signal propagation could not exist, including Maxwell's equations. The important element in this figure is the representation of capacitance *C1-C4*.

Capacitor *C1* represents power or an energy source. This could be a battery or AC mains voltage. Without a source of power nothing electrical would work or exist. Capacitor *C2* represents the input terminal of an electrical device. Elements *L1* and *L1'* represent the physical interconnect between energy source and input terminal. This interconnect could be a wire, battery terminals, a soldered connection between energy source and the printed circuit board, or any other means required for application of use.

Capacitor *C3* is where all work is performed. The purpose of having electrical products is to provide a useful function or intrinsic value to a user. This could be the motor driving a shaft, a central processing unit (computing devices, appliances, machine tools), transducers, anything! Between the input connector section and work area (*C3*) is another transmission line pair represented by *L2* and *L2'*.

Capacitor *C4* is the load or output of the work performed by *C3*. What is the purpose of having electrical products if they do not provide value or perform a function? This output could be a display screen, light bulb, heater element or antenna, for example. Again, like previous sections the interconnect between the work area and output is *L3* and *L3'*.

In theory, this structure works without degradation since resistance is generally very low and insufficient to cause a voltage drop or loss. This is why a lossless transmission system works best at lower frequencies.

The impedance of a transmission line is provided by (2.1). Depending on how one visualizes any transmission line, we describe it using inductance (L_o) and capacitance (C_o) (frequency domain aspects of Maxwell's equations). Digital design engineers prefer to use volts (*V*) and amps (*A*) as the metric unit of measurement (time domain or Ohms law).

$$Zo = \sqrt{\frac{Lo}{Co}} = \frac{V(x)}{I(x)} \tag{2.1}$$

In (2.1), there is subscript *(x)* for both voltage and current, representing the fact that time-variant propagation of a signal is present.

The ratio of line voltage to line current is constant with respect to length, regardless of source driver or load impedance. Equation (2.1) illustrates impedance in both the frequency domain $\left(Zo = \sqrt{Lo/Co}\right)$ and time domain $\left(Zo = V(x)/I(x)\right)$. The *(x)* subscript indicates that variations in *V* and *I* will exist along the line with respect to time (an AC sine wave), except for special cases. As a digital signal transitions from a logic low to a logic high state, or vice-versa, the impedance it encounters (voltage to current ratio) is equal to the characteristic impedance at that finite point within the transmission line and at a specific period in time.

2.6.2 Lossy Transmission Line

For most design configurations the lossless transmission line model does not exist, except at lower frequencies (<< 1 MHz). Most printed circuit boards operate at higher frequencies, generally above 1 MHz. Because of both inductance and capacitance some form of energy loss will occur. The magnitude of this loss is the magnitude of undesired common-mode current developed. The lossy transmission line model is shown in Fig. 2.9.

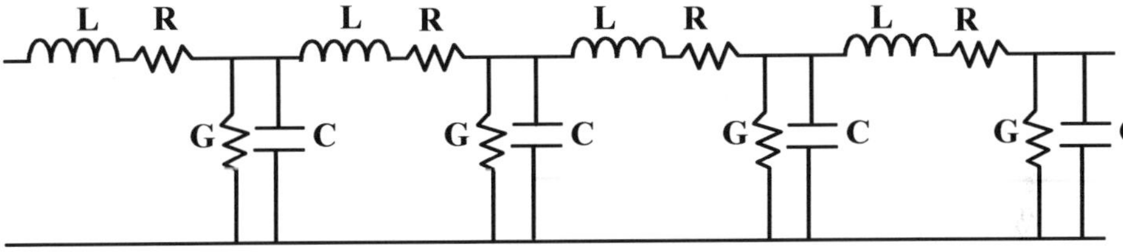

Figure 2.9 Lossy transmission line equivalent circuit.

Notice that there is significant difference between the lossy and lossless configuration, namely in the return path and the addition of conductance, *G*. With a multilayer printed circuit board, the RF return path is generally a very large metal plate or plane (compared to the signal trace size), usually copper. The impedance of this large plane is extremely small while the signal trace using a narrow width has higher impedance generally by several orders of magnitudes. This can potentially set up an unbalanced transmission line system. As presented in Chapter 1, any imbalance in a transmission line carrying differential-mode signals creates common-mode RF current. Visualizing the difference between a theoretically perfect transmission line (lossless) that is balanced, and a lossy line with an excessive amount of inductance (imbalance) in one path, helps us understand how these configurations appear to a propagating electromagnetic wave (actually the signal is within the dielectric surrounding the transmission line) and its effect on both signal integrity and development of undesired common-mode current.

Every transmission line has fixed impedance which consist of inductance (*L*) and resistance I that is in the ohms range. Planes have impedance that can be in the milliohm or nano-ohm range. At higher frequencies, inductive reactance can approach hundreds of ohms ($X_L=2\pi fL$).

Resistance prevents the flow of electric current through an element or device. The inverse of resistance is conductance, or the ease at which an electric current propagates down the transmission line. The SI unit of electrical resistance is the ohm (Ω) [$R=V/I$], while electrical conductance is measured in Siemens (S) [G=I/V]. Conductance and admittance are the reciprocals of resistance and impedance respectively, hence one Siemens is equal to the reciprocal of one ohm, and is also referred to as the *mho*.

With a difference in transmission line impedance between source and return path, it is easy to understand why lossy configurations increasingly exist in proportion to higher frequencies and why engineers need to remember that there will always be losses in every transmission line structure. The key to understanding this loss deals with how one designs a system. What may work for signal integrity in the time domain, and a "do not care" for digital designers is because computational analysis indicates that the system will be functional, however the circuit may not work in the frequency domain because common-mode RF energy is developed due to an imbalance that is oblivious to digital components, and which will somehow find a way to cause harmful interference. This is why one must evaluate all designs in both time and frequency domains in order to be a solid electrical engineer, not just a digital or analog designer working in just one domain.

There are several equations that describe lossy transmission lines (2.2). The impedance of a lossy transmission line is given by a complex equation that includes resistance, inductance, capacitance and conductance.

$$V(\omega, x) = Vo \exp(-\Gamma x) \exp(jt)$$
$$\Gamma = \alpha + j\beta = \sqrt{(R_L + j\omega L_L) + (G_L + j\omega C_L)}$$
$$Zo = \sqrt{\frac{(R_L + j\omega L_L)}{(G_L + j\omega C_L)}}$$
$$Zo = \textit{characteristic impedance}$$
$$L = \textit{line length}$$
$$R_L, G_L \quad \textit{may vary with frequency} \qquad (2.2)$$

2.7 Transmission Line Effects on Signal Propagation

Transmission lines can easily cause functional problems in addition to creating unwanted common-mode RF energy in both the time and frequency domains. Problems occur, as an example, when a signal on a printed circuit board encounters an impedance discontinuity or a change in the geometry of the transmission line path. We can easily analyze transmission line effects for high-speed, high-edge rate signals with computational analysis. When the output driver of a digital device switches logic states between 0 and 1 (or logic level low to logic level high), the voltage to current ratio must equal the characteristic impedance *Zo* of the transmission line as provided by (2.1 and 2.2), or a signal integrity problem will develop for that finite time period of discontinuity.

You cannot analyze transmission lines using Ohm's law except in very short segments. Kirchhoff and Ohm's law fail in large structures due to electromagnetic fields that propagate in free space and couples to parasitic return paths that Ohm's law does not consider. Ohm's law applies to transmission lines containing impedance which contin both resistance and reactance. DC analysis considers mainly resistance (steady state operation) while AC analysis use both resistance and reactance with a frequency dependent component ($2\pi f$). Any time variant signal such as a digital pulse with both rising and falling edges with respect to time is an AC signal. At higher frequencies it becomes impossible to use Ohms's law due to inductive reactance, therefore Maxwell's equations takes over requiring use of field solvers when performing computational analysis.

The focus of this book takes frequency domain signal propagation in a transmission line (electromagnetic field) and converts this conceptually to Ohm's law. If we understand signal propagation in one domain, functionality in the other is ensured especially when it comes to both signal integrity and EMC compliance.

If the impedance or any other characteristics of the transmission line never changes, the signal propagates without any alteration in parametric values that include voltage and current levels, timing, reflections, and numerous other items of concern that were listed in Section 2.1.

The goal of creating a system is to ensure high-quality signal integrity as well as reliable high-speed data transmission. In a digital system, a signal is transmitted from a driver to receiver in the form of logic 1 or 0, which is actually at specific reference voltage level. At the input a receiving device, the positive voltage potential to create logic state "1" (V_{ih}), is above a reference value such as 0V (a.k.a. ground). A lower voltage level above a reference to 0V, V_{il} is considered logic low ("0"). Figure 2.10 illustrates an ideal voltage waveform on the left side in a perfect digital world, whereas the illustration on the right in Fig. 2.10 shows how a signal appears in a real transmission line with losses due to discontinuities within the propagating path, or lack of proper termination for the signal.

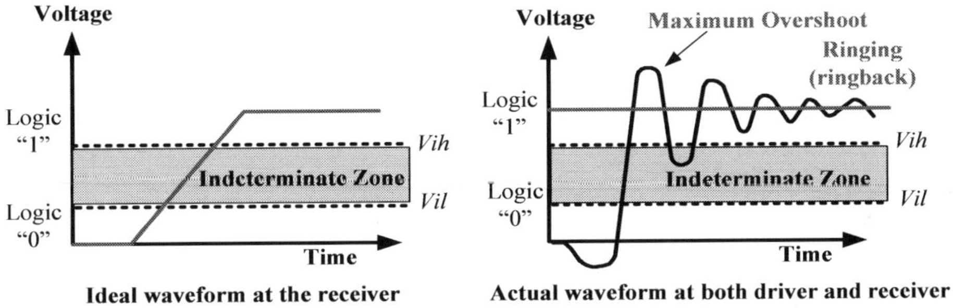

Figure 2.10 Reflection effect in a transmission line causing overshoot and ringing.

Complex data patterns are generally composed of a string of bits, 1 and 0s, and appear to be a continuous voltage waveform. The receiving device needs to sample the waveform in order to obtain binary encoded information. The data sampling process is usually triggered by the rising or falling edge of a clock signal as shown in Fig. 2.10. It is clear from the diagram that data must arrive at the receiving gate on time and settle to a non-ambiguous logic state to ensure the receiving component latches in the data stream. Any delay of the data due to propagation time or distortion of the data waveform from losses or impedance discontinuities will result in a failure of the data transmission image. If the signal waveform in Fig. 2.10 exhibits excessive ringing into the logic gray zone (indeterminate state) while sampling occurs, then either logic level cannot be reliably detected and the device may be destroyed by not knowing if the transistors internal to the device are either on or off; consuming infinite amount of current over a finite period of time that creates thermal runaway and destruction.

When dealing with transmission line effects, impedance becomes an important factor in designing for optimal performance. A signal that travels down a printed circuit board transmission line will be absorbed at the far end if, and only if, the transmission line is terminated in its characteristic impedance. If proper termination is *not* provided, most or a portion of the transmitted signal will be reflected back in the opposite direction of desired travel. If there is improper or lack of termination, reflections will occur resulting in a longer signal-settling time because of multiple overshoots and undershoots as the signal bounces between source and load, with total amplitude of the signal decreasing with each reflection.

Reflections occur every time there is an high impedance discontinuity in the transmission line path, but cannot occur when the impedance discontinuity is lower than the characteristic impedance of that transmission line. In high-speed systems, reflection noise can increase time delay and produces overshoot, undershoot and ringing. The root cause of reflection is impedance discontinuity along the signal transmission path. When a signal changes routing layers, and the impedance of the vias are not matched with the impedance of the transmission line (due to manufacturing variations, design considerations, etc.), a reflected signal will occur at the discontinuity boundary.

When a transmission line is routed over a solid plane with perforations at different locations (degassing holes, via holes, etc.), crossing a gap or split plane, having branches (T-stubs) or passing the proximity of another transmission line, an impedance discontinuity may occur and a reflection established. When the signal finally reaches the receiving end, if the load is not matched to the transmission line characteristic impedance a reflection again may occur depending on the load impedance, providing it is higher than the characteristic impedance of the transmission line path. To minimize any potential reflected event a common practice includes controlling trace characteristic impedance (through trace geometry and dielectric constant), eliminating stubs, choosing appropriate termination (series, parallel, RC, Thevenin, etc.), and always using a solid metal plane as a reference for RF return current physically adjacent to the source path.

The signal bouncing back and forth between source and load repeats until all energy in the transmission line structure is absorbed or dissipated. Absorption is from dielectric loss within the board material (called loss tangent-δ). This bouncing signal between devices, when viewed on an oscilloscope, is call ringing with each sine wave beginning at 0 degrees to 359 degrees of the analog waveform.

Ringback is a signal continually crossing a desired voltage threshold (both above and below) before settling to a steady state value such as 5/3.3V or 0V (a.k.a. ground). Depending upon the magnitude and duration of the re-crossing, settling time may need to be calculated from the final crossing, both for logic high and low. For a signal carrying a periodic signal such as clock, ringback is typically allowed as long as the signal settles to a steady-state value prior to satisfying setup timing requirements of the receiver.

The voltage that is reflected at the *end* (load side) of the transmission line can be greater in magnitude than the initial voltage level if termination impedance of the receiver is significantly *greater* than the characteristic line impedance. Conversely, the reflected voltage level will be *less* than the initial transmitted voltage level if termination or load impedances at the receiver is *less* than the characteristic line impedance. The amplitude of the reflected voltage at the end of the transmission line is easily calculated by (2.3).

$$V_r = V_i\left(\frac{R_t - Z_o}{R_t + Z_o}\right) = \rho V_i \qquad (2.3)$$

where:
V_r = reflected voltage at the far end
V_i = input voltage source level into the transmission line
R_t = load or termination resistance
Z_o = characteristic impedance of the transmission path, and
ρ = reflection coefficient.

If the load impedance is not matched to the characteristic impedance of the transmission line, a voltage waveform is reflected back toward the source. This occurs if no termination is provided such as series, Thevenin, RC or parallel. The value of this reflected voltage (V_r), and the percentage of the signal reflected back towards the source (%) is calculated by (2.4):

$$\%\,reflection = \left(\frac{Z_L - Z_o}{Z_L + Z_o}\right) \times 100 \qquad (2.4)$$

Equation (2.4) is the percentage of the transmission line voltage reflected back to the source by the impedance mismatch at the load. If $R = Zo$, reflection coefficient $\rho = 0$ and there is no reflected signal bouncing between source and load. The voltage level is constant.

If $R_t = \infty$, then $\rho = +1$. This means that 100% of the line voltage level is reflected back to the source. This reflected voltage signal may double in value since the observed measured voltage is the sum of the initial source voltage added to the reflected voltage level depending if phasing of the signals matches perfectly. For example, if the source driver sends +3.3V to a load with a high impedance that is several magnitudes greater than the characteristic impedance of the transmission line, 100% or +3.3V will be reflected back to the source. Since signal propagation of an electromagnetic field that exists in both directions within the dielectric of the board material, and depending on phasing of the outgoing and reflected wave, there may be a physical location somewhere in the transmission line route where the two waves will be coincident and add in phase, or subtract. This means that is it possible to have a voltage level of +4.7V maximum. If a digital component happens to be physically located somewhere on the net which is not at either the source or load location, this high voltage level may will exceed operational margins of that device and could cause the component to become non-functional or destroyed by high voltage levels. This occurs for both source and sink currents. Having proper termination in the transmission line to prevent a reflected wave is thus critical for optimal product design.

If $R_t = 0$, a short circuit exists and $\rho = -1$. This means there can never be a reflected signal. *The greater the mismatch in impedance between the transmission line and load, a proportional amount of reflected voltage will be established.* If both ends of the transmission line have mismatches, ringing will be created by the reflected wave bouncing between driver and receiver decreasing in amplitude with each reflection until absorbed within the board material until steady state is achieved. These bouncing waves can cause both signal integrity problems and development of undesired common-mode EMI.

An example of various waveforms due to impedance mismatches in a transmission line network is shown in Fig. 2.11. The item to note is that the words "High" and "Low" are several magnitudes in difference, such as $Z_0 = 20$ ohms (output impedance) and Z_{High} is >100 Kohms. If the low impedance nomenclature is significantly less than the transmission line characteristic impedance, a capacitive effect is observed by the propagating electromagnetic field. If there is too much capacitance in the transmission path or within the input capacitance of multiple loads on the net, the signal when viewed in the time domain may have its edge transition rounded or slowed down which could violate timing margins.

A circuit can be considered as a sequential collection of lumped elements with capacitive and inductive components. This condition occurs when a signal path segment is small compared to the wavelength of a signal's highest frequency spectral component propagating down the trace. As signal frequency increases, the circuit must be treated as a distributed transmission line. For this situation, controlled impedance, matched termination and radiated emission effects must be considered.

Depending on the input impedance of the of load, trace reflections if any may diminish or become a nonissue related to signal integrity when the quiescent state is achieved, yet EMI may still occur due to the development of common-mode RF currents due to an imbalance in the transmission line loop circuit. Once the signal achieves quiescent (DC) level, the transmission line now behaves as a typical wire operating at lower frequencies.

The velocity of propagation (t_{pd}) of a signal, or how fast it propagates between elements, is described by (2.5) and is dependent on "per unit length of both inductance and capacitance."

$$t_{pd} = \sqrt{L(per\ unit\ length)\ C(per\ unit\ length)} \qquad (2.5)$$

Source Z	Load Z	EMI Results	Waveform at load
Z_0	Z_0	None	
Z_0	High	Trace-trace coupling	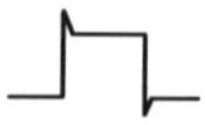
Z_0	Low	Edge rate changes	
Low	High	Trace coupling, EMI and crosstalk	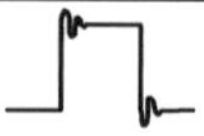

Figure 2.11 Transmission line effects based on source and load impedance ratio.

To illustrate transmission line effects as they relate to velocity of propagation, assume a printed circuit board transmission line has a propagation delay of 150 ps/inch (381 ps/cm) one way from source to load. Round-trip delay (including the image reflection time) is thus 300 ps per inch (762 ps round trip). If a clock driver has an edge rate of 2 ns, the transmission line characteristic of a short trace (t_{pd} is short compared to the rise time) is not a concern since the signal will reflect back to the source (762 ps total flight time), long before the next edge triggered event occurs (2 ns). If the transmission line is properly terminated, the transmitted signal will have all possible reflections absorbed at the load and dissipated within the network. Thus, a clean signal is available for optimal signal integrity. If the clock trace is however 10 inches (25.4 cm) long, a serious problem could occur (the round-trip length of the trace is now 20 inches [50.8 cm]). Total signal propagation time using the number above means the signal will return to the source after a period of 3 ns. Under this situation, both outgoing and reflected wave exist at the same time, which could cause a significant signal integrity problem based on (2.5).

When a clock or strobe line drives multiple integrated circuits using a single trace, additional distributed capacitance and inductive elements are encountered because of the additional components on the net. Each component has several pF of input shunt capacitance. This input capacitance increases total capacitance of the transmission line, thus increasing propagation delay of the signal. The increase in propagation delay occurs because total propagation delay is proportional to the square root of the capacitance per unit length (2.5). With a 2-ns and faster edge rate signal, transmission line effects become important for lead lengths of no more than a few inches or centimeters.

2.7.1 Conditions That Create Ringing

Figure 2.12 illustrates a typical circuit using lumped components. A source driver with internal series resistance, *Rs*, is shown along with inductance of the trace, *L*, (includes component lead wires) and distributed capacitance from trace to ground, *Xc*. This X_c elements is in addition to the input capacitance of the receiver.

Assume there is significant capacitive reactance in the trace plus input load capacitance [$Xc = 1/(2\pi fC)$] which is, for example, much less than load resistance (R_L) at higher frequencies. When a routed trace is physically short in length, inductance within the semiconductor package and decoupling capacitor lead-length interconnect loop inductance become the dominant cause of ringing. This is analyzed in terms of lumped circuitry where the damping of a simple *RLC* series circuit applies for this example.

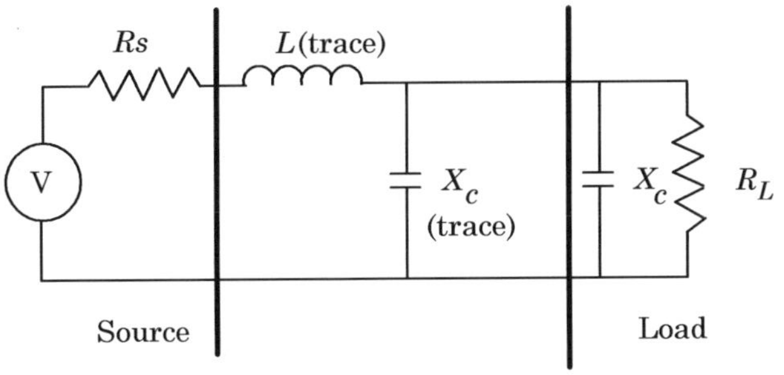

Ringing = $R^2 Xc/4 > 1$ (Underdamped)

Rounding = $Xc > 4L/R^2$ (Overdamped)

Figure 2.12 Equivalent circuit for ringing and rounding of signal propagation.

The condition for ringing is caused by having the transmitted signal under-damped at the load, described by (2.6).

$$\text{Ringing} = R^2Xc/4 > 1 \tag{2.6}$$

For rounding of the signal or over-damping at the load (2.7):

$$\text{Rounding} = Xc > 4L/R^2 \tag{2.7}$$

Figure 2.13 illustrates transmission line effects in a typical printed circuit board as it relates to signal integrity, namely reflections and ringing. Assuming +5V in the left figure, an overshoot of up to nearly 7V may occur somewhere in the system, *but only if* there is a 100% positive reflection caused by having a high impedance at the load. If overshoot does occur, this voltage level may exceed the operational margin level of a component on the net causing overstress or non-functionality, if not damage to the device itself.

In Figure 2.13, Plot A, a smooth signal pulse is propagated from source to load. Ringing may occur depending on whether the transmission line is properly terminated, which is insignificant in relation to the signal waveform that could affect signal integrity. Active components may exhibit some reflections and ringing generated by the output switching transistors, and in most cases this is acceptable. These output drive transistors are, in probability, generally non-ideal or have non-uniform drive characteristics due to their manufacturing process and design of the circuit. The behavioral models used for signal integrity or computational analysis are usually considered ideal or perfect. In actual usage, behavioral models used for simulation purposes often may not describe real, important transmission line effects.

In Figure 2.13, plot B, a poorly terminated transmission line results in significant overshoot and ringing. Ringing, if severe enough can cause false triggering of components if the voltage level exceeds transition logic levels (V_{IH} or V_{IL}). If the length of the transmission line is physically long with respect to the propagation delay of the signal (source-to-load and return), or if there are long distance intervals between devices, the signal will bounce back and forth between end points. Each ringback shown is a reflection from either the source or load, and will decrease in amplitude due to dielectric losses in the board material.

Figure 2.13, plot C, indicates what is observed at the source driver of an "un-terminated" transmission line. Backward reflections can cause noise margin upset and corrupt the quality of the desired signal if another clock transition occurs before the reflected pulse reaches the source driver at the wrong point in time. These reflections are also created by an electrically long signal trace (or long loading intervals), similar to Plot B. When backward reflections occur, the edge rate desired for proper operation is reduced to a slower time period. This signal degradation may be sufficient to prevent other sections of the printed circuit board from functioning at the intended speed of operation. Hence, performance is degraded, or the circuit may become unstable or broadly nonfunctional.

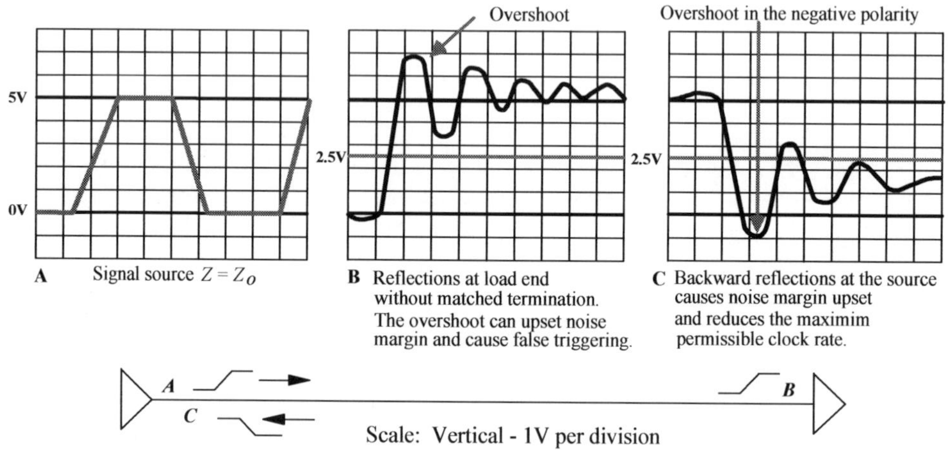
Overshoot
Overshoot in the negative polarity
5V
0V
2.5V
2.5V
A Signal source Z = Zo
B Reflections at load end without matched termination. The overshoot can upset noise margin and cause false triggering.
C Backward reflections at the source causes noise margin upset and reduces the maximim permissible clock rate.
A
C
B
Scale: Vertical - 1V per division

Figure 2.13 Ringing effect in a transmission line.

2.8 Transmission Line Termination Overview

When does transmission line impedance matching become required and how difficult is it to achieve? In Fig. 2.14 we see two systems connected by a transmission line (Z_o). For this configuration, do we make Z_L=Z_o or Z_s=Z_o? It may be necessary to match both source driver *and* load if there is any significant magnitude of impedance difference. Exceptions always exist and are detailed in [1, 2, 3 and 4]. If signal flow is bidirectional, both ends should be terminated. For a single-ended circuit (i.e., a clock signal from an oscillator to load), only one end of the trace requires termination, either source termination (series resistor) or load (Thevenin, parallel or *RC*). Implementing termination is easy. Numerous resources exist on the Internet or application notes on how to choose a proper termination methodology and calculate the value of components used.

If Z_L is fixed, the load is generally designed to match the transmission line characteristic impedance, making Z_L=Z_o. This transmission line connection usually exists when preexisting equipment or components is used with a specific drive impedance requirement. If the load impedance is not known, and this is often the situation when interfacing to a peripheral device, optimal value for Z_L must be chosen so that the load is matched as closely as possible to the transmission line's characteristic impedance ($Z_o \approx Z_L$). If Z_L is not known, termination pads in the layout should be provided on the printed circuit board to determine correct component values for impedance matching if computational analysis *is not able* to be performed. However, understanding the methodologies in [1, 2, 3 and 4] makes the process of figuring an optimal value of termination components easy without the need for trail-and-error attempts. Digital designers using SPICE or other modeling techniques can easily determine an optimal termination method along with component values.

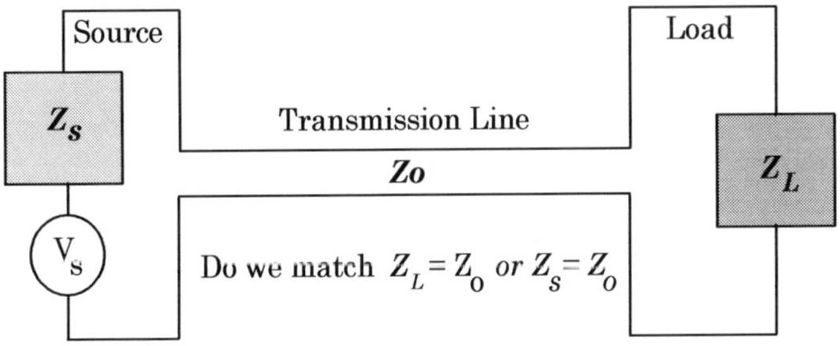

Figure 2.14 Impedance matching requirements of a circuit.

Example

Refer to Fig. 2.15. For a voltage pulse of amplitude V, driving transmission line Z_o, there is a drive current of $I = V/Z_o$. Assuming $V = 5$ volts and $Z_o = 5\ \Omega$ (an unrealistic value), we observe that the drive current would be 1A. If Z_o is now 50 Ω, drive current of 100 mA is now required. It is *not* advisable for both functional purposes (power supply loading and EMI) to use source drivers with 100mA drive capability for an application that generally requires only a few milliamps of drive current. For this reason, most components are designed to drive a typical trace impedance of 30–65Ω.

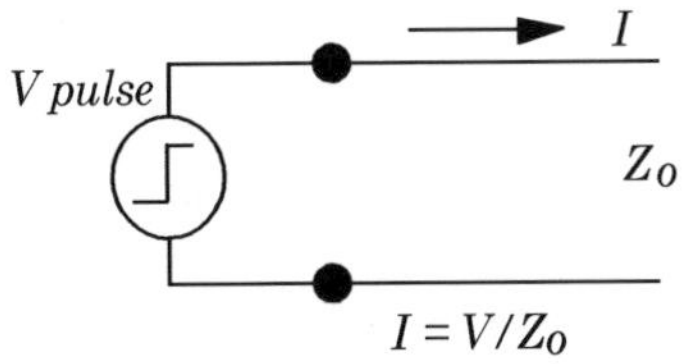

Figure 2.15 Simplified network for drive current example.

2.9 RF Current Distribution

A 0V reference plane allows RF current to return as an image current to its source to satisfy Ampere's law. This 0V plane establishes closed-loop circuit requirements discussed in Chapter 1. Current distribution on microstrip traces tends to spread out within a return plane, illustrated in Fig. 2.16 and exists in both forward direction and return, thus sharing a common impedance between trace and plane (or trace-to-trace), resulting in mutual coupling between these two transmission lines. Peak current density lies directly beneath the trace and falls off sharply from each side due to current spread distribution. This is observe as the bell shaped curve in Fig. 2.16 with the *top of the bell at maximum value* and *i(d)* is the magnitude of magnetic field current which is significantly less at distance "D" away from the center of the transmission line.

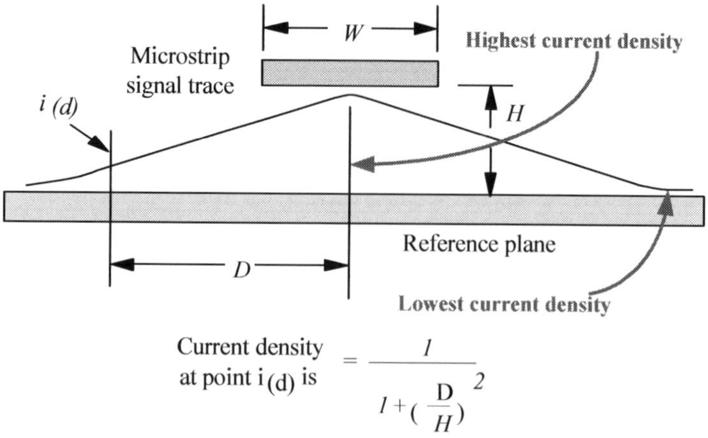

Figure 2.16 Current density distribution from trace to reference plane Courtesy Ref. [1].

When physical distance spacing is large between trace and plane, the loop area between the forward and return path increases (Chapter 3). This increased return path raises the inductance of the circuit proportional to the size of the loop area. Equation (2.8) calculates current distribution optimal for minimizing total loop inductance for both the forward and return current path. The current described in (2.8) also minimizes the total energy stored in the magnetic field surrounding the signal trace [1].

$$i(d) = \frac{I_o}{\pi H} \cdot \frac{1}{1+\left(\frac{D}{H}\right)^2} \tag{2.8}$$

where:
i(d) = signal current density, (A/inch or A/cm) as a flux boundary distribution
Io = total current (A)
H = height of the trace above the ground plane (inches or cm), and
D = perpendicular distance from the center line of the trace (inches or cm).

The mutual coupling factor is partially dependent on frequency of operation in addition to frequency-related skin depth effect (similar to resistance) as a component of the plane impedance. As skin depth decreases to ever diminishing values, the resistive portion of the plane impedance I also increases (per $Z=R+jX_L$). This increase will be observed with proportionality at relatively higher frequencies due to higher inductive reactance, X_L ($X_L=2\pi fL$).

2.10 Analysis of RF Return Paths

In digital systems, especially those using high-speed (fast) edge rate components, a low-impedance (low-inductive reactance) RF return current path must be present to minimize radiated electromagnetic interference (EMI) and to cancel out or minimize undesired common-mode magnetic field currents established in the closed loop circuit. A closed-loop circuit is required per Ampere's law. This applies to both the time and frequency domain aspects of circuit operation.

RF currents propagate using any transmission line path possible, generally the one with the lowest impedance which includes transference of the field due to parasitic capacitive coupling to a nearby metallic structures or system enclosure. If there is no low-impedance conductive path (<377Ω), free space (377Ω) becomes the transmission line. Free space is exactly what we do not want, especially with regulatory compliance concerns.

Examining Fig. 2.17, RF return currents do not have an optimal or low-impedance return path since there is no reference plane within close physical proximity of the signal trace to achieve mutual coupling or flux cancellation of the magnetic field present. Assume all components are tied to a common voltage and return reference and are at a significant physical distance away from signal traces. In the time domain functionality is ensured since there is optimal loop flow for DC voltage levels but not for RF energy (the electromagnetic field) that exist in the transmission line, generally above 10 kHz. Why should digital designers worry about frequency domain aspects or magnetic field cancellation when the circuit works with DC voltage levels of logic high and low, and if timing margins are acceptable for circuit operation?

As discussed in Chapter 1, time and frequency domain aspects of a circuit must be considered simultaneously. Return currents (DC voltage levels) travel through either the power and/or 0V return (ground). While single-sided printed circuit boards may be a cost-effective implementation for a simple design, the probability is high that this simple configuration will not pass various EMC test requirements if digital components are utilized (unless the digital rise and fall times are very slow, such as >1 microsecond with a very short transmission line route between source and load. Adding a 0V reference plane, or making the stackup a four-layer assembly will enhance overall performance related to signal integrity and EMC compliance. While cost is definitely low using a single-sided design, alternative methods of EMC compliance may be required which includes adding expensive metal covers or a metalized plastic enclosure for shielding that far exceeds the cost of using one more layers, along with administrative overhead to solve a design deficiency that adds weight (from the shield) along with design and manufacturing expense, plus time spent to solve a problem that could easily be fixed by using an extra plane in the printed circuit board.

In addition to developing common-mode RF current between circuits due to lack of a properly design RF return path, the printed circuit board can be sensitive to ESD events.

A high-current ESD pulse, along with its effective radiated field, will find a much lower impedance path provided by board copper than that observed by space (377 ohms). A radiated field can also be injected into traces by direct contact from a high energy pulse into the circuit. Component failure could occur or a functional glitch may develop degrading system performance or possible permanent damage to ESD sensitive components that do not incorporate internal diode protection against an ESD event. These ESD protection diodes are generally rated at 2kV.

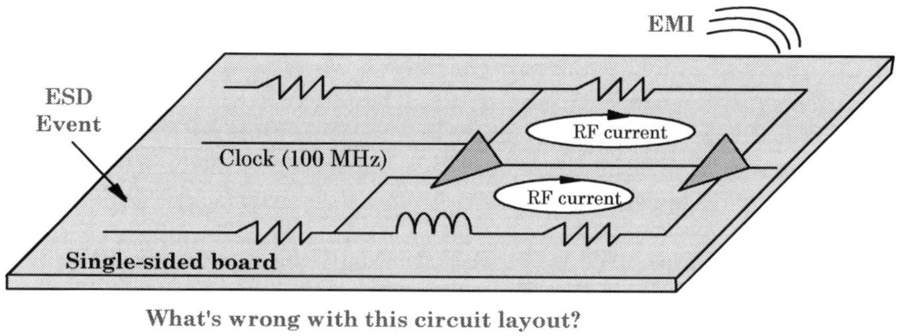

Figure 2.17 Typical PCB design without an optimal RF return current system.

A high-quality, 0V reference is the foundation of every digital printed circuit board design. If the 0V reference system is poor it becomes difficult to isolate or solve a radiated EMI problem. Having a poor 0V reference may also be the actual cause of common-mode development. Adding two more layers to a double-sided printed circuit board to make it a four-layer assembly requires minimal work by the board designer (essentially modifying the stackup assembly and layout definition file). These two additional layers can achieve significant reduction in radiated emissions as documented in numerous EMC publications, textbooks and technical papers (far too many to list in the Reference section).

There may be an infinite number of parallel return paths when many interconnects are provided. Since a large number of RF returns may be present, we can take advantage of this feature and reduce an unlimited number of possible RF return paths (essentially infinity) down to one ($\infty \rightarrow 1$). The number one (1) is called an image plane. Any physically large copper or metal laminate in a multilayer stackup provides a return path with minimal impedance for RF return currents. Since RF currents flow on the laminate in the first level of the metal through the process called skin effect, the voltage potential of the plane (i.e., +3.3V, +5V, +12V) is not a major concern, except under certain operating conditions. One disadvantage of using a multilayer board sometime lies in cost. For many applications, use of an extensive multilayer design is not economically possible. For this situation an alternate return path for RF currents must be established through use of a gridded ground system (not highly effective) or ground traces routed physically next to the signal [source] trace.

The discussion on skin effect refers to current flow that resides in the first skin depth level of metal at higher frequencies. At higher frequencies, time variant (RF) current does not and cannot flow in the center of a transmission line, (only DC or non-time variant electrons can), and is predominately observed on the outer surface of the conductive media. Different materials have different skin depth values. The skin depth of copper is extremely small above 30 MHz. Typically, this is observed at $6.6*10^{-6}$ (0.0017 mm) at 100 MHz. RF current present on a ground plane does not typically penetrate 1 oz. thick copper plane-0.0014” (0.036 mm) at frequencies of 100 MHz or above. Table 1.1 and Equation 1.23 (Chapter 1) provides details and the mathematics on calculating skin depth numbers. As a result, both common-mode and differential-mode RF currents flow only on the top (skin) layer of the plane. No significant current flows internal to the image plane or on the opposite side of the copper away from the signal line.

2.11 Creating an Optimal RF Return Path

An image plane is a layer of copper (either at voltage or return potential) physically adjacent to a signal routing plane. Use of an image plane provides a low impedance path for RF currents to return to their source (flux return) and reducing EMI emissions by cancelling out undesired common-mode currents due to losses within the system.

This return path is for both the signal and power/return distribution network. The term image plane was popularized by Ref. [5] and is now used as industry standard terminology.

An image plane also reduces ground noise voltage in addition to allowing RF currents to return to their source (mirror image) in a tightly coupled, nearly 100% relationship. The reason why coupling approaches 100%, but can never achieve this value, is because there must be physical distance separation between layers (planes) or traces in a printed circuit board assembly. Tight coupling between both signal and return paths provides enhanced magnetic field flux cancellation or minimization between these two transmission lines, which is another reason for use of a solid return plane without splits, breaks or oversized drilled through-holes.

As shown in Fig. 2.18, a signal propagates from source to load through a transmission line with a fixed impedance illustrated as a resistor. Technically, these resistors should be shown as inductors (inductance) if signal propagation is above the 10 kHz-100 kHz range. Due to the impedance of the transmission line at lower frequencies, impedance is mainly resistive and DC return current will find its way back to the source through the path of least resistance, or the power and return pins/planes between components and power supply.

At higher frequencies when inductive reactance starts to become noticeable ($X_L=2\pi fL$), the RF (AC or time variant) current will start to mirror image against the return or image path. What we now have, in this configuration, are two return paths, one for DC current the other for AC. Detailed discussion on inductance and AC current flow in a transmission line is presented in Chapter 3.

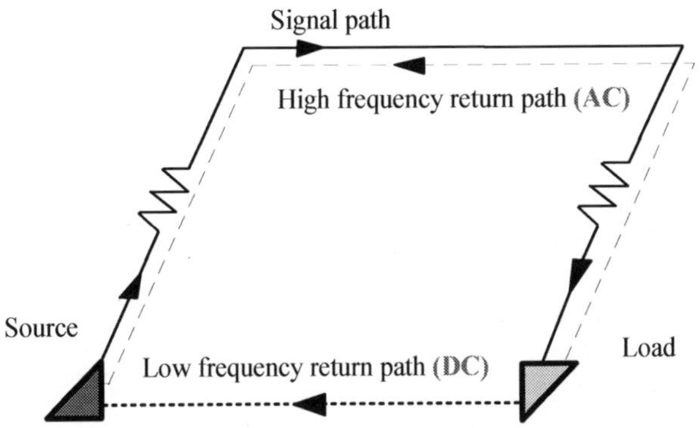

Figure 2.18 Current return paths – DC and AC.

Regarding image plane theory, the material presented herein is based upon an infinite-sized plane. All printed circuit boards are finite in dimension. Image planes cannot be relied upon for reducing or removing RF currents that exist on I/O cables and interconnects due to signals traveling across a boundary condition. This restriction is owing to approximating finite-sized conductive planes when dealing with signals that travel off the printed circuit board. When I/O interconnects are provided, the dimensions of the source and load impedance are important parameters to remember [7, 8, 9].

With every printed circuit board layout there will always be a loop area between components mandated by Ampere, illustrated in Fig. 2.19. The DC return current is through the power and return pins of the components connected by traces or a plane. As for RF (time variant) current flow, do we really know which path or where this current will propagate, either through a trace, plane or free space? If this was a multi-layer assembly, the AC return current would mirror image against the solid plane on the next physical layer, provided the distance spacing between planes is within a reasonable distance to couple the magnetic fields between source and return path. The DC return path however remains the same, path of least resistance and is generally a completely different path than what the RF current takes.

In most printed circuit boards, primary emission sources are established from currents flowing between components. Radiated emissions are modeled as a small-loop antenna carrying interference current, shown in Fig. 2.19. A small loop is one whose dimensions are smaller than a quarter wavelength *(λ/4)* at a particular frequency of interest. For most printed circuit boards, loops exist with small dimensions for frequencies up to several hundred MHz, if not GHz. When a dimension approaches λ/4, RF currents within the loop will appear out of phase at a distance such that the effect causes the field strength to be reduced at any given point.

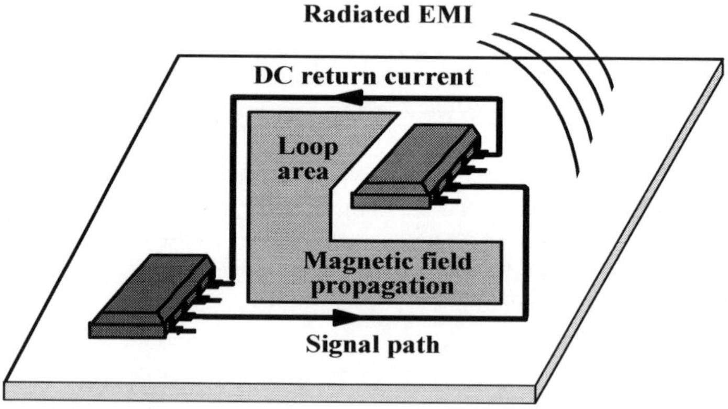

This configuration is for a single or double sided PCB
For multi-layer, loop area is in the plane directly below signal path

Figure 2.19 Loop area between components–single/double sided assemblies.

2.12 How RF Return (Image) Planes Work

In Chapter 1 we examined the need for flux cancellation or minimization. Image planes provide this function by allowing RF return currents to image back along its source path differentially (from source to load, then to the differential of the image return) thus cancelling out undesired common-mode current created from losses within the network. Here, the term *differentially* describes the phase relationship between signal and its return image. When an RF return path is placed in close proximity to a wire or trace, magnetic flux opposite in polarity cancels each other out. We now examine the physics of this discussion.

When time variant current propagates down a printed circuit board transmission line (trace), an electromagnetic field is generated. Maxwell's equations describe the development of both fields. Depending on the physical length of the transmission line, radiated EMI (RF energy) may be established. Both trace and copper planes have finite inductance. This inductance inhibits current buildup and charge energy whenever a voltage potential is applied.

Research [6] shows that if a two-wire transmission line is slightly unbalanced, the trace will radiate as an asymmetrical dipole antenna. This unbalanced structure will create common-mode radiated emissions at levels much greater than differential mode radiation within the closed loop circuit

Before examining how an image plane works within a printed circuit board, the following briefly summarizes differences between various types of inductance [8, 9], each discussed in greater detailed within Chapter 3.

- *Self-inductance*: The inductance that exists in a wire or printed circuit board trace.
- *Partial inductance*: The inductance from one wire segment relative to an infinite segment.
- *Mutual inductance*: Magnetic flux coupling from one transmission line to another.
- *Mutual partial inductance*: The effects that one inductive segment has on a second inductive segment.

2.12.1 Image Plane Implementation and Concept

Figure 2.20 illustrates an image plane within a printed circuit board along with mutual partial inductance. Comparing Fig. 2.20 to Fig. 2.19, the majority of RF current within the signal trace will return on the plane located directly below the signal trace within the boundary relationship previously illustrated by Figure 2.16. Within this return "image" structure, RF return current will encounter finite impedance (inductance). This return current thus produces a voltage gradient which is referred to as ground-noise voltage. Ground-noise voltage will cause a portion of the signal current to flow through the distributed capacitance of the ground plane.

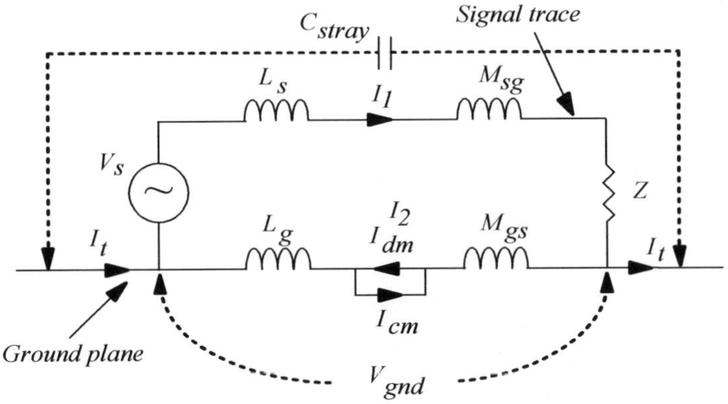

The currents in the return path may also be identified as I2

Figure 2.20 Schematic representation of a return path within a printed circuit board.

Common-mode current, I_{cm}, is typically much less than differential-mode current I_{dm} by many orders of magnitude. However, common-mode currents traveling in the same direction (I_l and I_{cm}) produce higher emissions than those established by differential-mode currents propagating in opposite directions (I_l and I_{dm}). This is because common-mode RF current fields are additive whereas differential-mode fields tend to cancel [5, 6, 10].

To reduce ground-noise voltage, it is necessary to increase the mutual partial inductance between the source path (transmission line) and its nearest image plane. Doing so provides for an enhanced return path for signal return current to mirror image back to its source. We can easily calculate ground-noise voltage V_{gnd} using Eq. (2.9):

$$V_{gnd} = L_g \frac{dI_2}{dt} - M_{gs} \frac{dI_1}{dt} \quad (2.9)$$

where (as shown in Fig. 2.15 and included in Eq. 2.9):
V_{gnd} = ground plane noise voltage
L_g = partial inductance of the ground plane
L_s = self-inductance of the signal trace
M_{sg} = partial mutual inductance between signal trace and ground plane (return path)
M_{gs} = partial mutual inductance between ground plane and signal trace
C_{stray} = distributed stray capacitance of the ground plane
I_t = total amount of RF current in the network, and
dI/dt = time variant current propagation (both in common- and differential-mode) within the transmission line.

To reduce I_t currents (Fig. 2.20), ground-noise voltage *(V_{gnd})* must be minimized. This is best accomplished by reducing distance spacing between signal path and ground plane (return path). In most cases, there is a limitation on ground-noise reduction since physical spacing between a signal and image plane must be at a specific, finite distance to maintain constant impedance of the transmission line for functionality and manufacturing reasons. Hence, there are limits to making the distance separation between two planes any closer than physically manufacturable. Ground-noise voltage can also be reduced by providing an additional low impedance path for RF currents to propagate. This additional return path may include ground traces or other conductive structures nearby that shares the total amount of flux that must return back to its source, including free space to satisfy Kirchhoff's law.

Since mutual partial inductance minimizes production of radiated RF currents, we now examine how differential-mode, I_{dm}, and common-mode, I_{cm}, currents are affected at the same time. Use of image planes significantly reduces these currents as was illustrated in Fig. 2.21. As discussed in Chapter 1, and as will be reaffirmed later in this chapter, desired differential-mode RF currents are propagated cleanly when equal and opposite currents exist within the signal trace and the RF return current path, or a properly balanced circuit with no losses. If cancellation of unbalanced currents is not 100%, the amount of current that is left over becomes undesired common-mode noise. It is this common-mode current that functions as an excitation source and develops the majority of EMI. This is because the un-cancelled RF current in the return path is added to the primary current in the signal path if both currents are traveling in the same direction. To minimize common-mode current development, we must maximize the mutual partial inductance between signal trace and image plane to "capture the flux," hence canceling unwanted RF energy or common-mode currents.

When an RF return plane or path is provided within a printed circuit board assembly, optimal performance may result when the return path is connected to a reference source such as chassis ground. This reference source must also be connected to the reference pins of components physically located at both the source and load ends of the transmission line [11]. For all digital and analog devices, the power and return pins inside the component's package (wafer or die) are connected to potential sources; both power and 0V return. Certain geometrical features can impact these connections and add inductance to the circuit. Only when the RF return path is connected to both the power and return pins of a component will a real image plane exist.

An image plane containing differential-mode signaling can produce common-mode currents through the process known as mode-conversion. Depending on distance spacing between trace and image plane, common-mode currents will be reduced by increased mutual partial inductance. The amount of common-mode current that travels in the return plane is thus dependent on the minimized distance separation between two conductive surfaces.

Image planes function because active devices must be connected to both a power and 0V return path. The return connection internal to the device package, connected to the return plane, provides for this reference and is the reason why image planes work.

When the image plane is removed, or distance spacing between source path and image plane becomes physically significant, a phantom image return path has to be created between trace and phantom plane as shown in Fig. 2.21. The RF image associated with these currents will now not cancel out in an efficient manner, thus increasing radiated RF energy. For an image plane to perform as desired, the plane should be infinite in size and not contain disruptions, slots or cuts [9] relative to the source transmission line placement.

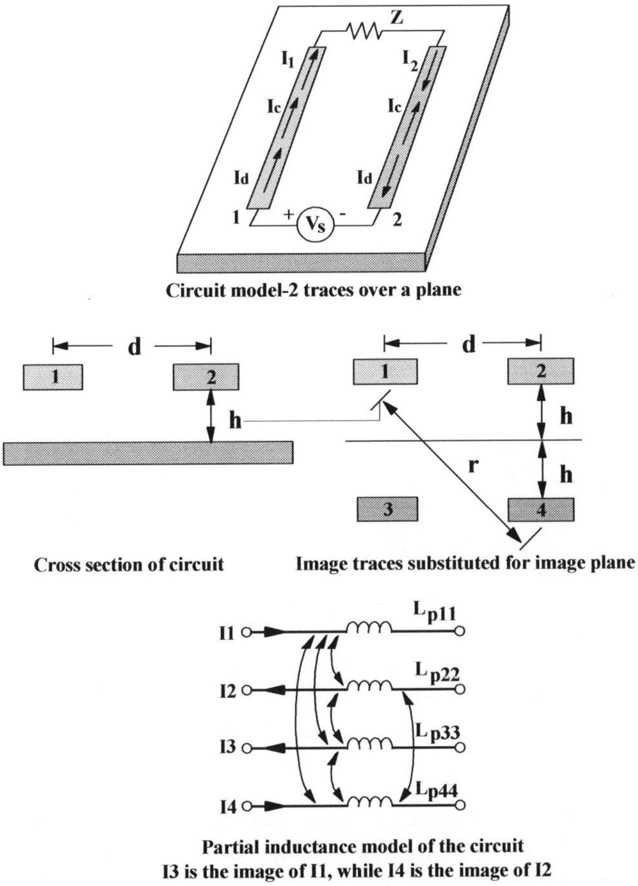

Figure 2.21 Image plane related to partial inductance and phantom current propagation.

2.13 Image or RF Return Path Violations

For an image plane to be effective, all signal traces must be located adjacent to a solid or metal return path/plane, generally copper, and must not cross an isolated or open area such as slots or gaps in the adjacent plane where the copper has been etched away. Exceptions are permitted using advanced, specialized trace routing techniques discussed later and elsewhere including differential pair routing. If a signal, or even a power/return trace is routed within or internal to any solid plane regardless of potential, this solid plane is thus fragmented and increases the physical loop area for RF return current flow from the adjacent signal routing layer. RF loop currents created can then couple to the adjacent signal layer across this absence of copper and create an EMI or signal integrity problem.

Figure 2.22 illustrates a violation of the image plane concept. The return plane can now no longer function as a solid, undisturbed RF return path. Vias placed in an image plane do not degrade imaging capabilities, except where continuous row of ground slots are provided discussed in the next section. The signal trace (transmission line) in Fig. 2.22 is the solid line going horizontal and the RF return path with a longer route is the dashed line. The percentage of current flow is directly proportional to distance spacing from the impedance discontinuity. The vertical lines are slots located in the adjacent return plane.

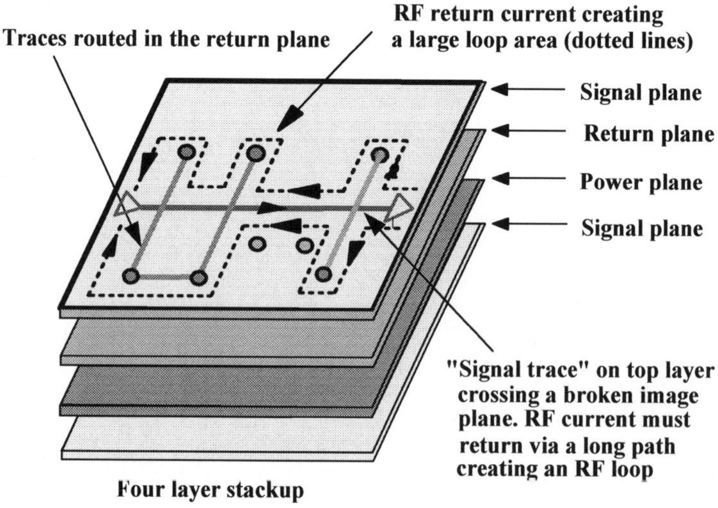

Figure 2.22 Image plane violation with traces.

Another area of concern that lies with plane discontinuities is use of through-hole components. An excessive number of through-holes in a power or return plane creates what is called the Swiss Cheese Syndrome. The copper area within the plane is reduced because many holes overlap (i.e., oversized drilled through-holes commonly found with press-fit connector) leaving large areas of discontinuities. This is shown in Fig. 2.23, left side. Return current flows on the image plane (an adjacent layer) around the oversized through-hole pattern while the signal trace crosses the discontinuity [1]. The percentage of RF current flow is dependent on the distance spacing relative to the location of the trace route and the last slot opening. Although there is 100% flux return, one path may have 70% and the other side 30%, as an example, which means an unbalance in transmission line return current due to some current taking a longer return path. This imbalance creates undesired common-mode current.

As for the right side of the signal trace in Fig. 2.23, there are no plane discontinuity directly under the trace, hence optimal flux cancellation.

This longer return path adds inductive reactance in one-half of the closed loop network creating common-mode voltage, *E*, easily calculated by (2.10). With increased inductive reactance in the return path, there is reduced differential-mode coupling between signal trace and the RF return path (less flux cancellation). For through-hole components that have a space between pins (non-oversized holes), optimal reduction of signal and return current is achieved through less inductance in the signal return path and the existence of the solid RF return plane.

$$E = -L(dI/dt) \tag{2.10}$$

where:
E = Common-mode voltage (volts/time period)
L = Inductance in the transmission line path (ohms)
dI = amount of time variant current in the transmission (Amps), and
dt = time rate of change of the digital signal (seconds).

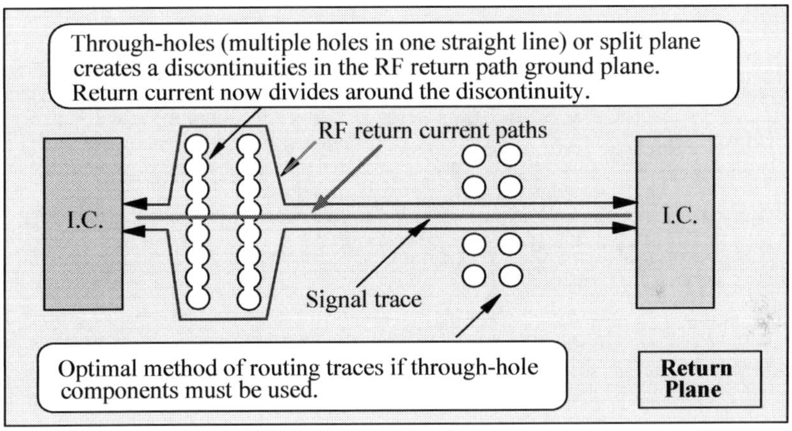
Through-holes (multiple holes in one straight line) or split plane creates a discontinuities in the RF return path ground plane. Return current now divides around the discontinuity.
RF return current paths
I.C.
I.C.
Signal trace
Optimal method of routing traces if through-hole components must be used.
Return Plane

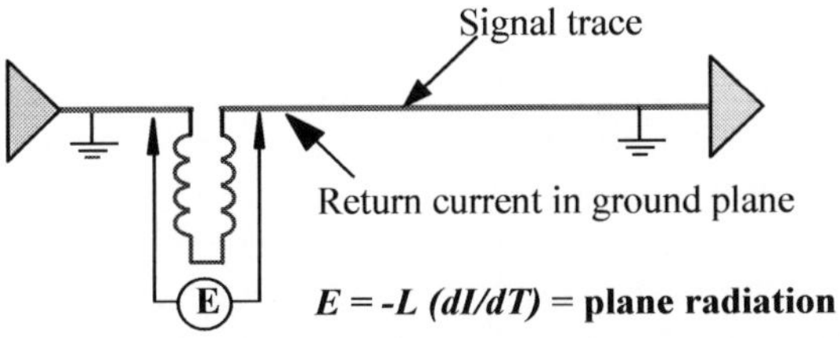
Signal trace
Return current in ground plane
E
E = -L (dI/dT) = plane radiation

Equivalent circuit showing inductance in the return path. This inductance is approximately 1 nH/cm.

Figure 2.23 Ground "loops" using through-hole components (slots in plane).

As was shown in Fig. 2.16, the image plane under the signal route should be a total of three times (3) the width of the trace for optimal flux cancellation.

Generally, any slot in a plane will cause RF problems for signals that route directly over the discontinuity. For higher-speed, high-threat signals[3], alternative methods of routing traces between slots, opening or absence of copper in a solid plane must be implemented using computational analysis and advanced layout techniques.

An example of a trace that must traverse across a slot or partition due to functionality reasons is shown in Fig. 2.24. In order to establish a continuous RF return path across a split plane, use of a capacitor in bypass mode whose self-resonant frequency is calculated to be the same as the switching noise frequency provides significant benefit however, these bypass capacitors are very limited in effectiveness over a narrow range of frequencies, unless a capacitor with high *ESR* is used to increase the bandwidth of operation. Chapter 4 provides details on use of capacitors for various applications such as bypassing and how to calculate an optimal value.

Improvement of up to 20 dB has been observed during functional testing and has been reported in many technical papers not listed in the Reference section. The capacitor must be chosen for optimal performance calculated for the *self-resonant frequency* of the signal present on the trace and not based on a common value such as 100 nF (0.1 μF). It is cautioned however, that this technique may result in reactance-based phase shifts in the current relationships between the traces and their images, impacting the magnitude of flux cancellation or minimization. Again, details on use of AC bypass capacitor implementation is found in Chapter 4.

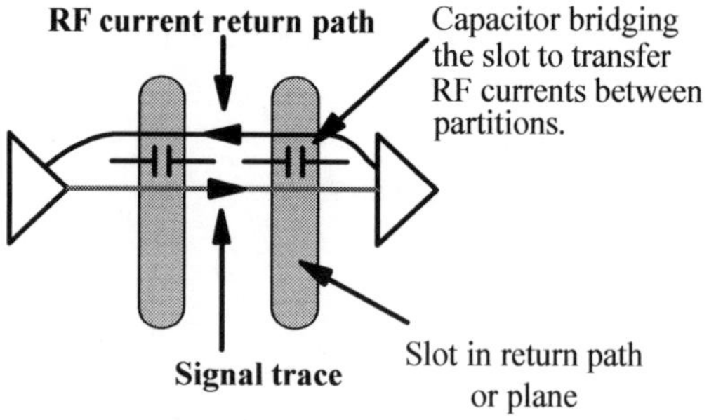

Figure 2.24 Passing RF return current across a split plane using a bypass capacitor. Figure courtesy: Elya Joffe

2.14 Layer Jumping–Use of Vias

It is generally assumed that if each and every printed circuit board trace is routed adjacent to a solid plane there will be tight coupling of common-mode RF currents along the entire trace route. In reality, this assumption is partially incorrect. When routing a clock or high-threat signals, one must transition this trace to another signal routing layer while maintaining a continuous RF return path physically adjacent to the via where the

---

[3] High-threat refers to high-bandwidth, RF spectral components that propagate as an electromagnetic field down a transmission line or trace. These signals include clocks, video, address lines, analog circuits and the like. All of these sensitive circuits may either radiate RF energy or be susceptible to an externally induced field disturbance requiring a mandatory low-impedance RF return path to complete the closed-loop circuit.

layer jump occurs if a multi-layer assembly is used. This return path via must be tightly coupled to the signal trace for maximum magnetic field flux cancellation.

As a signal jumps from one routing layer to another, RF return current must follow the trace route and closely couple to it in differential-mode to prevent development of undesired common-mode current. When a trace is routed internal to a printed circuit board between two planes (power and return) or two planes at the same potential, RF return current is shared between these planes in proportion to the distance spacing between the trace and plane.

If an asymmetrical stackup exist in stripline configuration, the majority of the RF return current will image against the plane physically closest to the transmission line. The remaining amount of return current that is not coupled to this closely spaced plane will occur on the other plane located physically further away. If the planes are at the same potential (both power or 0V or return), we maintain optimal flux capture and cancellation. If the planes *are at different potentials*, such as power and return or 0V, we may now have a situation to where undesired common-mode current may be developed since RF return is not able to couple to its correspondence source trace in an efficient manner. Not that this will be cause for system failure, but can be a contributing effect to the creation of common-mode EMI.

If both planes are at the same potential (i.e., 0V reference or ground) the RF return current jump will occur at any via connecting all planes of the same potential, which essentially creates a coaxial based transmission line and ensures enhanced flux cancellation.

2.14.1 Layer Jumping Concerns

Layer jumping requires a low inductance path between planes of the same potential. When layer jumping must occur between horizontal and vertical routing layers containing high-speed or critical signals of interest, the designer should incorporate ground or stitch vias at “each and every” location where signal axis jumps is executed. The ground via is always at 0V potential. This is illustrated in Fig. 2.25.

A ground via is placed directly adjacent to each signal route jump from a horizontal to a vertical routing layer. Ground vias can be used only when there are more than one 0V (ground) reference planes. This via is connected to all ground planes (0V reference) and serves as the RF return path for the signal jump currents, creating a coaxial based transmission line configuration.

When using ground a via, a continuous path is now present to maintain a constant RF return physically located 100% adjacent to a signal route.

For stripline topology:

- If two planes are of the same potential, interconnect vias can be used.
- If both planes are at different potential, interconnect vias *cannot* be used and an alternate means of layer jumping of critical traces must occur. This alternate means may include use of a bypass capacitor (multi-layer) or RF return trace for double-sided assemblies.

When only one 0V reference plane is provided and the alternate plane is at voltage potential, commonly found in a four-layer printed circuit board, maintaining a constant RF return path for RF currents becomes challenging. There is a common misconception that decoupling capacitors performs as an optimal means of jumping RF current between planes of different potential. Capacitor used in bypass mode, not decoupling mode (providing charge), transfers RF return current in a specially designed low impedance transmission line structure when jumping layers and whose value must be calculated based on the expected frequency that the signal in the corresponding trace operates at. The impedance of decoupling capacitors used for energy replenishment in power distribution networks may exceed the impedance of the via and might not provide optimal performance, if there is excessive inductive reactance in the transmission line relative to the via, and vice-versa.

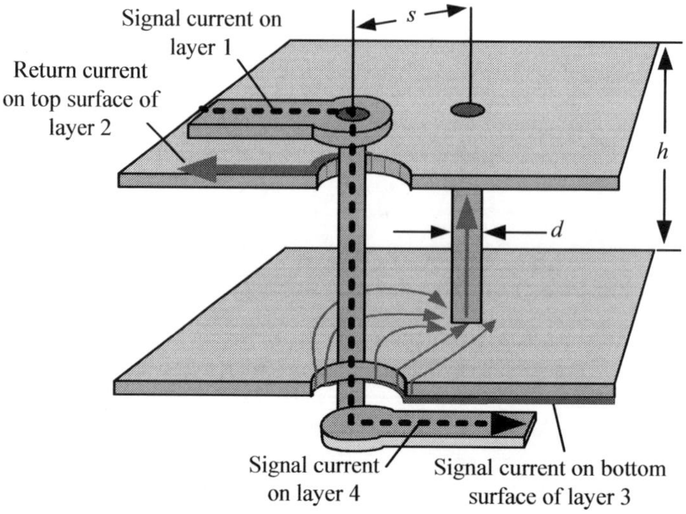

Figure 2.25 Using ground stitch via during layer jumping.
Source: Short-Term Impedance of Planes-Online Newsletter; Vol. 6, Issue 5:
Dr. Howard Johnson

There are two ways to ensure a coaxial based transmission line routing is present (Fig. 2.26); using a reference trace at an appropriate potential to cancel magnetic flux, generally the power or return (ground) pin of a component; or making the layer jump physically next to a bypass capacitor used in decoupling mode. The RF reference trace must be located no more than one-trace width distance away from the signal line to ensure tight magnetic field coupling.

How can we minimize use of multiple ground vias or reference traces when layer jumping and routing density is tight? In a properly designed printed circuit board, the layout designer generally routes all high-threat traces manually before use of an autorouter, which speeds up the routing process. Much freedom is permitted in routing these initial traces (i.e., clocks and high-threat signals). A layer jump for any high-threat trace must first be located by the layout engineer [manually placed] physically adjacent to the via of an existing *ground pin* of any component, assuming there are multiple ground planes and the trace is routed stripline between two ground planes. The transmission line carrying a signal with high levels of RF energy now co-shares this component's ground via ensuring a coaxial based transmission line is created (Fig. 2.27).

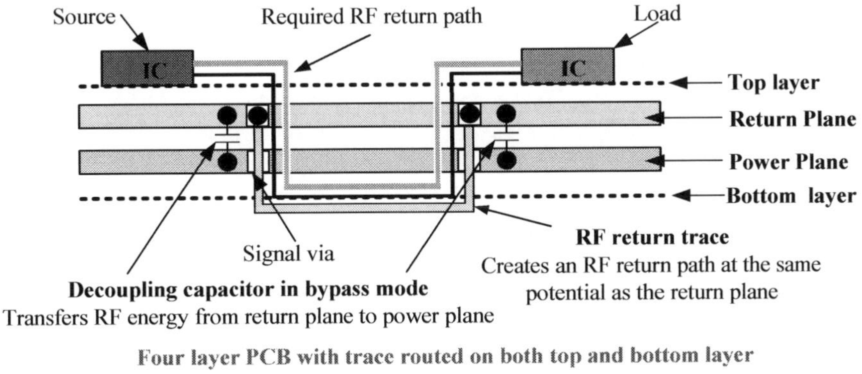

Figure 2.26 Creating an RF return path on a four-layer printed circuit board.

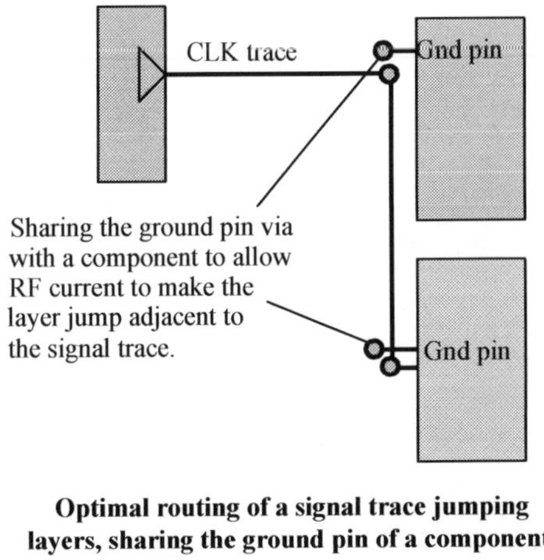

Figure 2.27 Manual layer jumping to create an optimal RF return path next to a via.

2.15 Split Planes and Their Effect on RF Return Path Discontinuity

A split plane is a solid plane segmented into two or more partitions. With regard to split planes and RF return path continuity, earlier sections in this Chapter discuss the need to ensure a coaxial based transmission line is present at all times. Should a trace cross a split, development of common-mode current will be established, which in turn may cause operational disruption.

The only time one generally splits a plane is when using mixed logic devices such as analog-to-digital converters or processors that require an analog partition such as Ethernet. The major concern lies with traces routed on an adjacent signal plane that crosses the split plane. Crossing a split in a plane under any condition is not recommended especially microstrip routing however, differential-pair signaling can cross a split if both traces are routed physically close to each other since the impedance discontinuity that will occur is equal and opposite and cancels out undesired magnetic flux. Single-ended signals are the primary transmission lines of concern when crossing splits. This section discusses split planes with regard to analog and digital partitions.

A simple example of split plane topologies are shown in Fig. 2.28 for mixed signal applications. The top left drawing has a continuous ground plane. Two circuits have been partitioned with separate functional power planes (analog and digital) connected by a ferrite bead that prevent digital switching noise from entering the analog partition.

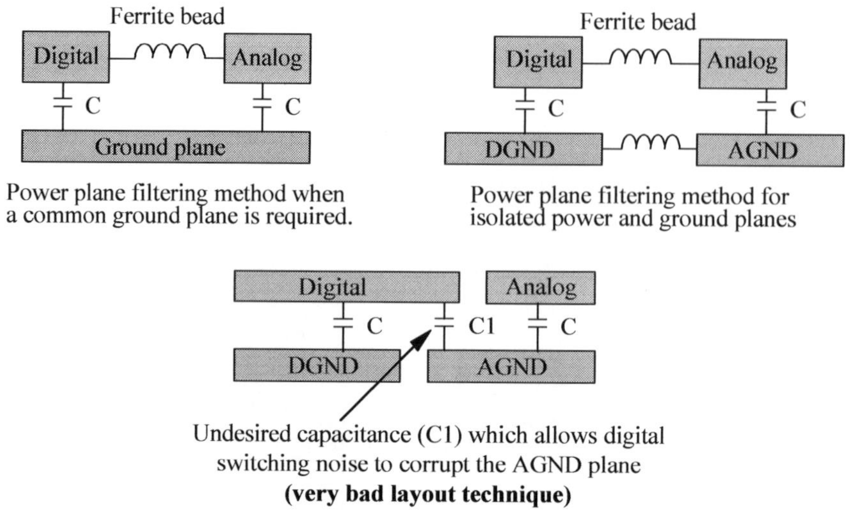

Figure 2.28 Variations on split plane configurations.

Sometimes use of a ferrite bead in this configuration may present a minor voltage drop that could, under certain operating conditions, cause the analog device to not have optimal input power and can create functional problems. When using this layout technique with component mounting pads installed in the artwork to determine if the bead is needed or not, or to change the impedance of the bead for a particular frequency range of performance, the ferrite bead may have to be converted to a 0-ohm resistor. This ferrite bead is used when only one or a few pins of a device needs analog power not sourced by a local power plane at a specific voltage potential regardless of ground potential.

The digital and analog circuits in the upper right illustration are connected by a ferrite bead for both power and ground (true isolation of between partitions). Like the above paragraph, there is now a voltage potential difference between the planes. It is "not advisable" to put any impedance in the ground (return) plane between analog and digital unless the analog device mandates this type of circuit layout, but may be highly effective under certain operating condition discussed below.

The primary reason to not split the ground plane is that the traces routed on the adjacent layer will have to cross this split creating potentially serious EMI problems due to large *L(dI/dt)* in the return path; this breaks up the coaxial based transmission line routing requirements mandated for optimal EMI and flux cancellation. If the ground plane is not split in a mixed signal application, then routing of traces is simplified along with enhanced signal integrity and EMI performance without the need to reroute traces around the isolated partition.

The bottom illustration is another configuration totally isolated for both power and ground between digital and analog but has an unfortunate overlap between the two partitions where part of the digital power partition overlaps the analog ground plane.

This example *maximizes* undesired RF field coupling from the digital section to analog through plane capacitance, and is never to be implemented in a real printed circuit board layout!

> *CAUTION: Various integrated circuit devices may require that a specific partition design be implemented using ferrite beads, and some devices might not be compatible with the examples provided below. Verify use of ferrite beads for power and return (ground) with the component manufacturer first to determine if they are acceptable and under what specific conditions of use.*

With multilayer printed circuit boards, power and return planes are required for functional reasons. An example is separation of analog and digital voltage reference areas (i.e., +3.3 VDD from +3.3 ADD). A primary design and layout concern is to guarantee that overlaps from split plane implementation does not occur, especially between analog and digital partitions. Due to overlap interplane capacitance between the two segments (Fig. 2.28, bottom configuration), identified as *C1*, will allow RF switching noise to traverse from the noisy digital ground plane to the analog ground plane. If noise coupling does happen, we may end up with both a signal integrity and EMI problem within the analog section. The DC voltage potential of the planes remains between the two partitions, yet higher frequency common-mode noise is now developed and which must be avoided across this inductive interconnect.

If additional high-frequency isolation is required, we can isolate one plane (power) or both potentials (power and return) with ferrite beads (one or two, circuit functionality dependent). It is critical that a physical inductor never be used! Extreme caution is advised if this design technique is used as in "most" cases, a ferrite bead or inductance used in this application causes more harm than good!

Inductive reactance from a physical inductor will create a significant amount of common-mode RF across the physical device due to $E=-L(dI/dt)$. If L is increased to a large value, so is development of common-mode current. We must however be careful when using this design technique which is utilized for only specialized circuits and functional requirements.

If both analog and digital planes contain high-frequency switching noise, it is usually better to keep both power and return planes isolated to prevent crosstalk from occurring between the two. If a common analog-to-digital reference plane is required, a ferrite bead should be placed only across the analog to digital power partition to create AVDD from DVDD such as the top left illustration on Fig. 2.28 and use a solid plane for ground.

2.15.1 Digital-to-Analog Partitioning (Split Return Plane)

Concerns exist for proper partitioning of digital-to-analog circuits, especially the return (ground) plane. A common ground reference between digital and analog may or may not be required. How do we know if a split plane should occur? The answer is easily learned by studying the application notes of the circuit device with regard to how the manufacturer designed the silicon substrate. If the component vendor designed its silicon circuit (the wafer die) to have a common digital/analog ground partition, then we do not split the ground plane between analog and digital. Depending on how the silicon is partitioned within its package, its utilization on the printed circuit board must match that of the silicon.

Question: Do we, or do we not, split the return (ground) plane with mixed logic devices.
Answer: *It depends!*

Analog/digital mixed signal devices can be described, by definition, as a printed circuit board shrunk down to microscopic size. There are both power and return planes built into the die. The question we must ask: "Is the return plane inside the die partitioned with a moat or split plane." If the answer is yes, then a ferrite bead will be required on the printed circuit board. Most analog/digital devices so not have a partitioned return plane in the die. In this case, splitting the return plane in the printed circuit board is highly discouraged. It is mandatory to have a single ground plane and not a split plane!

Read the application note carefully. If there is the statement saying "connect analog ground to digital ground in one location, then in reality there is only one return plane into the silicon (non-split plane in the silicon), and making a narrow connection at single point on the printed circuit board adds extra inductance when in fact, why is there any split incorporated at all when everything is referenced to the same potential? Under this condition, do "not" split the plane. When looking at the big picture, if analog ground is to be connected to digital ground at one location, why make routing difficult for an adjacent signal plane when there is only one 0V reference required.

If a split partition is however required between analog ground and digital ground, all analog discrete components and traces must reside within the analog section as shown in Fig. 2.29, while digital traces and component remain in the digital partition.

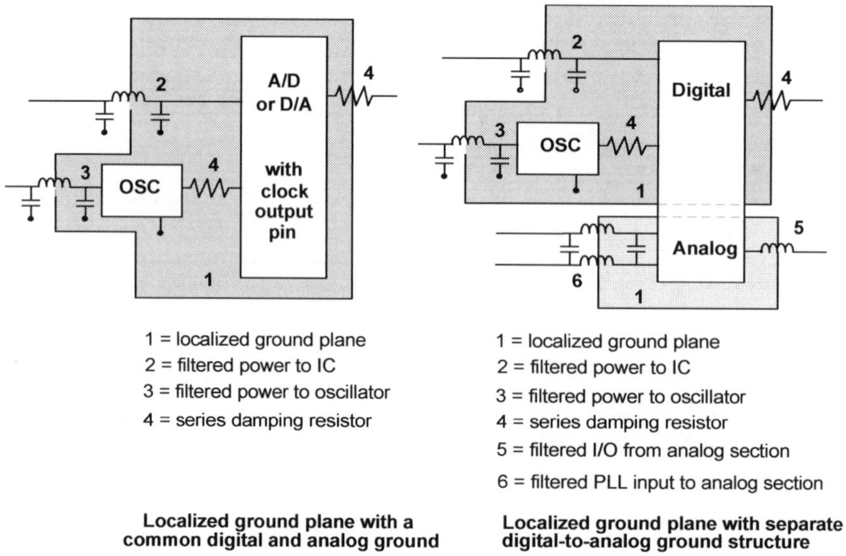

Figure 2.29 Localized ground plane–digital-to-analog partition.

During printed circuit board component placement, we may sometimes have to create a very convoluted shape that zigzags between the pins of a component if the component manufacturer did not provide an optimal pinout configuration that allows for ease of digital-to-analog partitioning.

For most design applications, the analog power input to the device is filtered with a ferrite bead as well as a localized decoupling capacitor. The capacitor is located within the "quiet" analog plane or localized return plane.

2.15.2 Using a Ferrite Bead versus Inductor in Split Plane Configurations

The reason not to use inductors as a filter between digital and analog partitions is easily visualized and best shown in Fig. 2.30.

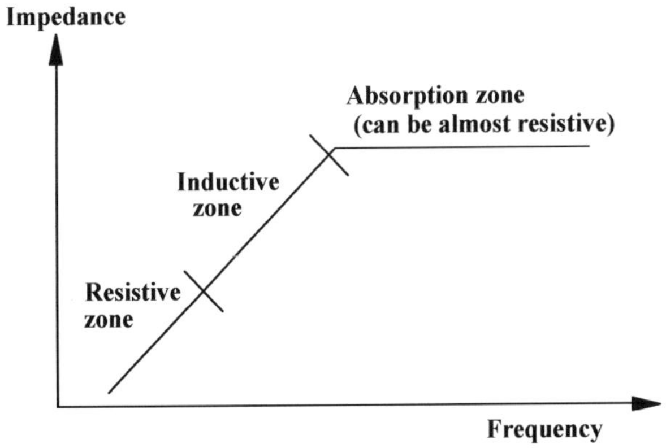

Figure 2.30 Ferrite material performance characteristics.

Ferrite material has practically zero impedance or extremely small DC resistance at lower frequencies. The ferrite bead is thus essentially transparent to DC levels and acts as a small inductor having little effect due to small X_L (X_L=$2\pi fL$) At higher frequencies, RF currents within the power distribution structure now starts to see both resistive and inductive elements. At these higher frequencies, inductive reactance starts becomes high preventing RF current from propagating, making the resistive characteristics dominate (low resistive impedance at high frequency). It is the resistive element that prevents RF current propagation with this energy dissipated as heat internal to the ferrite material.

Basically, a ferrite component is a ferromagnetic device that behaves like a large frequency dependent resistor that keeps RF energy from traveling between locations with minimal DC voltage loss in the transmission line. An inductor, on the other hand, has large inductive reactance, (X_L=$2\pi fL$) that is frequency dependent. Inductive reactance is exactly what we *do not want* within a transmission path under any condition!

Occasionally an inductor is substituted for a ferrite bead with the belief that operational performance is enhanced at a particular frequency. This belief actually causes more harm than good. The reason why we "never" use an inductor in lieu of a ferrite bead, except under special design considerations, is that parasitic capacitance is present between the two terminals of the inductor, plus parasitic capacitance between each inductor winding. With both *inductance* and parasitic *capacitance*, a resonant circuit is created somewhere in the frequency spectrum. Depending on the values of L and C, we may be allowing undesired RF current at a particular frequency to pass between the isolated areas which is what we do not want when the inductor starts to behave as a capacitor, described in detail within Chapter 1, Section 1.13 (*Hidden Schematic or Parasitics of Passive Components*). Once RF current passes through the inductor, undesired RF currents may cause harmful disruption to other functional circuits, which were supposed to be operated from a clean filtered power supply or transmission line routing but now contain dirty, or switching noise.

If an isolated ground plane contains low-frequency circuits (i.e., analog), and another isolated ground plane has high-frequency switching components (digital), it sometimes becomes mandatory to isolate both planes with a ferrite bead depending on the device's function and the manufacturer's requirements for power and/or return plane isolation.

Functional requirements of the analog circuitry may also be a reason for analog plane isolation because of low CMMR input protection or low-level voltage transitions in a noisy environment.

Ferrite bead isolation is required, only if, no high-frequency switching energy is permitted between the two areas. If both areas contain only low-frequency components and there are no high-frequency RF energy threats (high-edge rate switching noise), a filter may not be required. A single-point bond connection is thus preferred between the two planes which means no split plane under any condition.

Another method to minimize loop inductance when a split plane is required, and a trace must cross the split containing high-frequency or high-bandwidth signal, is use of a capacitor in bypass mode across the split (i.e., ground-to-ground or power-to-power).

A capacitor in bypass mode allows RF energy (the electromagnetic field present) to propagate between functional partitions while maintaining DC isolation. An example of this layout technique is shown in Fig. 2.31. The advantage of using this bypass capacitor is to minimize the RF return loop current path and development of common-mode current due to excessive inductance in that path. Instead of the RF return path jumping from top to bottom layer as shown in Fig. 2.31, the RF return path remains on the same layer, properly coupled to the signal trace. It does not matter what the plane potential is (DC value) as the capacitor is not providing charge to keep the circuits functional but to pass an electromagnetic field (RF return) with the smallest loop area possible.

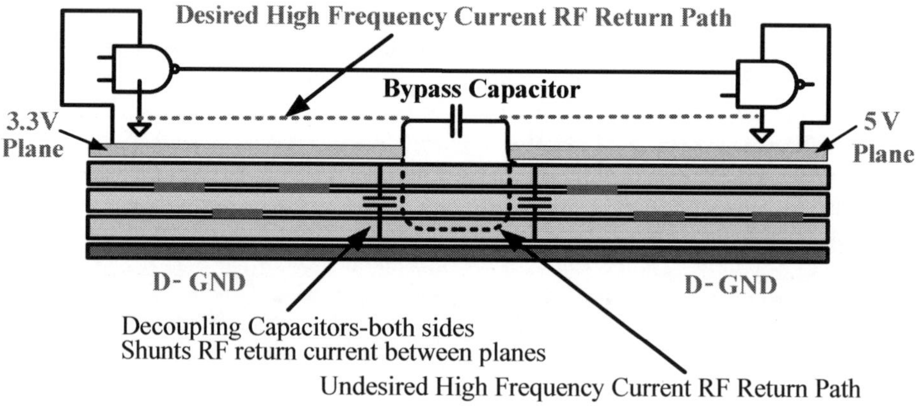

Figure 2.31 Crossing a split plane using a bypass capacitor.
Figure courtesy: Elya Joffe

2.16 Flux Cancellation Concepts (Optimizing RF Current Return)

The concept of implementing flux cancellation is simple. However, one must be aware of many pitfalls and oversights that may occur when implementing flux cancellation or minimization techniques. With one small mistake many additional problems can develop creating more work for the EMC engineer to diagnose and debug.

The easiest way to implement flux cancellation is to use *image planes* [5] or an RF return path from load to source which includes implementation of a coaxial based transmission line. Regardless how well we design a printed circuit board, magnetic and electric fields will always exist due on Maxwell's equations. If we cancel or minimize undesired magnetic flux propagating in both source and return path, the probability of having a signal integrity and/or an EMI event is ensured.

How do we cancel or minimize magnetic lines of flux during layout? Various design and layout techniques are available. A brief summary of some of these techniques is presented below. Not all techniques are involved with flux cancellation/minimization but are still provided for completeness. Due to the focus of this book (*EMC Made Simple*), some of the items below are not discussed herein but thoroughly presented in detail within [3 and 4].

- Proper stackup assignment and impedance controlled transmission lines routing within multilayer boards.
- Routing a signal trace carrying high levels of RF energy (i.e., clock trace) as physically close as possible to an adjacent RF return path, thus creating a coaxial based transmission line structure. This RF return path can be a solid plane (power or return-it makes no difference), a ground grid, a return trace located closely adjacent to the source trace (single- and double-sided boards), or other means to achieve a coaxial-based transmission line routing topology.
- Capturing magnetic flux created internal to a component's housing package into the 0V reference system to reduce component radiation using a localized plane directly under the device which is at the opposite potential from the potential on the adjacent layer. Example: If layer 2 is a ground plane, then the localized plane on layer 1 is at power potential. This alternating plane configuration provided additional decoupling capabilities while also capturing magnetic flux that is absorbed within the copper shield barrier.
- Carefully choosing logic families to minimize RF spectral distribution from silicon and trace radiation (use of slower edge rate devices if possible). Sometimes changing the slew rate of a driver using software can be highly effective since most components come in only one speed-fast!
- Reduce RF common-mode currents developed on transmission lines by minimizing the RF drive voltage from a high-speed driver using a series resistor.
- Provide for a high quality power distribution network for components that consume significant amount of current switching all signal pins simultaneously under maximum capacitive load (Chapter 4).
- Properly terminating clock and signal transmission lines to prevent ringing, overshoot, undershoot and other potential signal integrity problems.
- Using data line filters (i.e., ferrite beads) and common-mode chokes on selected nets, especially differential-mode signals when crossing a boundary, such as I/O interconnects (Chapter 6).
- Proper use of bypass capacitors when interfacing to I/O cables to shunt common-mode currents to chassis ground.

As seen in the above list, undesired magnetic flux is only part of the reason on how EMI is created within a printed circuit board. Other areas of concern follows.

- Creation of common-mode and differential-mode currents between circuits and I/O cables due to impedance discontinuities.
- Ground loops creating a propagating magnetic field, which in turn develops common-mode currents if not properly cancelled out.
- Component radiation due to poor package design (i.e., the silicon radiating by itself regardless of how it is implemented on a printed circuit board)
- Lack of, or having a stable power distribution network, causing plane bounce which in turn causes functional disruption to other components.
- Impedance mismatches in transmission lines.

REFERENCES

1. Johnson, H. W., & M. Graham. 1993. *High Speed Digital Design.* Englewood Cliffs, NJ: Prentice Hall.
2. Bogatin, E. 2009. *Signal Integrity-Simplified.* Englewood Cliffs, NJ: Prentice Hall.
3. Montrose, M. I. 1999. *EMC and the Printed Circuit Board-Design, Theory and Layout Made Simple*, Hoboken, NJ: Wiley/IEEE Press.
4. Montrose, M. I. 2000. 2nd ed. *Printed Circuit Board Design Techniques for EMC Compliance*, Hoboken, NJ: Wiley/IEEE Press.
5. German, R. F., H. Ott, & C. R. Paul. 1990. "Effect of an Image Plane on Printed Circuit Board Radiation." *IEEE International Symposium on Electromagnetic Compatibility,* pp. 284–291.
6. Dockey, R. W., & R. F. German. 1993. "New Techniques for Reducing Printed Circuit Board Common-Mode Radiation." *IEEE International Symposium on Electromagnetic Compatibility*, pp. 334–339.
7. Ott, H. 1988. *Noise Reduction Techniques in Electronic Systems.* 2nd ed. Hoboken, NJ: John Wiley & Sons.
8. Ott, H. 2009. *Electromagnetic Compatibility Engineering.* Hoboken, NJ: John Wiley & Sons.
9. Paul, C. R. 2006. *Introduction to Electromagnetic Compatibility*, 2nd ed. Hoboken, NJ: John Wiley & Sons.
10. Paul, C. R., K. White, & J. Fessler. 1992. "Effect of Image Plane Dimensions on Radiated Emissions." *IEEE International Symposium on Electromagnetic Compatibility*, pp. 106–111.
11. Montrose, M. I. 1996. "Analysis on the Effectiveness of Image Planes within a Printed Circuit Board." *IEEE International Symposium on Electromagnetic Compatibility*, pp. 326–332.
12. King, W. Michael, "*EMCT: Electromagnetic Compatibility Tutorial*" Module One, Co-branded by IEEE.

3 Inductance Made Simple

Inductance is defined as the ratio of total magnetic flux that couples (passes) through a closed-loop path to the amplitude of the current that produced the flux. Inductance is described mathematically by (3.1).

$$L = \frac{\psi}{I} \quad (Henries) \tag{3.1}$$

where:
L = total loop inductance
ψ = magnetic flux (Webers), and
I = total current flowing in the transmission line (must have a closed loop circuit for current to flow).

For a closed-loop circuit, inductance is a function of loop geometry as well as the shape and dimension of the wire or transmission line, which includes printed circuit board traces. To describe inductance within a closed loop circuit, we must examine the effects of various forms of inductance which includes *self-inductance, mutual inductance, partial inductance* and *mutual partial inductance.*

Inductance is the ability of a coil, wire or loop to store energy created as a result of a time variant magnetic field that propagates through the transmission line. Inductors oppose voltage present in proportion to the time rate of change of the current present. This property to store energy is called self-inductance.

The term 'inductance' was coined by Oliver Heaviside in February 1886 with the symbol L. The SI unit of inductance is Henry (H), named after American scientist and magnetic researcher Joseph Henry.

The important points to remember about inductance, as discussed herein is that 1) Inductance is everywhere associated with conductors; and 2) Loop area is a critical aspect of defining inductance.

3.1 Types of Inductance

3.1.1 Self-Inductance

Consider a coil with of N turns carrying current I in the counterclockwise direction (Fig. 3.1) as described by the Right Hand Rule. If current flow is steady then magnetic flux through the loop remains constant. However, suppose current I changes with respect to time. According to Faraday's law, an induced EMF (Electromotive Force) will be developed to oppose the change in current flow. The induced current will flow clockwise if $dI/dt > 0$, and counterclockwise if $dI/dt < 0$.

The property of a loop circuit with a magnetic field tends to oppose any change in current. This is called "self-inductance," self-induced EMF or back EMF which we denote as ε_L. All current-carrying loops exhibit this property. An individual inductor is a circuit element with a large value of self-inductance.

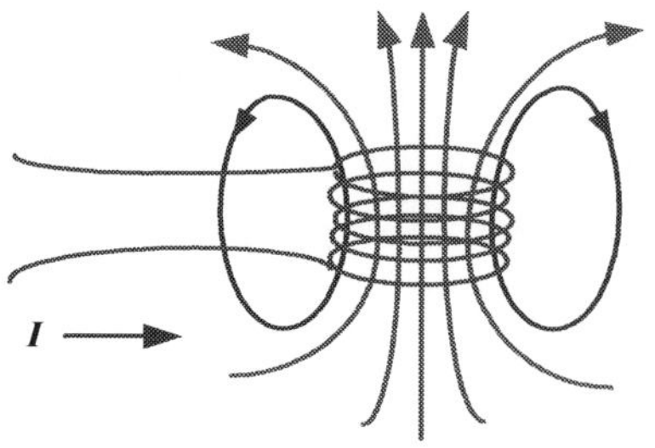

Figure 3.1 Magnetic flux through a current loop with time variant current.

Inductance is fundamentally related to the number of magnetic field lines (flux) that exist around any transmission path. We count the number of magnetic field lines around a current carrying conductor in SI units of Webers, a quantitative definition of self-inductance, *L*, or Henries (Webers/ampers-Equation 3.1). For most interconnect structures, the value of inductance is typically a small fraction of a Henry, such as nanoHenry (*nH*).

The simplest solution of this equation is constant current with no voltage, or a current changing linearly in time with constant voltage present (3.2).

$$v = -L\frac{dI}{dt} \quad (volts) \tag{3.2}$$

where:
v = voltage present
L = total inductance in the loop circuit, and
dI/dt = change in current with respect to time.

Self-inductance is also defined by the number of turns (N) of magnetic flux (Ψ) linking each other divided by the current (I) producing the flux (3.3).

$$L = \frac{N\psi}{I} \quad (Henries) \tag{3.3}$$

Inductance, described by Eqs (3.1, 3.2 and 3.3) is also determined by the magnetic permeability μ of the medium (transmission line) and the size, length and spacing of the conductors that form a current loop path. Inductance in a path cannot be determined until a loop circuit exists that includes both output and input, permitting current to flow.

Voltage induced in a wire depends on how fast the current changes states, which is called switching noise, delta *I* noise, or ground bounce. Delta *I* noise across a conductor does not necessarily dictate which current created the field lines; whether it is the field lines from its own current source or another conductor's current. If there is adjacent time variant current near a wire, some of the field lines from the second wire may interact with the first wire. If the time variant current in the second wire changes states, it will cause a voltage to be created across the primary wire.

When coupling occurs between two transmission lines, generally set up (for example) as traces on a printed circuit board, we call the induced noise in the adjacent conductor "crosstalk" or "cross-coupling." To be able to analyze real world problems that involve multiple conductors it is necessary to be cognizant of all currents which produce field lines. The effects are the same yet more complex when there are many conductors sharing flux paths.

Figure 3.2 illustrates how inductance is defined based on loop area. The smaller the loop area, a smaller amount of total self-inductance due to linkage of magnetic flux propagating in both source and return path as a result of both physical layout and mutual coupling which is detailed later in this Chapter. In Fig. 3.2, both circuits have the same physical transmission line lengths but different self-inductance values.

$\uparrow$ L as the loop area size $\uparrow$

$\uparrow$ L as the wire diameter $\downarrow$

$$L = \frac{\textbf{Magnetic Flux } (\Psi)}{\textbf{Current } (I)}$$

Figure 3.2 Inductance and loop area physical dimensions.

When the diameter of the transmission line wire becomes smaller inductance increases, and when opposing (out-of-phase) differential mode currents in a circuit are brought spatially together, inductance may also be reduced. As for physical size, the larger the loop area between source and return (note that there is no load resistance in the drawing), total inductance will also increase.

Rules of thumb:

- Inductance goes up as the physical loop area increases or the physical diameter of the transmission line (a.k.a. wire) decreases.
- Inductance goes down as wire size diameter increases, or the physical distance between source and RF return path is minimized.

Within a printed circuit board, total loop area includes the distance from a silicon die through its mounting pads to a routed breakout trace on the assembly, the physical routed distance in all axis (x-, y- and z [propagation through vias]) all the way to the load and its return path. Total inductance of a printed circuit board loop can become substantial, enough to cause a signal integrity or EMI problem through the process of common-mode current development across this inductance creating a loss in the transmission line.

3.1.2 Mutual Inductance

A variant of self-inductance is mutual inductance, which describes how time variant voltage present in one electrical circuit induces an electric current in another circuit in the opposite direction, described by Lenz's Law.

To distinguish the source of the field lines, we use the term self-inductance to refer to field lines from its own currents, and mutual inductance to field lines created from linking two or more transmission lines each coupling flux to one another. In other words, the self-inductance of a wire is set up by the number of field lines from its own charge (of current) independent of the presence of another conductor's current.

Before examining an enhanced description of mutual inductance, a discussion on coupled inductors must occur first.

3.1.2.1 Coupled Inductors

When steady-state current flows in a coil, a magnetic field (B) is produced in a second coil (Fig. 3.3). However, when the magnetic field is not changing (steady state), Faraday's law instructs that there will be no induced voltage in the secondary coil. If the switch is opened to stop current flow, there will be a change in the magnetic field in the left side of the illustration. Coil voltage will then be induced in the second coil from the discharge effect.

A coil is a reactionary device, which does not change inductive value with respect to a time variant magnetic field in a given dynamic spectra. Induced voltage present in one coil will cause current to flow in a second coil, which tries to maintain the magnetic field.

The fact that the induced field always opposes the change is described by Lenz' law. Once current flow is interrupted and the circuit re-energized to cause the current to flow again (as in an alternating wave current), an induced current in the opposite direction will oppose that buildup of that magnetic field. This is the basic principle on how transformers work. This change of current in one coil affecting both current and voltage in the second coil is the basis of mutual inductance.

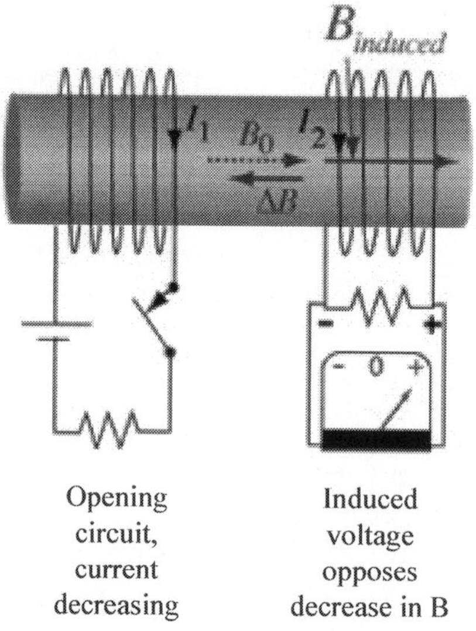

Figure 3.3 Concept of mutual inductance as a transformer model.
Artwork provided courtesy: Georgia State University.

3.1.2.2 Describing mutual inductance

Mutual inductance, M, is defined as the proportionality between EMF generated in a coil to the change in current from another coil that created it. The most common application of mutual inductance is the transformer shown in Fig. 3.3. Excessive mutual inductance is generally the cause of unwanted coupling between conductors in a circuit.

Mutual inductance has the relationship of (3.4):

$$M_{21} = N_1 N_2 P_{21} \tag{3.4}$$

where:
M_{21} = mutual inductance (the subscript specifies the relationship of the voltage induced in coil 2 due to the current in coil 1
N_1 = number of turns in coil 1
N_2 = number of turns in coil 2, and
P_{21} = permeance of the space occupied by the flux.

Mutual inductance also has a relationship with a coupling coefficient. This coupling coefficient is always between 1 and 0 and is a convenient way to specify the relationship between certain orientations of inductors that are linked together by (3.5):

$$M = k\sqrt{L_1 L_2} \tag{3.5}$$

where:
k = coupling coefficient and $0 \le k \le 1$
L_1 = inductance of the first coil, and
L_2 = inductance of the second coil.

Once mutual inductancc, M, is known it can be used to predict the behavior of a circuit using (3.6):

$$V_1 = L_1\frac{dI_1}{dt} - M\frac{dI_2}{dt} \tag{3.6}$$

where:
V_1 = voltage across the inductor of interest
L_1 = inductance of the inductor of interest
M = mutual inductance
dI_1/dt = derivative, with respect to time of the current through the inductor of interest, and
dI_2/dt = derivative, with respect to time of the current through the inductor coupled to the first inductor.

The circuit diagram of Fig. 3.4 is a representation of mutually coupled inductors: a transformer or two transmission lines. This transmission line structure can also be extrapolated to be two traces on a printed circuit board in close proximity, and which can cause inductive crosstalk to occur between the two. The two vertical lines between the inductors indicate a *solid core* that the wires are wrapped around (the same schematic would apply even if the wires were wrapped around an air core). The symbol "n:m" indicates the ratio between windings of the left inductor to the windings of the right inductor, also showing the dot convention which indicate the direction of current flow.

The minus sign arises because of the direction of current flow of I_2. With both currents going into the dots, the sign of M will be positive.

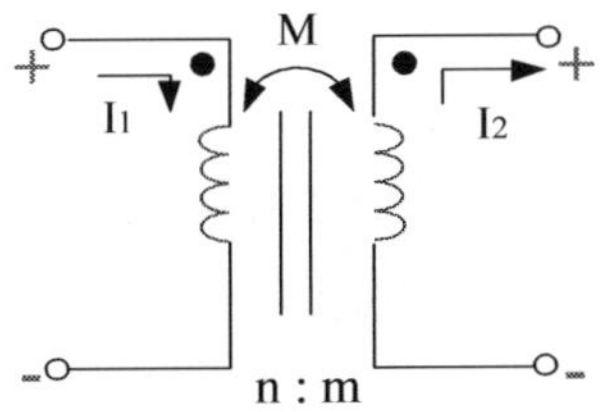

Figure 3.4 Mutual coupling between two inductors or transmission lines.

3.1.3 Partial Inductance

Partial inductance is technically defined as the internal inductance of a conductor due to magnetic flux that is present within the conductor [2], but this definition is not entirely true.

Inductance is defined normally for closed-loop circuits. To simplify the need to study partial inductance, we investigate separate sections of a current loop. This approach allows visualization of overall effects that a transmission path has in a circuit. To lower overall inductance, it is first necessary to reduce the inductance of the section that has the greatest amount of inductance. Reduction may occur by shortening the transmission line length, removing vias, increasing the width or thickness of the conductor, collapsing the distance from one opposing phase of a circuit to the other, or other methods that include trace reorientation.

Partial inductance is useful for estimating the voltage drop across part of a circuit due to the inductance of that particular section. Care must be taken since the voltage drop or potential difference is not uniquely defined in the presence of time-varying fields.

The total partial inductance of a closed-loop segment is the sum of all sections. This is describe by (3.7), which is a way to break overall loop inductance into pieces in order to find total inductance of the transmission line.

$$L_{total} = L_{partial\ segment1} + L_{partial\ segment2} + \ldots + L_{partial\ segment\ n} = \sum_{i=1}^{n} L_i \tag{3.7}$$

With Eq. (3.7), the static current within each segment is identical. L_{total} is the total flux of current in the loop. With this information, we can define partial inductance for a particular segment as the ratio of flux coupling to the current within a particular segment per (3.8).

$$L_{partial\ segment\ i} = \frac{\psi_i}{I} = \frac{flux\ due\ to\ segment\ "i"\ that\ couples\ the\ loop}{amplitude\ of\ the\ current\ in\ segment\ "i"} \tag{3.8}$$

The concept of partial inductance is for a single loop only. Obviously, different loops with different symmetries will have different values of partial inductance.

The total internal inductance of a conductor will decrease when the perimeter or circumference of the loop area becomes physically smaller. Internal impedance (inductance portion of the impedance equation) will still increase with the square root of frequency owing to skin effect. Because of skin effect at higher frequencies (generally greater than 20 MHz), inductance that exists within the center portion of the conductor plays a minor role in the overall inductive performance of the conductor as frequency increases. The parameter of interest is thus partial inductance, which is frequency independent, not the total inductance of the trace or transmission line.

3.1.4 Mutual Partial Inductance

Mutual partial inductance [2, 3, 4] is the key element that allows an image or return plane to provide for flux cancellation. Flux cancellation occurs by having magnetic lines of flux link between two transmission lines create an optimal return path for RF currents to propagate from. Based on the Right Hand Rule (Faraday's Law), if magnetic flux in the source path is counterclockwise, and the return path clockwise in relation to the source path, flux cancellation occurs but only within the minuscule boundary of distance separation.

Self-partial inductance applies to a given segment of a loop independent of location or orientation to any other loop segment. Given current within a wire or trace, a nominal loop is now defined, bounded by a wire segment on one side and infinity on the other. These two perpendicular wire segments extend from the ends of the segments into infinity. This is illustrated in Fig. 3.5. Since self-partial inductance is present between a wire segment and an infinite structure, we can develop the concept of *mutual partial inductance* [4].

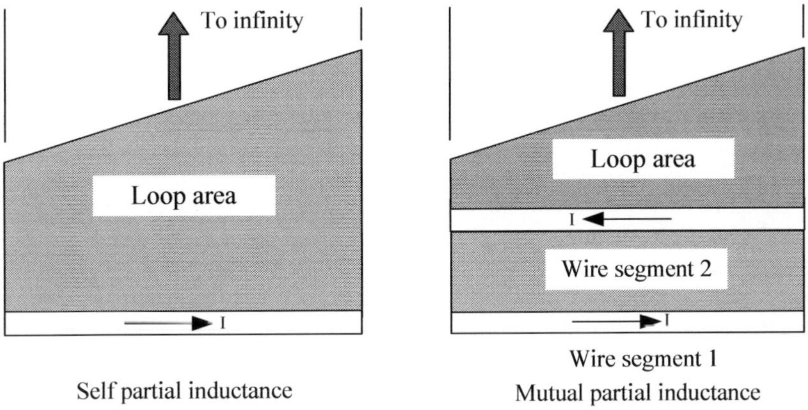

Figure 3.5 Loop area defining self and mutual partial inductance.

Consider an isolated conductor (or trace) with length L carrying current I. The *self-partial inductance* of conductor, Lp, is the "ratio of net magnetic flux generated by a current I passing through the loop (or between a conductor and through infinity beyond the trace), divided by current I within the wire segment" [2].

Self-partial inductance is theoretically independent of the proximity to adjacent conductors. Closely spaced conductors, however, can alter the self-partial inductance of one or both of the transmission lines. This is because one conductor will interact with the other and cause current distributions over the entire length to deviate from a uniform condition. This typically occurs when the ratio of wire separation to radius is less than approximately 5:1. A separation radius of 4:1 for two identical wires means that a third wire may fit between the two original wires, if the wire radius is identical [2].

Between two conductors closely spaced, *mutual partial inductance* exists. Mutual partial inductance, Mp, is based on the distance spacing between parallel traces or wire segments. The distance, s, is the ratio of "magnetic flux due to current in the first conductor that passes between the second conductor and into infinity" to "the current in the first conductor that produced it." Mutual partial inductance is easily observed in Fig. 3.5 with the electrical schematic shown in Fig. 3.6. The voltage developed across the conductors from this configuration is described mathematically by (3.9) [2].

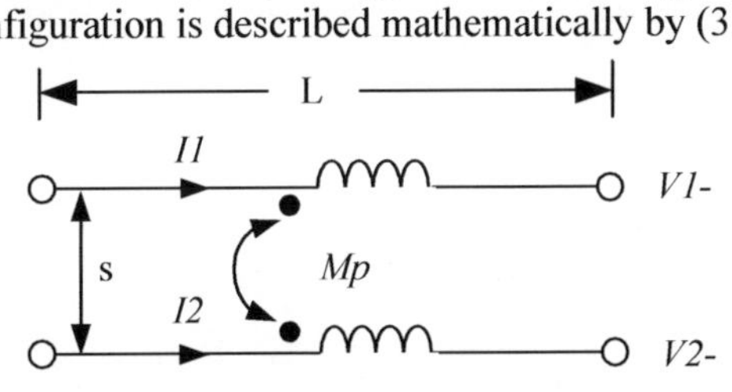

Note: Every trace or conductor contains inductance

Figure 3.6 Mutual partial inductance between two conductors.

$$V_1 = L_{p1}\frac{dI_1}{dt} + M_p\frac{dI_2}{dt}$$

$$V_2 = M_p\frac{dI_1}{dt} + L_{p2}\frac{dI_2}{dt} \qquad (3.9)$$

When performing computational analysis in either the time or frequency domain, we must consider both lumped and distributed inductance, capacitance, and resistance of traces, vias and planes simultaneously. The most difficult parameter to quantify is inductance. Unlike capacitance and resistance, inductance is a dynamic property of a closed-loop current path and the spatial geometry of that path.

With the concept of mutual partial inductance having been presented, consider both traces in Fig. 3.6 are now carrying a signal of interest. The trace identified as $V1$ is the signal path and $V2$ is RF current return. Assume now that both conductors constitute a signal path and its associated return so that $I_1=I_2$ and $I2=-I_1$. If there is no mutual coupling between the two conductors the circuit cannot function since it is not closed-loop (refer back to Chapter 2 for a detailed explanation). If the circuit ends up being a complete loop, voltage drop can be described by (3.10).

$$V_1 = (L_{p1} - M_p)\frac{dI}{dt}$$

$$V_2 = -(L_{p2} - M_p)\frac{dI}{dt} \qquad (3.10)$$

According to (3.10), in order to reduce the voltage drop across a conductor, we must *maximize* mutual partial inductance between that conductor and its associated conductor within the same circuit. The most direct way to maximize mutual partial inductance is to provide a path for RF return current as close as physically possible to the signal trace. The most optimal design technique is use of a plane located adjacent to the signal trace with the smallest distance spacing possible. An alternative way to maximize mutual partial inductance for single- and double-sided printed circuit boards is to provide a return path (trace) adjacent to the signal trace in the same axis with a distance spacing that is also as small as possible.

To view effects of both partial and mutual partial inductance consider two parallel traces or trace over a plane. Partial inductance will always exist in a conductor (by default). Inductance will create an antenna structure at a specific resonant frequency based on physical dimensions. Mutual partial inductance minimizes the effects of partial inductance. By locating two conductors close together, individual partial inductance becomes minimized, which is a desired design requirement for EMI compliance within the boundary of an "image" between the conductors.

To optimize mutual partial inductance, the currents in the two conductors must be equal in magnitude and opposite in direction and phase. This is why image planes (and RF return tracks) work as well as they do. Because mutual partial inductance exists between two parallel wires, a certain amount of inductance must be present. Table 3.1 provides details on mutual partial inductance between two parallel wires with various spacing between the two [2].

Since mutual partial inductance was examined for signal traces above, how does inductance relate to power and return planes separated by a dielectric? The mutual partial inductance between planes is maximized when distance spacing between the two is also minimized. In addition to minimizing mutual partial inductance, inter-plane capacitance is increased, desirable for enhancing the quality of the power distribution network.

Table 3.1 Mutual partial inductance between two parallel transmission lines.

Conductor Separation	Common Length		
	1 inch (2.54 cm)	10 inches (25.4 cm)	20 inches (50.8 cm)
½ inch (1.25 cm)	3.23 nH	137.9 nH	344.9 nH
¼ inch (0.63 cm)	6.12 nH	172.4 nH	414.7 nH
1/8 inch (0.32 cm)	9.32 nH	207.3 nH	484.8 nH
1/16 inch (0.16 cm)	12.7 nH	242.2nH	1.1 nH

3.2 Impedance and Transmission Line Behavior Related to RF Return Current

3.2.1 Typical transmission line configuration

As discussed in Chapter 2, *Transmission Line Theory Made Simple*, every transmission lines consist of resistance, inductance and capacitance. A simplified version of this configuration is illustrated in Fig. (3.7). This topology describes every transmission line structure that exist (wire, coax, cables, printed circuit board traces, etc.) except free space, leading us to discussing the path of least impedance and how both electric and magnetic fields propagate within a dielectric. From here, we apply this principle to real-life applications in a simplified manner.

Understanding electromagnetic theory in a simplified manner is necessary before applying theory to real-world applications.

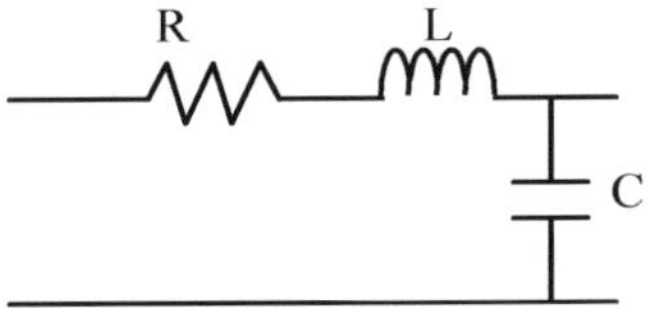

For wiring only: 2 π fL > R for frequencies > 10 kHz

Figure 3.7 Simple example of a transmission line topology.

3.2.2 Path of Least Impedance

Current always take the path of least impedance. Below approximately 10 kHz, current travels through the path of least resistance. Above 10 kHz, the path starts to become inductive due to a higher frequency being propagated within the transmission line (3.11). It is important to note that the complex portion of the impedance equation includes inductance (*L*) and frequency (*f*). As frequency goes up, so does the complex portion of the equation becomes very large quickly, and swamps out the value of resistance.

Impedance equation: $Z = R + j X_L$ (3.11)

Inductive reactance portion: $X_L = 2\pi f L$

- As frequency increases, X_L increases in proportion to frequency
- Resistance and inductance is always fixed due to physical constraints
- The value of R is generally very low, usually a few ohms or less
- X_L starts to become significantly than R for frequencies > 10 kHz

---

Example: A piece of wire (antenna) is in a 10 cm loop configuration.

- Resistance is only a few ohms however the antenna has inductance of 1 μH/cm.
- At 100 MHz, the 10 cm of a loop antenna has $X_L = 2\pi$ (100 MHz) (10 μH) = 6.28 KΩ, which significantly exceeds the value of resistance by many orders of magnitudes.

Most wire, as an example, is approximately a few ohms or less, resistively. This value is not frequency dependent (except for skin effect and related skin depth at very high frequencies). As a result, it is easy to calculate total transmission line impedance of the circuit. At lower frequencies, total impedance will typically be low in the order of a few ohms. For printed circuit board implementation, we seek to have a fixed transmission line for impedance control, which for most applications is a desired 50 ohms (single-ended signal with respect to the return). As the operating frequency approaches 1 MHz, 100 MHz, 1 GHz and above, the resistance portion of the impedance equation becomes irrelevant because jX_L has become very large, sometimes by many orders magnitudes.

Remember, impedance of space is 377 ohms. If the impedance of a transmission line is any value larger than 377 ohms, RF return current will take any path of least impedance, including air to satisfy both Kirchhoff and Ampere as we are dealing with an electromagnetic field and not DC potential in the return path.

One must control the loop inductance for each and every transmission line. It is noted in (3.11) that current propagation within the transmission line is not a variable in the equation.

3.2.3 RF Return Current Travel in a Transmission Line

RF current, regardless of frequency, always takes the path of lead impedance using the following order:

- Least Resistance → Least Reactance → Least Inductance → Smallest Loop
- Below 10 kHz (in typical copper structures) RF current takes the path of least resistance. Above 1 MHz (in typical copper structures) RF currents take the path of least inductance. Between 10 kHz and 1 MHz, current propagation is shared between the two paths with the path having lowest impedance dominating.
- If the load impedance is much greater than the wiring shunt capacitance, wire inductance becomes dominant. With many conductors, the path of least inductance is the one with the smallest loop area.

To enhance visualization of current flow, both at low and at higher frequencies, examine Fig. 3.8. In this figure, we observe RF return current flow in this simple transmission line using a coax connected to an oscillator sweeping the frequency range from DC to 2 MHz. The end of the coax is terminated into a resistor with characteristic impedance that matches the impedance of the coax, in this case 50 ohm. A copper strap connects the shield at both source and load and is long enough to put a current clamp over the strap. The current clamp measures the amount of magnetic flux present, or low frequency return current.

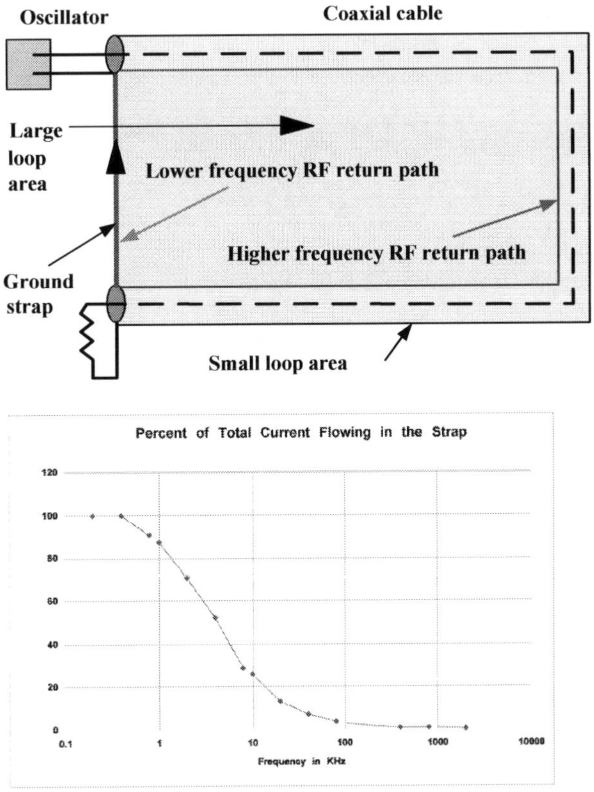

Figure 3.8 RF current flow based on the path of least impedance.
Source: Public domain

The oscillator is varied from DC to 2 MHz. It is observed that at DC level, or thereabouts, 100% of the RF return current flows in the strap. This is due to the "path of least resistance." As frequency increases, the amount of RF return flux observed by the current clamp decreases, with the RF return current mirror imaging against the center conductor of the coax. This mirror image is a result of mutual partial inductance between source and return. Once we approach approximately 100 kHz, nearly all RF return current flow back to the source is in the braid. At 1 MHz, almost no current can be measured by the clamp. The reason for this is due to current taking the "patch of least impedance."

To help visualize RF current flow, Figs. 3.9 and 3.10 is provided [7] showing the RF return path for two signals; 1 kHz and 1 MHz. The configuration is identical to graphic artwork of Fig. 3.8; coax with oscillator and termination.

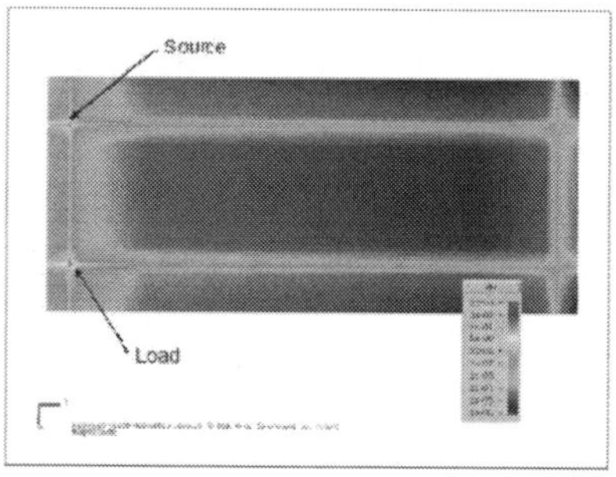

Figure 3.9 Method of moments results for current density at *1 KHz* (least resistance).

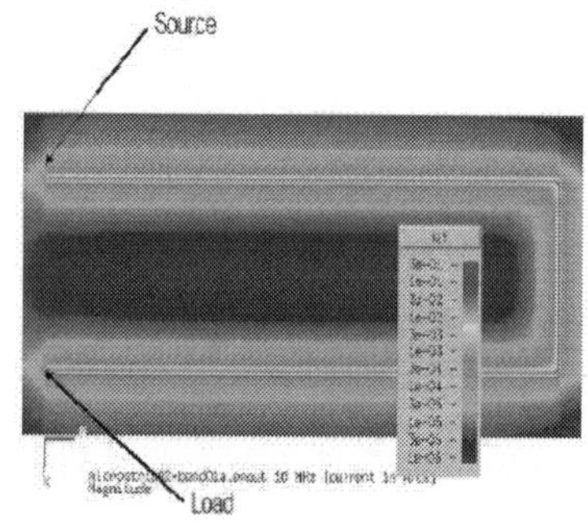

Figure 3.10 Method of moments results for current density at *1 MHz* (least impedance).

3.3 Inductance Concerns Related to Printed Circuit Board Layout and Trace Lengths

3.3.1 Loop Inductance

Consider a transmission line (printed circuit board trace) using the simplified configuration of Fig. 3.11. Both source and load elements are deleted for purpose of this discussion. The source path (top) wraps back on itself (bottom). The current from the source produces field lines that couple to return, but in the opposite direction. We can determine the field lines in the entire loop circuit by adding all lines that exist in both paths. With both wire segments, current in the source is in the opposite direction of return at the same magnitude assuming no losses in the transmission line routing. Field lines from the return path are based on direction of travel from the source and will be in the opposite direction. These flux lines will subtract from the total field lines in the bottom segment.

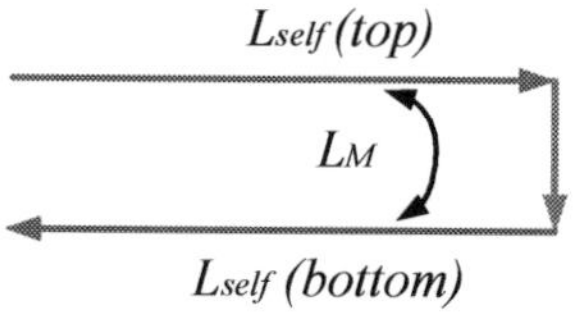

Figure 3.11 Loop inductance.

In a printed circuit board, routing a signal trace as close as possible to a solid plane, or adjacent trace carrying RF return current in the opposite direction is exactly what we want in order to minimize development of common-mode RF current (differential-mode signaling). Flux cancellation, or minimization, is the key to success since we have minimized loop inductance. Loop inductance was described by Eq. (3.2) [$V = -L(dI/dt)$]. By minimizing total [loop] inductance, common-mode voltage that may be present (V) becomes smaller. With less voltage drop across a fixed impedance (e.g., transmission line), less RF current will be developed but not enough to drive antenna structures present within board layout. The mechanism that describes this complex aspect of circuit layout is easily understood using "*Ohms Law*" or "*Maxwell Made Simple®*" (Chapter 1).

3.3.2 Loop Mutual Inductance

Loop mutual inductance is actually the self-inductance of a loop. If there are two independent loops, there is mutual inductance between the two which is the number of field lines from RF current in one loop that transfers a magnetic field to the other. When current in one loop changes states it also changes the number of field lines around the other loop and induces noise in that loop. The amount of noise coupling is described again by Eq. (3.2) [$V = -L(dI/dt)$].

One of the coupling paths in the two adjacent transmission lines is actually the mutual inductance between loops. When the return path is not a uniform plane, mutual inductance between two adjacent signal and return paths can be large as in the case of an IC package or a connector. This is commonly called simultaneous switching noise or SSN. The most important way of reducing SSN is to reduce the mutual inductance between loops by moving them farther apart or decreasing the physical area of each loop, for example.

When return currents of two signal lines share the same conductor, loop mutual inductance is dominated by the partial self-inductance of the overlap region. This inductance can be very large, and is the primary reason to have independent return path pins in a connector, or a separate pin for each signal path.

3.3.3 Decoupling Capacitor Mounting Related to Lead Inductance

As shown in Fig. 3.12, two common mounting dimensions are provided for both a 0603 and 0402 surface mount capacitor. Although the package size is basically the same, lead inductance is still identical based on manufacturing requirements to have a 10 mil (0.254 mm) breakout trace from pad to via. Even this small distance route can be highly inductive, enough to inhibit the capacitor from performing to its maximum capability of providing charge when used in a decoupling application.

Table 3.2 and 3.3 [7] provides numeric values on how inductive breakout traces are on a printed circuit board.

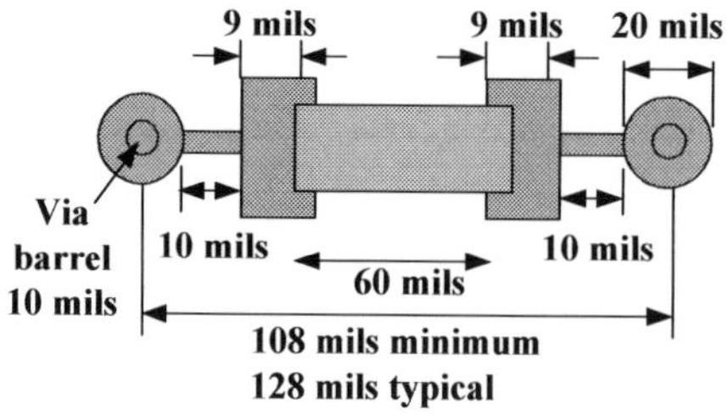

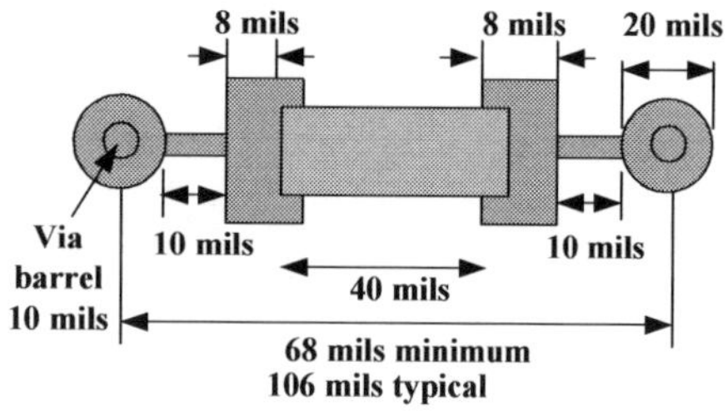

Figure 3.12 Capacitor mounting dimensions.
[Ref. 7]

Table 3.2 Mounting inductance for typical capacitor configurations.

Distance into planes (mils)	0805 typical/minimum 148 mils (3.76 mm) between vias)	0603 typical/minimum 128 mils (3.22 mm) between vias)	0402 typical/minimum 106 mils (2.69 mm) between vias)
10	1.2 nH	1.1 nH	0.9 nH
20	1.8 nH	1.6 nH	1.3 nH
30	2.2 nH	1.9 nH	1.6 nH
40	2.5 nH	2.2 nH	1.9 nH
50	2.8 nH	2.5 nH	2.1 nH
60	3.1 nH	2.7 nH	2.3 nH
70	3.4 nH	3.0 nH	2.6 nH
80	3.6 nH	3.2 nH	2.8 nH
90	3.9 nH	3.5 nH	3.0 nH
100	4.2 nH	3.7 nH	3.2 nH

Table 3.3 Mounting inductance of capacitors
50 mils (12.7mm) total distance from capacitor pad to via pad.

Distance into planes (mils)	0805 typical/minimum 208 mils (5.28 mm) between vias	0603 typical/minimum 188 mils (4.8 mm) between vias	0402 typical/minimum 166 mils (4.2 mm) between vias
10	1.72 nH	1.6 nH	1.4 nH
20	2.5 nH	2.3 nH	2.0 nH
30	3.0 nH	2.8 nH	2.5 nH
40	3.5 nH	3.2 nH	2.8 nH
50	3.9 nH	3.5 nH	3.1 nH
60	4.2 nH	3.9 nH	3.5 nH
70	4.5 nH	4.2 nH	3.7 nH
80	4.9 nH	4.5 nH	4.0 nH
90	5.2 nH	4.7 nH	4.3 nH
100	5.5 nH	5.0 nH	4.6 nH

3.3.4 Via Configuration and its Effect on Lead Inductance

Figure 3.13 illustrates, in conceptual form, various mounting configurations commonly found on a printed circuit board. Trying to put specific inductance values for these configurations provides no value for the purpose of this discussion since inductance is dependent on loop size, dimensions of the trace, via and mounting pads, the interconnect distance to planes internal to the printed circuit board (z-axis), and other parametric values, however the range provided below is fairly accurate and give a visualization of what may occur in a printed circuit board.

The typical inductance for interconnects (vias) are on the order of:

- Pair of surface traces 10–15 nH/inch (25-38 nH/cm)
- Pair of vias for decoupling capacitor 0.40–1 nH and 200–500 pH each
- Plane inductance 0.1 nH

These numbers indicate that trace and via inductance are significantly greater than planar inductance, which is usually in the picoHenry range.

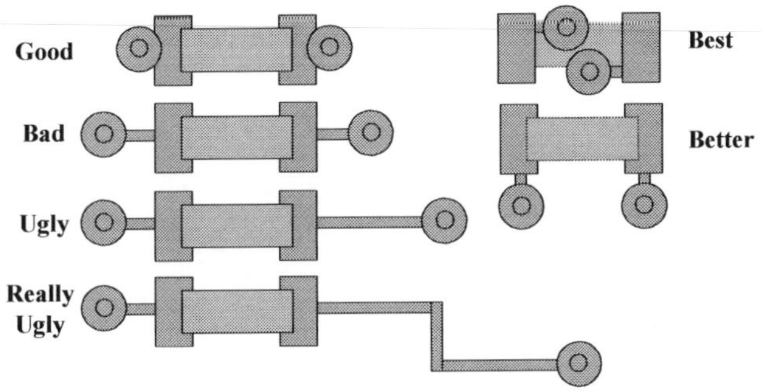

Figure 3.13 Typical via configuration and their relationship to lead inductance. [Ref. 7]

Another method on how to enhance decoupling capacitor placement on a printed circuit board is to minimize lead inductance (Fig. 3.14). There is less inductance as we move from the left to right in the illustration due to having a smaller loop area, thus smaller loop inductance. Smaller loop inductance in turn leads to higher self-resonant frequency operation of the component, along with enhanced performance related to signal integrity and minimizing development of undesired common-mode RF energy.

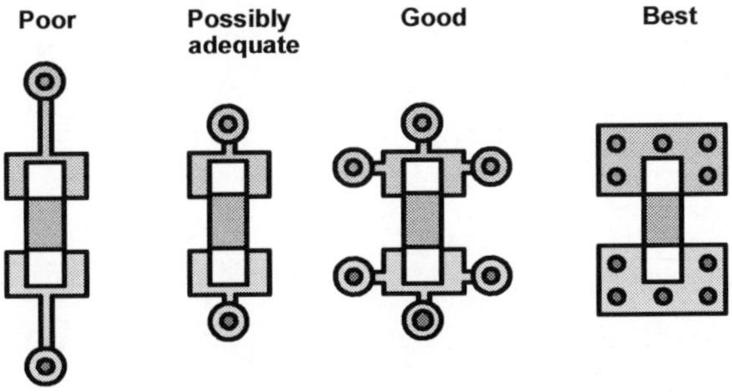

Figure 3.14 Improvement on minimizing lead inductance for capacitor placement.

REFERENCES

1. Montrose, M. I. 1999. *EMC and the Printed Circuit Board Design-Design, Theory and Layout Made Simple.* Hoboken, NJ: John Wiley & Sons/IEEE Press.
2. Paul, C. R. 2006. *Introduction to Electromagnetic Compatibility*, 2nd ed. Hoboken, NJ: John Wiley & Sons.
3. Dockey, R. W., & R. F. German. 1993. "New Techniques for Reducing Printed Circuit Board Common-Mode Radiation." *IEEE International Symposium on Electromagnetic Compatibility,* pp. 334–339.
4. Hubing, T. H., T. P. Van Doren, & J.L. Drewniak. 1994. "Identifying and Quantifying Printed Circuit Board Inductance." *IEEE International Symposium on Electromagnetic Compatibility,* pp. 205–208.
5. Ott, H. 1988. *Noise Reduction Techniques in Electronic Systems.* 2nd ed. Hoboken, NJ: John Wiley & Sons.
6. Ott, H. 2009. *Electromagnetic Compatibility Engineering.* Hoboken, NJ: John Wiley & Sons.
7. Archambeault, B. *Inductance and Partial Inductance – What's it all mean?* Retrieved from http://web.mst.edu/~jfan/slides/Archambeault3.pdf.

4 Power Distribution Made Simple

For every electrical product regardless of application, having an optimal power distribution network (PDN) is critical. Components on printed circuit boards are sensitive to fluctuations in the power and return distribution system; electric motors function properly when a specific voltage and frequency is present. Telecommunication systems require an electromagnetic field to propagate uninterrupted energy between source and load. Without a stable power distribution network, signal integrity is not ensured. These are only an example of electrical products that require a power system that must not have under any condition switching noise, power fluctuations, spikes, droops, blackouts along with a host of other items that may cause functional disruption.

In this Chapter we begin by defining the need for a power distribution system and how it relates to transmission line theory. From here, discussion occurs on various applications of capacitor use and their relationship to maintaining a stable power delivery network. Other aspects of capacitor use in PDNs includes resonances, energy storage, impedance, using paralleling discrete devices, lead inductance, physical placement, hazards associated with capacitor implementation and buried capacitance, along with additional topics not listed in this long list.

4.1 The Need for Optimal Power Distribution

An optimal power distribution system is required for:

- Providing a stable voltage and return reference between components to ensure functional operation.
- Distributing optimal power to components in order to minimize plane bounce or switching noise from creating common-mode currents on internal antenna structures.

A stable voltage and return system allows components, both analog and digital, to operate within certain tolerance levels without functional degradation. If the voltage rail changes by several millivolts, sensitive or critical analog circuits may stop functioning properly. Digital devices are more tolerant to voltage swings however, if the period of the voltage variation exceeds switching logic transition levels of digital devices (rising and/or falling edge rate) signal integrity problems may occur. Essentially, what we would like to have is an unlimited amount of pure uninterrupted DC voltage appearing at the power and return pins of components without noise spikes or degradation of voltage and current levels.

A key item of concern when designing a PDN is to:

1. Use a low impedance connection between circuits to minimize voltage drops in the transmission path.
2. Between components there is always a finite physical distance to transfer charge which means a finite voltage drop will occur across this distance. Depending on the inductance of this particular interconnect significant fluctuations of power may occur.

In Fig. 4.1, two simplified circuits connected to a DC power source are presented. Interconnects between components are represented schematically as an inductor. Current must propagate through the transmission line to charge the input capacitance of the component connected to the PDN without significant loss of parametric values such as voltage and current levels, or timing margins. Inductors cause a voltage drop to occur, which could be significant based on various transmission line parameters.

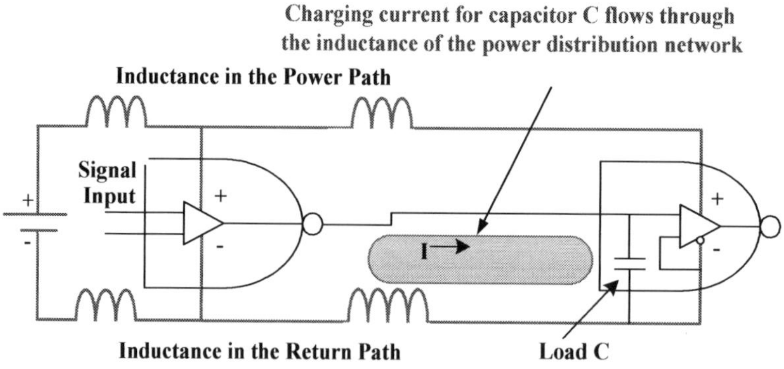

Figure 4.1 Simplified visualization of a power distribution system.

Each circuit device has a both an operating voltage maximum and minimum specification in which functional operation is guaranteed. To inhibit any change in supply voltage that otherwise would occur due to varying current demands, it is necessary to place charge reservoirs of current adjacent to circuits and to minimize series inductance of the interconnect (prevent voltage drop from occurring). A capacitor functions as an optimal charge reservoir because, as opposed to an inductor, the voltage across a capacitor (neglecting parasitic elements) will not change abruptly during a surge of current consumption.

4.2 Power Distribution Network as Transmission Lines

Power distribution networks can be represented as a two conductor transmission line with a defined characteristic impedance and propagation delay. Planes within a circuit board are in reality transmission lines, identical to signal traces that are routed between components. The difference between a plane and trace is physical dimensions only. Planes are extremely wide in both x- and y-axis while signal traces are very narrow. Both transmission lines propagate a signal (e.g., power or data) between devices. A return path must also be present (Figure 4.2).

In Fig. 4.2, lower illustration, there is inductance shown only in the power rail (i.e., physically small trace with high impedance) but little to no inductance in the return path (wide plane and low impedance). There is always a finite amount of inductance present. The reason why minimal inductance in the return path is described in greater detail later in this Chapter.

4.3 Primary Requirements for Enhanced Power Distribution

The amplitude of a power supply transient response is directly proportional to the characteristic impedance of the power distribution system, Z_0. It is thus critical for stability to exist in order to reduce the impedance of the PDN to the lowest value possible. This is easily explained using Ohms law, $V=IZ$. With a fixed voltage level, high impedance in the PDN will minimize the amount of available current. If we have very low impedance (Z), a larger amount of current (I) is available for component operation and plane bounce is minimized.

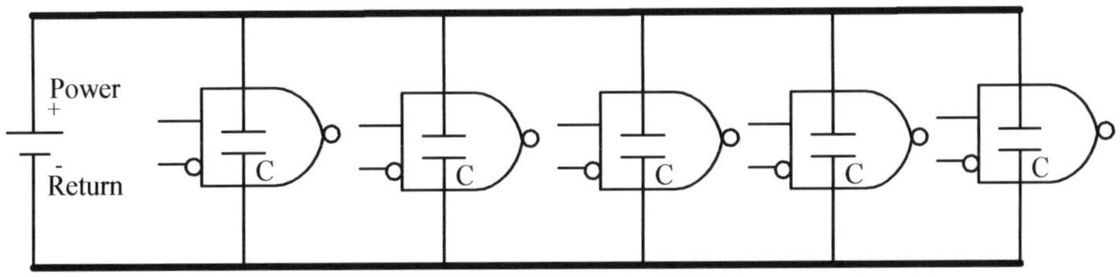

Ideal Power Distribution Network – Multiple Loads

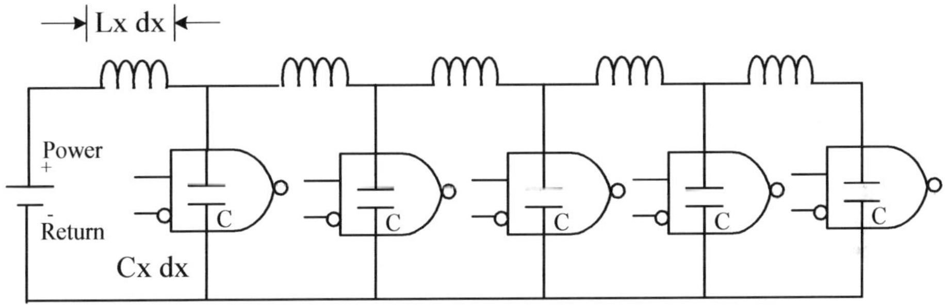

Real-Life Power Distribution Network – Multiple Loads

Figure 4.2 Power distribution network represented as a transmission line.

The impedance of every transmission line is described by (4.1) in two ways. One way is using characteristic inductance and capacitance in the frequency domain (refer to Chapter 2 for details), and the other Ohms law, where (x) represents change in voltage or current with respect to time. Both are exactly identical and differ only in the manner of consideration during the design stage. With the need to have low impedance for the PDN, we do so by minimizing inductance (numerator) or increasing capacitance (denominator). This can be achieved on a printed circuit board through the process of:

- Reduction of loop area between conductors (less inductance)
- Placing conductors as close together as possible (greater capacitance and spatially cancelling inductance)
- Increase of conductors' width (less inductance and greater capacitance, with respect to another conductor)

$$Zo = \sqrt{\frac{Lo}{Co}} = \frac{V(x)}{I(x)} \qquad (4.1)$$

4.4 Defining Capacitor Usage on Printed Circuit Boards

It is generally assumed that a capacitor has only one function, to provide energy charge. This is an incorrect assumption. The technical definition of a capacitor is "two conductors separated by a dielectric." Storing energy is only one application of use. In this section we discuss three primary functions, although there are other applications such as filtering and wave shaping in analog circuits.

We need within a PDN a means of overcoming physical and time constraints caused by digital circuitry switching logic levels. Digital logic usually involves two possible states, "0" or "1." The setting and detection of these two states are achieved with switches internal to components that determine whether the device is to be at logic LOW ("0") or logic HIGH ("1"). There is a finite time period for the device to make this determination. Within this window, a margin of protection is provided to guarantee against false triggering, but with limits. Moving the logic switching state near the trigger level creates a degree of uncertainty. If we add high-frequency switching noise, the degree of uncertainty increases and false triggering may occur.

Importantly, the three most common applications for capacitors when used in a printed circuit board are: bulk, bypass and decoupling. Other functional applications are discussed later in this chapter.

> *Note: It is common practice for engineers to interchange the term bypass with decoupling, assuming they perform the same function. This assumption is only partially "correct." There is a significant difference in application and use. Under certain conditions, a capacitor that is located between power and return to minimize switching noise (decoupling function) may also transfer RF energy from one plane to another (bypassing).*
>
> *A capacitor in the picofarad (pF) range provides minimal to no value for decoupling due to limited energy storage capacity, yet it will easily shunt switching noise in the upper MHz range to the 0V reference. A bypass capacitor can be provided in the same-axis (i.e., ground to ground), whereas decoupling is always used in the z-axis (between power and return planes) within a printed circuit board stackup.*

Optimal implementation is achieved using appropriate capacitors for a specific application: bulk, bypass, and decoupling. All capacitor use must be engineered for a specific function and not left to chance: including self-resonance, equivalent series inductance (*ESL*), equivalent series resistance (*ESR*), packaging dimensions, temperature and voltage ratings. In addition, the dielectric characteristics and properties of the capacitor must be properly selected and not be left to random choice from past usage, experience or rules-of-thumb.

4.4.1 Bulk Capacitor Description

A bulk capacitor is typically a large value capacitor (e.g. 1 μF to 1000 μF) that helps maintain constant DC voltage and current levels within a PDN. Bulk capacitors ensure there is sufficient power (voltage and current) available to all components within a specific radius of operation. With every binary edge transition, components consume current for functional operation as well as driving signal transmission lines. Components receive power from planes (power/return) or specially designed traces. This source of energy needs to be replaced quickly, which is performed by localized power sources generally called decoupling capacitors. After decoupling capacitors deliver their charge, they themselves need to be recharged. Bulk capacitors located in the vicinity of the decoupling devices provide this recharge through the interconnect impedance of the planes, which is very small when implemented carefully.

If a printed circuit board has many large power consuming devices, the voltage drop from lumped combination of current and resistance usage (*IR* drop) may be excessive. Bulk capacitors brings the voltage level back to where it is supposed to be.

To illustrate voltage drop in a printed circuit board with many components consuming current at the same time (i.e., +5 VDC) with fixed transmission line impedance at a distance from the power connector, the IR drop inherent in the PDN may now have dropped to a steady state value of +4.5 V. Digital components are not reliable if they are rated for +5 V and input voltage drops below +4.75 V. A bulk capacitor brings the voltage level back up to +5 V, thus ensuring functional operation and stability.

There are many types of capacitors used for bulk applications. They are typically identified by their dielectric component such as; electrolytic, tantalum, lithium ion, mica and ceramic, then by the nature of the dielectric: polarized on non-polarized.

4.4.2 Bypass Capacitor Description

A capacitor used in a bypass application transfers RF energy from one area to another, such as RF fields on the shield of cable to chassis ground; transferring RF return current across a split plane to minimize loop area and development of *L(dI/dt)*; or shunting high-frequency switching noise from a power plane to 0V or chassis ground. Use of bypass capacitors is a primary element in removing unwanted RF energy from causing harmful disruption, in addition to providing other functions of filtering (bandwidth limited).

Bypass capacitor selection is often based on the self-resonant frequency of operation versus other critical parametric items. Details on calculating this value is found in Section 4.6. An important application for bypass capacitors, generally forgotten or unknown by designers, is to spread them all over the printed circuit board even in areas where there are no components to help lower the impedance of the power/return plane pair. This may include capacitors in the pF range, depending on the physical distance between the planes. If the impedance of the PDN is high at a particular physical location, significant plane bounce may occur causing functional disruption to components located elsewhere on the printed circuit board. This is due to reflective wave switching developed as a result of an impedance discontinuity, such as a via.

4.4.3 Decoupling Capacitor Description

Decoupling capacitors minimizes RF energy from being injected into the power distribution network from components consuming DC current at the speed of the switching device. Decoupling capacitors also provide a localized source of DC power and is particularly useful in reducing peak current surges propagated across the board.

Since the voltage level specification for components is fixed, changing power demands are manifested during each state transition as current is demanded. The power supply must accommodate for variations in current draw with as little change as possible in the PDN, especially the reference voltage. When current demand within a device changes rapidly, the PDN may not be able to respond to delivery of charge instantaneously. As a consequence, the voltage level may drop for a brief period before the power supply can reestablish itself back to full value. Per Ohms law, with a fixed impedance (interconnect between silicon and the PDN), and an increase in current draw, the voltage level drops concurrently with the period of time that current is being consumed. The DC power supply must quickly adjust to provide optimal current delivery but can only effectively maintain the output voltage for events at frequencies from DC to a few hundred kHz due to the switching speed of the voltage regulator used in the power supply. For transient events that occur at frequencies above this range, there is a time lag before the power supply can respond to new current demand level. This is where use of bulk and decoupling capacitors become mandatory.

Decoupling capacitors work as a local energy storage device directly at the input pins of components to minimize plane bounce. This capacitor cannot provide *continuous DC power* because it stores only a small amount of energy, however this energy can respond very quickly to changing current demands. These capacitors effectively maintain power-supply voltage at frequencies from hundreds of kHz to several hundred MHz (in the milliseconds to nanoseconds range). Because of inherent inductance and resistance in components, decoupling capacitors are less effective for switching events occurring above or below this range, and in reality provide less benefit above 200 MHz when contrasted against the impedance structure of a well-developed PDN. Above this frequency range, use of an internal decoupling capacitive structure is required, meaning the power and return planes become the primary source of dynamic charge. Buried capacitance is discussed in Section 4.22.

4.5 Review of Resonance (Basic Circuit Analysis)

Resonance occurs when the reactive value difference between the inductive and capacitive vector is zero. This is equivalent to saying that the circuit is purely resistive in its response to AC voltage. Three types of resonant circuits are common, although more types of basic circuit configurations exist.

- Series resonance
- Parallel resonance
- Parallel *C*–series *RL* resonance

Resonant circuits are frequency selective because they propagate RF current at specific frequencies. A series *LCR* circuit will pass this frequency (measured across *C*) if *R* is high and source resistance is low. If *R* is low and source resistance high, the circuit will reject the signal that happens to be present at that calculated frequency. A parallel resonant circuit placed in series with a load will thus reject a specific frequency, the opposite of series.

4.5.1 Series Resonance

The overall impedance of a series *RLC* circuit is defined by $Z = \sqrt{ESR^2 + (X_L - X_c)^2}$. If an *RLC* circuit is to behave resistively, values can be calculated where ω $(2\pi f)$ is called the *resonant angular frequency* (Fig. 4.3).

$$X_L = X_C$$

$$\omega L = \frac{1}{\omega C}$$

$$\omega = \frac{1}{\sqrt{LC}}$$

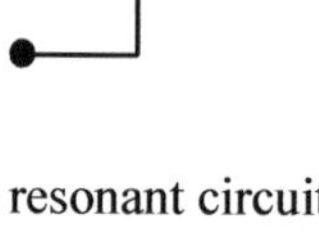

Figure 4.3 Series resonant circuit.

With a series *RLC* circuit at resonance:

- Impedance is at a minimum
- Impedance equals resistance
- The phase angle difference is zero
- Current is at a maximum
- Power transfer (*IV*) is maximum

4.5.2 Parallel Resonance

A parallel *RLC* circuit is shown in Fig. 4.4. The self-resonant frequency is the same as for a series *RLC* circuit.

$$\omega = \frac{1}{\sqrt{LC}}$$

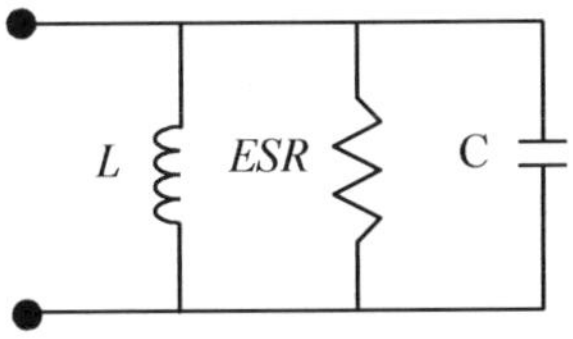

Figure 4.4 Parallel resonant circuit.

With a parallel *RLC* circuit at resonance:

- Impedance is at a maximum
- Impedance equals (equivalent series) resistance
- The phase angle difference is zero
- Current is at a minimum
- Power transfer (*IV*) is minimum.

This circuit configuration produces a resonance when inductance *L* equals capacitance *C at a specific frequency*. The resistor in this configuration can be of significant benefit under certain conditions to prevent parallel resonance.

If the equivalent series resistance becomes very large, it may exceed inductance at the resonant frequency. If resistance is just right, inductive impedance is equal to resistance, again at the resonant frequency which is what we desire. Resonance is the condition under which the inductive impedance is equal to the capacitive impedance.

4.5.3 Parallel *C*–Series *RL* Resonance (Anti-resonant Circuit)

Practical resonant circuits generally consist of an inductor and capacitor in parallel. The inductor will have some resistance that when combined with the X_L inductive value, changes the bandwidth of the resonance through a process known as "Q" or quality factor. Low-Q resonances have a wider and effective decoupling bandwidths. High-Q resonances produce "notch" bandwidths that are not broadly useful for energy storage but find valuable use in filter networks. The equivalent circuit of a parallel *C*-series *RL* is shown in Fig. 4.5. The resistance in the inductive branch may be a discrete, physical element or the internal resistance of the physical inductor.

At resonance, the capacitor and inductor trade the same stored energy on alternate half cycles. When the capacitor discharges the inductor charges, and vice versa. At the anti-resonant frequency, this tank circuit presents high impedance to the primary circuit current, even though the current within the tank is high. Power is dissipated only in the resistive portion of the network.

$$\omega = \sqrt{\frac{1}{LC} - \left(\frac{R}{L}\right)^2} \approx \frac{1}{\sqrt{LC}} \quad [R << \omega_0 L]$$

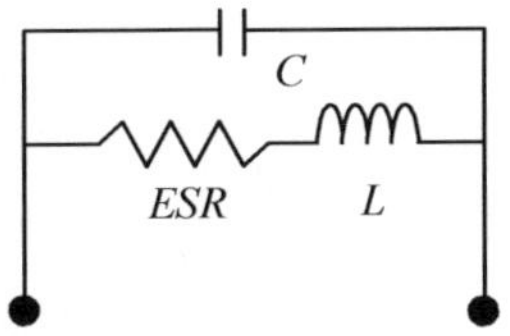

Figure 4.5 Parallel C–Series RL resonant circuit.

The anti-resonant circuit is equivalent to a parallel *RLC* circuit with resistance is Q^2R.

4.6 Physical Characteristic of Capacitors

An *ideal* capacitor has no losses in within its dielectric. Current is always present between the two plates. Because of this current, an element of finite inductance is associated with the parallel plate configuration with one plate charging while its counterpart is discharging. A mutual coupling factor is added to the overall performance and operation.

All capacitors can be modeled as an *RLC* circuit, where *L* (inductance) relates to lead length and body construction, *R* (resistance) in the leads and *C* (capacitance). A schematic representation is shown in Fig. 4.6, with variable "*d*" the physical distance spacing between the two plates.

At a calculable frequency presented later, the series combination of *L* and *C* becomes resonant providing very low impedance and effective RF energy shunting at that specific frequency. At frequencies above self-resonance, the impedance of the capacitor becomes increasingly inductive and performance, especially for use in decoupling becomes less effective. Hence, both bypassing and decoupling applications are affected by lead inductance (surface mount, radial, or axial styles) in addition to the physical trace length interconnect between capacitor and the PDN.

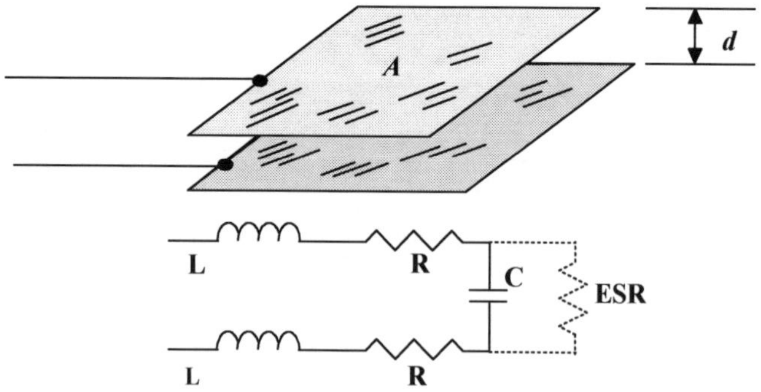

Interconnects internal to the capacitor plates contain both inductance and resistance – *ESR* is mainly the resistance inherent within the dielectric material

Figure 4.6 Capacitor model with resistance, inductance and capacitance.

As shown in Fig. 4.6, there will always be some series resistance in the lead interconnect from the capacitor pad to the printed circuit board, which is magnitudes less compared to the dielectric resistance (*ESR*) that exist between the two plates. The value of *R* (interconnect) may be pico-ohms and *ESR* in the milliohms range (internal resistance of the dielectric material). This means during simulation analysis we can generally ignore the *R* element, with *ESR* the primary parametric value of concern.

4.6.1 Capacitor Types

There are various types of capacitor families, all with different features and performance capabilities based on dielectric material used. A brief summary of "commonly" used capacitors are detailed in Tables 4.1 and 4.2.

Table 4.1 Summary of various capacitor types.

Electrolytic	Large capacitance value Large physical size Low ESR Low end of life capacitance value
Tantalum	Capacitance value from 1 μF to 1000 μF Medium to small package size Large range of ESL, some with low ESR
Ceramic	Very small capacitance values Small package size Very low ESR Lowest cost with the highest reliability
Capacitor arrays	Ceramic capacitors dielectric Multiple contacts per device package Very low ESL High cost

Table 4.2 Typical usage of capacitor families and operating range.

Power supply DC/DC converters	DC to 2 kHz
Large capacitors Electrolytic or tantalum	2 kHz to 1 MHz
Small capacitors Tantalum or ceramic	1 MHz to 50 MHz
Printed circuit board planes Between the power and ground planes	50 MHz and up
Integrated circuit package Between power and ground planes	Above 100 MHz
Inside the integrated circuit (the silicon die) Thin oxide capacitor	Above 500 MHz

4.6.2 Commonly Used Dielectrics

The following capacitor styles are commonly used, all based on the type of dielectric utilized. Regardless of dielectric family, the important item to remember is application and environment of use.

1. **Ceramics**: This is the most commonly used capacitor style on printed circuit boards. The main difference between ceramics is the temperature coefficient that affects capacitance and dielectric loss. Both are an important part of any power distribution network. The question one must ask when choosing a ceramic capacitor is "Which dielectric version should be selected"?
   - **C0G or NP0** (negative-positive-zero): This style is typically found in the 1 pF to 100 nF range, with ±5% tolerance maximum. NP0 dielectrics have the lowest losses of the ceramic family and are used in filters as timing elements and for balancing crystal oscillators with the lowest ESR possible along with outstanding temperature/voltage properties. They are physically larger in size and more expensive than standard ceramics. NP0 also refers to the shape of the capacitor's temperature coefficient graph (how much the capacitance changes

with temperature). NPO means that the graph is flat and the device is not affected by temperature changes.

- **X7R** capacitors have reasonable voltage and temperature coefficients and are the most popular capacitor used for decoupling applications. Typical values are from 100 pF to 22 μF, 10% tolerance, and are good for non-critical coupling and timing applications. Temperature values can go up to 125 °C.
- **X5R** is similar to X7R, with reduced temperature reliability and values extended to 100 μF.
- **Y5V** is used to achieve high capacitance values, but with very poor voltage and temperature characteristics.
- **Z5U** dielectric is used to achieve high capacitance but with much worse voltage and temperature characteristics than Y5V. It is the cheapest ceramic capacitor available, used in non-critical circuits and has typical values from 1 nF to 10 μF with 20% tolerance. Acceptable only for bypass or decoupling applications where there is no significant change in ambient temperature.

When choosing a ceramic capacitor, one will see a code that describes capacitor characteristics (Table 4.3). The parametric values in red are the most commonly used values for capacitors on printed circuit boards.

Table 4.3 Capacitor classification code (Class 2).

EIA Class 2 Classification					
Minimum Temperature		**Maximum Temperature**		**Capacitance Change Permitted**	
X	-55 °C	4	+65 °C	A	±1.0 %
Y	-30 °C	5	+85 °C	B	±1.5 %
Z	-10 °C	6	+105 °C	C	±2.2 %
		7	+125 °C	D	±3.3 %
		8	+150 °C	E	±4.7 %
		9	+200 °C	F	±7.5 %
				P	±10 %
				R	±15 %
				S	±22 %
				T	+22 % / -33 %
				U	+22 % / -56 %
				V	+22 % / -82 %

There are two basic ceramic capacitor families, Class 1 and Class 2, with Class 2 being the version used in nearly all applications and are highly popular. Class 1 capacitors are chosen by those requiring specific application of use for critical circuits or unique environments that are extreme such as satellite deployment.

- Class 1 ceramic capacitors are accurate, temperature-compensating capacitors and are the most stable version over voltage, temperature and to some extent frequency. These capacitors also have the lowest losses possible and therefore are especially suited for resonant circuit applications where parametric characteristics are essential, or where a precisely defined temperature coefficient is required. An example of use is in compensating temperature effects for a given circuit that must be stable over a wide range of temperature.

- Class 2 ceramic capacitors have a dielectric with high permittivity and therefore better volumetric efficiency than a Class 1, but with lower accuracy and stability. The ceramic dielectric is characterized by a nonlinear change of capacitance over the total temperature range and whose capacitance value additionally depends on the voltage applied. Class 2 capacitors are suitable for bypass and decoupling applications or frequency discriminating circuits where low losses and high stability are not of major importance. The nomenclature of dielectrics with regard to stability is provided in Table 4.3.

An inexpensive and commonly used dielectric material is Z5U (barium titanate ceramic). This material has a high dielectric constant that allows small capacitors to have large capacitance with self-resonant frequencies from 1 MHz to 20 MHz, depending on design and construction. Above this self-resonance frequency, performance of Z5U decreases as the loss factor of the dielectric becomes dominant which limits its usefulness to approximately 50 MHz.

Another dielectric material commonly used in critical circuits is NP0 (strontium titanate). This material has enhanced higher-frequency performance owing to its lower dielectric constant. Capacitors using this material are unsuitable for decoupling below 10 MHz. NPO is also a more temperature-stable dielectric. Capacitance value (and self-resonant frequency) is less likely to change when the capacitor is subjected to changes in ambient temperature.

A problem when Z5U and NP0 are provided in parallel is that the higher dielectric material, Z5U, can damp the resonance of the more frequency-stable, low-dielectric constant material NP0. For switching noise below 50 MHz it is preferred to use only a good, low inductance Z5U (or equivalent) capacitor. This is because Z5U combines excellent low-frequency decoupling with reduction in radiated emissions, if temperature tolerance is not a concern.

2. Electrolytic capacitors have a larger capacitance per unit volume making them valuable in relatively high-current, low-frequency electrical circuits, i.e., power-supply filters or as coupling capacitors in audio amplifiers.
   - **Aluminum electrolytic.** These capacitors have very large capacitance to volume ratio, inexpensive and polarized. Primary applications include smoothing and storage reservoir of capacitive energy for use in power supplies.
   - **Tantalum electrolytic.** These capacitors have large capacitance to volume ratio, smaller size, good stability, wide operating temperature range and long reliable operating life, extensively used in miniaturized equipment and computers. They are available both polarized and non-polarized.
3. Other less common capacitor families include the following. This is not a comprehensive list of all capacitor types available. Those not listed are used in very specialized applications beyond the scope of this book:
   - **Paper**: This style of capacitor found common use in antique radio equipment, consisting of a paper dielectric and aluminum foil layers rolled into a cylinder and sealed with wax. These capacitors had values up to a few µF and working voltage up to several hundred volts. They also consist of oil-impregnated bathtub types with a rating up to 5 kV for motor starting and high-voltage power supplies, and up to 25 kV for large oil-impregnated energy discharge.
   - **Polycarbonate**: These capacitors are good for filter circuits with low temperature coefficient and excellent properties, however they are expensive.
   - **Polyester**: This capacitor's dielectric ranges from about 1 nF to 10 µF and is used as signal capacitors and integrators in analog circuitry.

- **Polystyrene**: This style of capacitor is usually found in the pF range and used for ensuring a stable signal in analog circuits.
- **Polypropylene**: This is a low-loss, high voltage capacitor, resistant to breakdown voltages commonly used in special transmitting networks.
- **PTFE or Teflon**: These are higher performing capacitors and more expensive than other plastic dielectrics.
- **Silver mica**: These are fast and stable for HF and low VHF RF circuits, but cost considerably more than other dielectric families.
- **Film capacitor**: These capacitors have very low ESR and ESL and are primarily used in power circuitry with higher surge or pulse load capabilities over other capacitor families.

4.6.3 Impedance Plots Based on Dielectric Composition

An example of magnitude of impedance presented to a printed circuit board by a capacitor, based on dielectric material is shown in Fig. (4.7), illustrating why it is important to select an optimal dielectric for an intended application. When devices consume a significant amount of transient current, the impedance of the capacitor needs to be lower than a calculated target impedance over a spectral bandwidth of operation. This target impedance is easily calculated using Ohm's law. The lower the impedance of the power distribution network, a greater amount of current will flow.

As shown in Fig. 4.7, the curve achieves low impedance within a narrow frequency range. The sharpness of the curve which is in the shape of a big "V" is known as the Quality Factor (Q) of the circuit. This shape could covers either many frequencies (large bandwidth-low Q) or a narrow bandwidth-high Q. We calculate Q using (4.2).

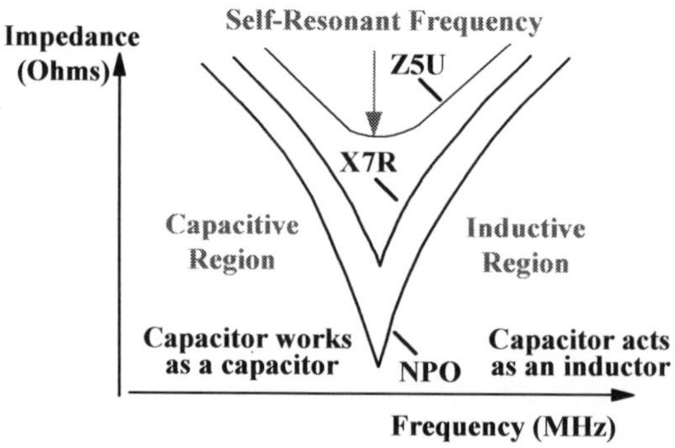

Figure 4.7 Impedance curves due to different dielectric material.

$$Q = \frac{ESL}{ESR} \tag{4.2}$$

If total resistance (*ESR*) increases, the spectral bandwidth of the curves becomes greater since *ESR* is in the denominator, but at the effect of creating a higher target impedance. There is a tradeoff on having a capacitor operate over a large range of frequencies or a little, relative to the amount of current charge required to keep components functional and operating within certain margins of performance. If use of a high *ESR* capacitor is required for a larger bandwidth of operation, ensure that the value of *ESR* is below the target impedance of the PDN.

4.6.4 Effective Range of Decoupling Capacitor Families Based on Target Impedance

Capacitors function over a specific bandwidth of frequencies for an intended application of use. In some situations we need to have a low Q (broad spectral coverage) or high Q (very narrow range). The bandwidth of operation for a capacitor is based on the dielectric material and manufacturing construction along with other factors such as *ESL*, *ESR*, and of course capacitance value. In order to achieve a large amount of capacitance for the power supply section of a printed circuit board (especially for the voltage regulator), a physically large value of capacitance is required. To maintain the power supply from collapsing in case of AC mains drop in voltage, or loss of several input cycles, an electrolytic capacitor with a value of 100 μF or more is commonly used to hold up the input voltage and current level until the abnormal or fault condition is removed. This large value capacitor also provides sufficient charge prior to energizing the DC portion of the printed circuit board. Electrolytic and tantalum dielectrics provide the greatest benefit for this application, but only at lower frequencies generally below 2 MHz, which is where most power supply operate.

In order for a capacitor to provide significant value to maintain stable power levels, the self-resonant frequency must be below a specified target impedance (Fig. 4.8). This figure also illustrates several capacitor families and their bandwidth of operation.

Ceramics, the most commonly used capacitor family for decoupling PDNs, cease to provide significant value above several hundred MHz. In order to provide essentially unlimited charge above this cutoff limit, internal decoupling structures must be utilized, detailed in Section 4.22 (Buried capacitance). Buried capacitive structures consist of power and return planes spaced close together. Figure 4.8 shows only an approximation on the range of frequencies that a capacitor family functions best at. It is therefore critical to choose a capacitor family for an intended application and not just based on total capacitance desired. Therefore, in almost every application there will be different types of capacitor families utilized.

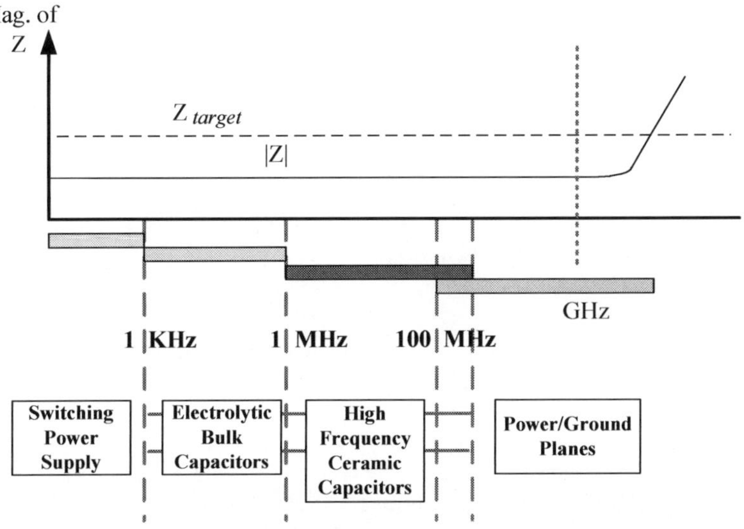

Figure 4.8 Effective range of typical decoupling capacitors. (Source: Ansoft Corporation)

4.6.5 Energy Storage Capabilities of Capacitors

If a capacitor is used for bulk or decoupling application it ideally should be able to supply necessary current during a state transition of a logic device to ensure component functionality. We can easily calculate how much capacitance is required based on current requirements with respect to time using (4.3). The difficult part in solving (4.3) is knowing total value of inrush surge current when all output pins source or sink current simultaneously under maximum capacitive load, a parametric value rarely provided by component manufacturers in their datasheets. This line charge current must be added to the quiescent current required for the core of the processor when in static or DC mode of operation. Most components have both core and I/O sections that use either the same or different voltage levels. Both sections must have optimal decoupling capacitance, calculated separately.

Also, manufacturers of digital components rarely specify the minimum or actual edge rate that a component operates at within the silicon die. The Δt requirement is usually unknown or not provided by the device manufacturer, given as "typical/maximum" but rarely do we find "minimum" edge rate, which is the primary variable of concern when trying to calculate an accurate target impedance. Devices consume transient current only during edge rate transition, and if we do not know the actual value or how fast the digital device really operates at, we are unable to calculate optimal capacitance for that specific time period.

In order to determine Δt with any level of accuracy, take the typical value provided in the datasheet and multiple by 0.6. In reality, the silicon operates approximately 60% faster than the datasheet value for most digital components. The reason for this is when a device is fabricated at the silicon level, it must be made faster than published specifications in order to meet functional targets. Semiconductor devices must have tolerance margins during production. If the component's internal switching speed is faster than a datasheet value, this is a better device for logic designers but at the expense of increase EMI and signal integrity problems.

$$C = \frac{\Delta I}{\Delta V / \Delta t} \qquad \text{for example,} \quad \frac{20\,mA}{100\,mV / 5ns} = 0.001\,\mu F \text{ or } 1000\,pF \tag{4.3}$$

where:

C = total amount of capacitance required
ΔI = current transient
ΔV = allowable power supply voltage change (ripple permitted by the component)
Δt = switching time of the event (take the typical value in the datasheet and multiple by 0.6 to get a more accurate number for calculation purposes)

The functional response of a decoupling capacitor is based on a sudden change in demand for current against the characteristic nature of the parameters that create the structure. It is often useful to interpret frequency domain impedance response in terms of the capacitor's ability to supply current. This charge transfer ability is also for the time domain function for which the capacitor is generally selected for. The low-frequency impedance between the power and return planes indicates how much voltage on the circuit board will change when experiencing a relatively slow transient. This response is an indication of the time-average voltage swing experienced during a fast transient. With low impedance (Z), more current is available to components under a sudden change in voltage requirement per Ohm's law. Higher-frequency impedance indicates how much current the board can initially supply in response to a fast transient. Boards with lowest impedance above 100 MHz can supply the greatest amount of current (for a given voltage change) during the first few nanoseconds.

Current requirements for each component must be identified to properly assess decoupling and power distribution requirements. Current requirements are as follows.

- *Quiescent*: The steady-state current level required for normal operation.
- *Output capacitive load*: Capacitive charging current that the driver must send to a load.
- *Transmission line load*: The current required to propagate an electromagnetic field from source to load. This requires a slower time period to keep the line charged until all transmission line reflections have subsided.
- *Device output charge*: The amount of output charge switching current available.
- *Cyclic switching*: Component decoupling capacitance recharge current.

The underlying component of the power distribution system is DC distribution. By definition, DC voltage is time invariant. When analyzing AC power requirements, the DC voltage drop or noise margin budget is usually considered *only* to determine the voltage drop budget allowance.

4.6.6 Impedance (Actual Self-Resonant Frequency)

The equivalent circuit of a capacitor was shown in Fig. 4.6. The impedance of a capacitor is described by (4.4).

$$|Z| = \sqrt{R_s^2 + \left(2\pi f L - \frac{1}{2\pi f C}\right)^2} \tag{4.4}$$

where:
Z = impedance (Ω)
R = equivalent series resistance – *ESR* (Ω)
L = equivalent series inductance – *ESL* (H)
C = capacitance (F)
f = frequency (Hz).

From (4.4), total $|Z|$ exhibits a minimum impedance (4.5) at a specific resonant frequency, fo, such that:

$$fo = \frac{1}{2\pi\sqrt{LC}} \tag{4.5}$$

In reality, impedance equation (4.4) includes hidden parasitics that are present when we take into account total *ESL* and *ESR*. A discussion of hidden parasitics was provided in Chapter 1.

Equivalent series resistance loss (*ESR* is a term referring to resistive losses in a capacitor, namely that contained in the dielectric) includes distributed plate resistance of the metal electrodes, the contact resistance between internal electrodes and external termination points, when compared to the resistance inherent in the dielectric composition which is generally several orders of magnitude lower. Skin effect at higher frequencies increases this resistive value in the leads of the component package and interconnect traces, thus total *ESR* is larger at higher frequencies than its value at DC.

Equivalent series inductance (*ESL*) is the loss element; however, by definition any (ideal) reactive element is lossless. Capacitors stores and returns energy but does not dissipate it. Energy dissipation (losses) only occurs in resistive elements. The value of *ESL* in a capacitor must keep to a minimum to prevent restriction of current flow within a device package. The tighter the restriction, the higher the current density resulting in higher *ESL*. The ratio of width to length of the capacitor plates must be taken into consideration to minimize this parasitic element.

From (4.6), we have a variation of the equation (4.4) with *ESR* and *ESL* substituted as variables.

$$|Z| = \sqrt{(ESR)^2 + (X_{ESL} - X_C)^2} \tag{4.6}$$

where:

$$X_{ESL} = 2\pi f(ESL)$$

$$X_c = \frac{1}{2\pi fC}$$

In order for an ideal capacitor to provide optimal performance, we need pure capacitance and no inductance. Inductance (*ESL*) increases total impedance per (4.6). Power and return planes are optimal in providing low-impedance decoupling (delivery of charge) versus discrete components for this reason.

For an ideal capacitive structure where current uniformly enters from one side and exits from another, inductance will be practically zero, a primary characteristic of power and return planes. With this configuration, *Z* will approach R_s at higher frequencies and will not exhibit an inherent resonance, which is exactly how power and return planes function within a printed circuit board. This is illustrated in Fig. 4.9.

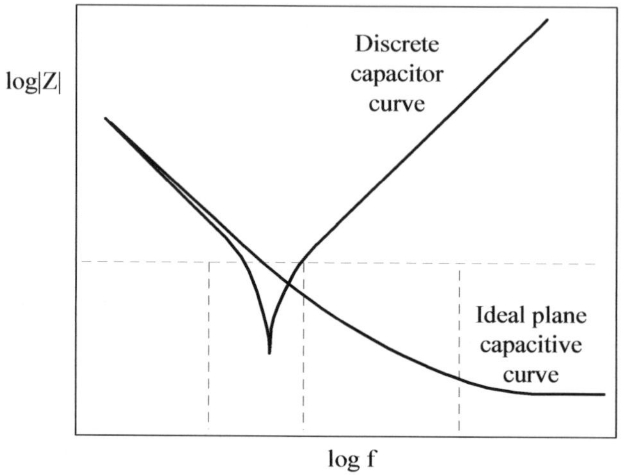

Figure 4.9 Theoretical impedance frequency response of ideal planar capacitors.

The impedance of an "ideal" capacitor decreases with frequency at a rate of -20 dB/decade. Because a capacitor has finite inductance in its structure that includes loop area interconnect, this inductance prevents the capacitor from behaving in a theoretically perfect manner.

It should be noted that power traces in two-sided boards that are not laid out for ideal flux cancellation (lowest loop inductance) and are in effect, extensions of lead interconnect inductance. Having extra inductance shifts the self-resonance of the power distribution system to a lower frequency while preventing charge energy from quickly passing to the device and maintaining optimal voltage levels.

Above self-resonance, the impedance of a capacitor becomes inductive and increases at +20 dB/decade as shown in Fig. 4.10. Above self-resonance, the capacitor ceases to function as a capacitor and behaves as an inductor. The magnitude of *ESR* is extremely small above 100 kHz and as such, does not significantly affect the self-resonant frequency due to the variable f in the reactance portion of the impedance equation ($X_L = 2\pi fL$).

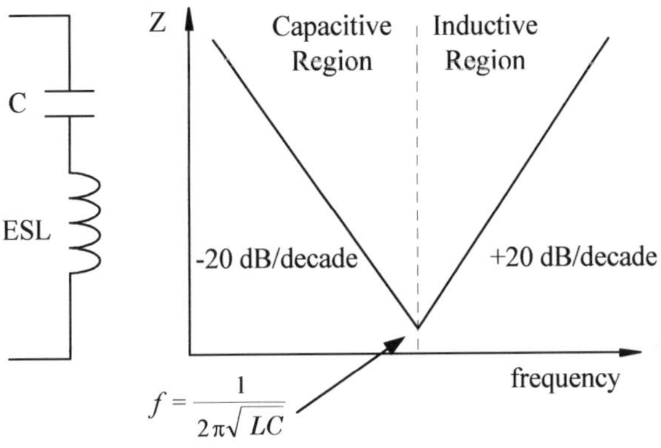

Figure 4.10 Effects of lead length inductance within a capacitor.

The effectiveness of a capacitor in reducing power distribution noise at a particular frequency of interest is described by (4.7).

$$\Delta V(f) = |Z(f)| \cdot \Delta I(f) \tag{4.7}$$

where:
ΔV = allowed power supply sag or ripple permitted
ΔI = current supplied to the device under maximum load conditions
Z = impedance of the PDN
f = frequency of interest.

To optimize power distribution by ensuring noise does not exceed a desired tolerance limit, $|Z|$ must be less than $\Delta V/\Delta I$ for the required current supply (Ohms law). Maximum impedance $|Z|$ should be estimated from maximum ΔI required. For example, if $\Delta I=1A$ and $\Delta V=3.3V$, then the impedance of the capacitor must be less than 0.3 Ω, if this device is to provide any value as a localized charge storage source.

In order for an ideal capacitor to work as desired, the device should have as much capacitance to provide required charge with minimal lead and loop inductance. In addition, the capacitor must have low internal resistance to obtain the least possible impedance. For this reason, power and return planes are optimal in providing low-impedance decoupling since inductance in planes is generally or negligible or have very low inductance. Planes can also have nearly unlimited charge storage potential if designed properly and closely spaced. An example of a plane structure with nearly unlimited charge is called buried capacitance, discussed in Section 4.22.

4.6.7 Resonance of a Capacitor When Installed on a Printed Circuit Board

When selecting a decoupling capacitor, storage capacity and resonant frequency must be calculated based on logic family used and edge rate transition of digital components. A capacitor remains capacitive up to its self-resonant frequency. Above self-resonance, the capacitor starts to function as an inductor due to both internal package and interconnect inductance. Inductance minimizes the ability of the capacitor to decouple or remove RF energy present between power and return by ensuring a stable power distribution network is available at all times to minimize plane bounce. Table 4.4 illustrates the self-resonant frequency for three types of ceramic capacitors, one with standard leads (radial or axial) and the other surface mount. The self-resonant frequency of surface mount (SMT) capacitors will always be higher, although interconnect inductance may obviate this benefit if installed on a printed circuit board using a interconnect trace and not via directly to planes.

The values in Table 4.4 were calculated using Eq. (4.5). Actual values will vary based on application of use and method of mounting, including lead interconnect loop inductance.

When performing SPICE analysis on various package-size SMT capacitors, all with the same capacitive value, the self-resonant frequency will change by only a few MHz between packages while keeping all other variables unchanged. SMT package sizes of 1210, 0805, 0603, 0402 and 0201 are common in today's products using various types of dielectric material. The primary different is mainly package inductance. The dielectric material used does not affect self-resonant frequency of operation. One can expect a change between package families to be approximately ± 2-5 MHz.

Table 4.4 Approximate self-resonant frequency of various packaging styles. (lead length dependent) – example only

Capacitor Value	Through-hole * 0.64cm (0.25 in.) leads	Surface mount ** (0805)	Surface mount *** (0402)
1.0 µf	2.6 MHz	5 MHz	5.3 MHz
0.1 µf	8.2 MHz	16 MHz	16.8 MHz
0.01 µf	26 MHz	50 MHz	53 MHz
1000 pF	82 MHz	159 MHz	168 MHz
500 pF	116 MHz	225 MHz	237 MHz
100 pF	260 MHz	503 MHz	530 MHz
10 pF	821 MHz	1.6 GHz	1.68 GHz

* For through-hole, L = 3.78 nH (15 nH/inch = 5.9 nH/cm)

** For surface mount, L = 1 nH *** For surface mount, L = 0.9 nH

Leaded capacitors are nothing more than surface mount devices with external interconnects. A typical axial or radial component has on the average approximately 2.5 nH of inductance for every 0.25mm (0.10 inch) of lead length. Surface mount capacitors average 1 nH total lead-length inductance. Total inductance is based on typical trace dimensions and will vary with width, thickness and length which are added to the fixed value from the capacitor's package.

An inductor does not change its resonant response like a capacitor; instead the *magnitude* of the device's impedance will change as frequency changes. Parasitic capacitance between windings of the inductor however can cause a parallel resonance to occur, which may alter desired response. When frequency increases, the magnitude of inductive reactance increases since the variable "*f*" frequency is in direct proportional in the equation ($X_L = 2\pi fL$).

RF current traveling through any impedance will causes an RF voltage potential difference between the input and output terminals of the inductive portion of the capacitor. Consequently, RF common-mode current is developed within the device described by Ohm's law, $Vrf = Irf * Z$ (Chapter 1). A critical design concern when implementing capacitors for decoupling deals with total lead and loop inductance. SMT capacitors perform better at higher frequencies than radial or axial because of lower lead and loop inductance. Table 4.5 details the magnitude of impedance of a 15 nH inductor versus frequency per ($Z = j2\pi fL$).

Table 4.5 Magnitude of impedance of a 15-nH inductor versus frequency.

Frequency (MHz)	Z (ohms)
0.1	0.01
0.5	0.05
1.0	0.10
10.0	1.0
20.0	1.9
30.0	2.8
40.0	3.8
50.0	4.7
60.0	5.7
70.0	6.6
80.0	7.5
90.0	8.5
100.0	9.4

Figure 4.11 illustrates the self-resonant frequency of a discrete capacitor with 6.4 mm (0.25 in.) leads, either radial or axial. Capacitors remain in the capacitive phase until they approach self-resonance (null point or lowest value of impedance) before changing phase and going inductive. Above self-resonance, they proportionally cease to function for RF decoupling; however, they may still be the best source of charge for the device even at frequencies where they are inductive, but only if there are lots of capacitors at the same value since the Thevenin equivalence of capacitor in parallel means *ESL* decreases overall. Inductance is what causes capacitors to become less useful at frequencies above self-resonance.

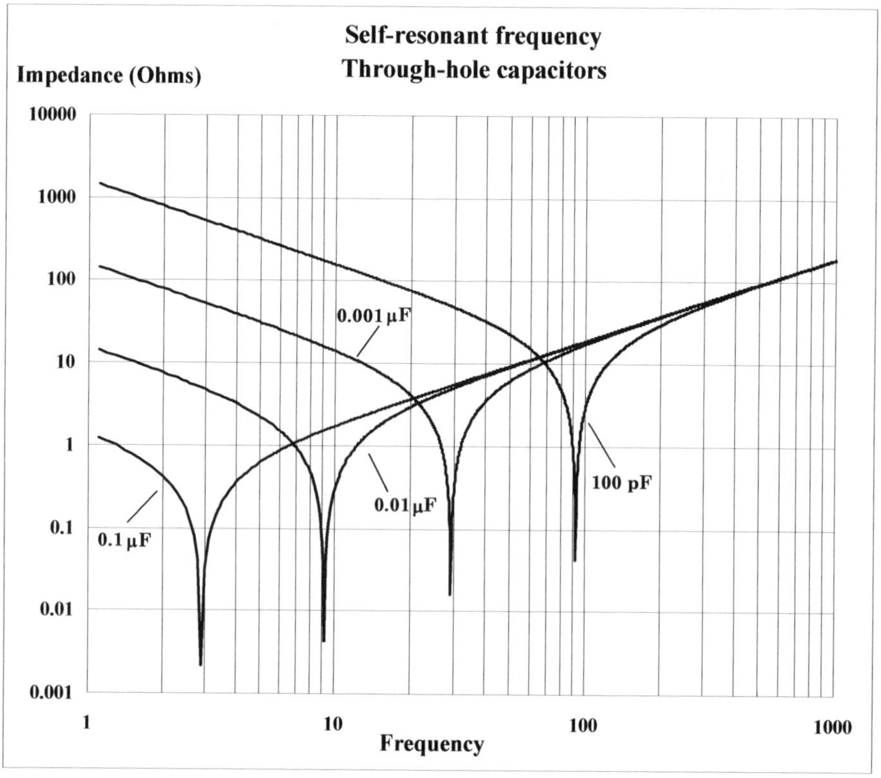

Figure 4.11 Self-resonant frequency of capacitors with radial or axial leads. (30-nH typical lead inductance, less interconnect inductance to a PDN)

Certain logic families produce a large amount of RF energy throughout the frequency spectrum. The RF switching noise established by components is generally higher in frequency than the operational bandwidth that a capacitor has below a target impedance. For example, a 100 nF (0.1 μF) capacitor may not provide adequate decoupling for higher speed devices. A 1 nF (0.001 μF) capacitor is a more appropriate choice for components operating at 150 MHz however, the total amount of energy storage is significantly less by two magnitudes. This means that if 1 nF capacitors are to be used, many more will be required in order to achieve the same amount of charge as a single 100 nF capacitor (100 capacitors that are 1 nF each will be required to achieve the same amount of charge as a single 1μF capacitor, although their self-resonant frequency of operation is much higher).

There will always be a finite amount of inductance associated with a capacitor due to the method of mounting on a printed circuit board. This includes the breakout trace and via interconnects to both internal power and return planes. Even a small amount of inductance may change the self-resonant frequency of operation significantly, making the capacitor ineffective for optimal or desired performance within a specific bandwidth of operation.

Although an engineer may take into consideration trace and lead inductance during the design cycle, a printed circuit board designer sometimes locate passive components where space is available rather than at a specific position for optimal effectiveness. In some cases, auto-placement of vias from the power and return planes to the capacitor occurs with a physical trace between pad and via, which is the worst design layout technique possible related to signal integrity and development of common-mode current.

We now compare the difference between capacitors with radial or axial leads and surface-mount (SMT). SMT devices have significantly less lead inductance by several orders of magnitude, thus their frequency of operation is significantly higher. Figure 4.12 shows a similar plot of the self-resonant frequency of various values of ceramic SMT capacitors. All capacitors have the same internal lead inductance for comparison purposes.

Note that in Fig. 4.12, compared to Fig. 4.11, the self-resonant frequency is shifted significantly higher and the magnitude of impedance (y-axis) is also much less. Having a SMT decoupling capacitor operate at a higher frequency with a larger bandwidth of operation, in addition to lower *ESR* provides enhanced performance in delivery of charge.

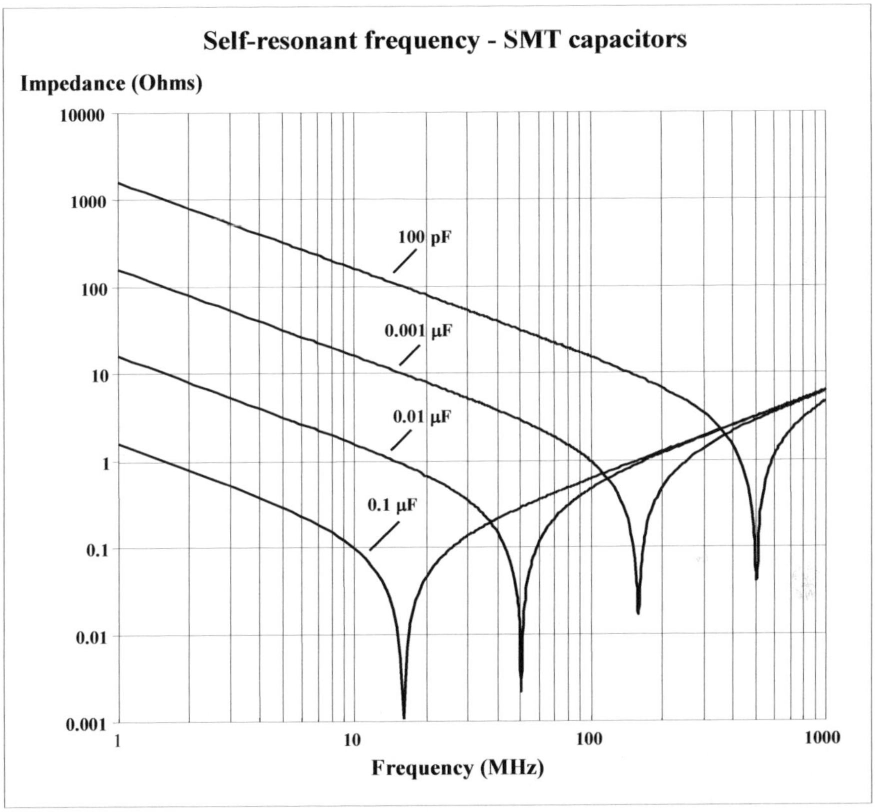

Figure 4.12 Self-resonant frequency of SMT capacitors.
(ESL = 1-nH, typical of most SMT capacitors, less interconnect inductance to a PDN)

4.7 Capacitors Placed in Parallel (Anti-Resonant Effect)

It is common practice to make provisions for parallel decoupling of capacitors with intent of providing a greater spectral distribution of performance while minimizing switching noise induced within any power distribution network. Board-level-induced noise, sometimes called "*Delta-I*," is a primary cause of common-mode EMI creation. This *delta-I* noise is generally referred to as ground bounce; do not forget that the power plane may also bounce which mean in reality we should refer to this as plane bounce, potential irrelevant.

Plane bounce is defined as a difference of potential within either the voltage or 0V from their reference level, all the way from the silicon die to the PDN through inductive interconnects. The majority of bounce occurs across the internal lead-bond wires of a component's package, assuming this style of packaging is used. To minimize bounce and interconnect inductance from bond wires, Ball Grid Array (BGA) and flip chip configurations are used. The best means of minimizing lead inductance from a silicon die is when the wafer is embedded internal to a multi-layer printed circuit board assembly where the interconnect is at the microscopic level.

Research on the effectiveness of multiple decoupling capacitors shows that parallel decoupling may not be significantly effective and that at higher frequencies, only a 6 dB improvement may occur over the use of a single larger value capacitor. Although 6 dB appears to be a small number for suppression of RF current, it may be all that is required to bring a noncompliant product into compliance with international EMI specifications [2].

> *Above the self-resonant frequency of the larger value capacitor where its impedance increases with frequency (inductive), the impedance of the smaller capacitor is decreasing (capacitive). At some point, the impedance of the smaller value capacitor will be smaller than that of the larger capacitor, and will dominate, thereby giving a smaller net impedance than that of the larger value capacitor alone* [2].

This example of 6 dB improvement occurs as a result of lower lead and device-body inductance provided by capacitors in parallel. Two parallel sets of leads have less inductance (by 50%) than that provided by one capacitor (Thevenin equivalent circuit). This reduced inductance is a reason why parallel decoupling works as wells as it does. However, one serious disadvantage does exist with parallel decoupling.

Figure 4.13 shows a plot of two bypass capacitors, 10 nF (0.01 μF) and 100 pF, both individually and in parallel. The 10 nF capacitor is resonant at 14.85 MHz. The 100 pF capacitor has its resonant frequency at 148.5 MHz. At 114 MHz, there is a large increase in the magnitude of impedance owing to this parallel combination. This is because the 10 nF capacitor has changed phase and is inductive while the 100 pF capacitor is still in the capacitive phase. There is now both L and C elements that meet each other at 114 MHz, causing an anti-resonant effect of the impedance. An anti-resonant frequency is exactly what we do not want! At this particular frequency, any harmonic of a clock or spurious transition will be observed as a powerful, transmitting signal. For this example, the third harmonics of a 38 MHz oscillator is 114 MHz, identical to the anti-resonant frequency of 114 MHz [2]. At this frequency we now have a serious radiated emission problem.

Figure 4.13 shows that at 500 MHz, impedance of individual capacitors starts to become identical. Also in the plot we see that the 10 nF capacitor at 15 MHz (standalone) shifts to 16 MHz (due to the Thevenin parallel inductance of the two capacitors physically adjacent to each other). This slight shift in frequency may improve EMI performance by 6 dB, detailed in [2]. The same frequency shift also applies to the 100 pF capacitor. This 6 dB improvement is only valid over a limited frequency range or bandwidth based on target impedance.

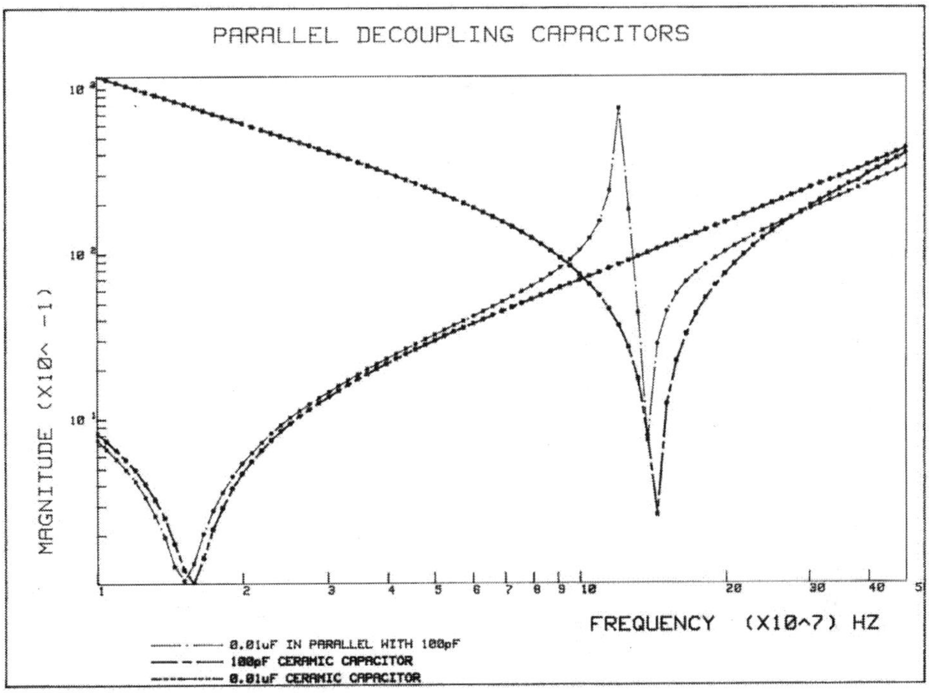

Figure 4.13 Two different capacitor values in parallel showing anti-resonant effect. (Courtesy – Dr. Clayton Paul [2])

To further examine what occurs when two capacitors are used in parallel, a Bode plot is shown in (Fig. 4.14). The frequency responses of the magnitude at various break frequencies are identified as C1, C2 and C3 (4.8) [3].

$$f_1 = \frac{1}{2\pi\sqrt{LC_1}} < f_2 = \frac{1}{2\pi\sqrt{LC_2}} < f_3 = \frac{1}{2\pi\sqrt{LC_3}} = 2f_2 \qquad (4.8)$$

By decreasing lead inductance of the larger sized capacitor, 10 nF (0.01 μF), we can obtain the same results by a factor of 2. For this reason, a single capacitor may be more optimal than two, especially if minimal lead and loop length inductance exists. Another reason to not use different value capacitors is that the smaller value one will have significantly less energy storage capability. By paralleling many larger value capacitors without the need to have a different value capacitor by 2 orders of magnitude, not only do we get greater energy charge but also higher self-resonant frequency of operation and lower impedance.

To remove RF current generated by components switching many signal pins simultaneously (and it is desired to parallel decouple for a specific reason), it is common practice to place two capacitors in parallel (i.e., 100 nF and 1 nF). Two orders of magnitude is required because these devices will form a resonant tank circuit (capacitor plates and interconnect inductance) and worry more about who is getting the charge than providing charge.

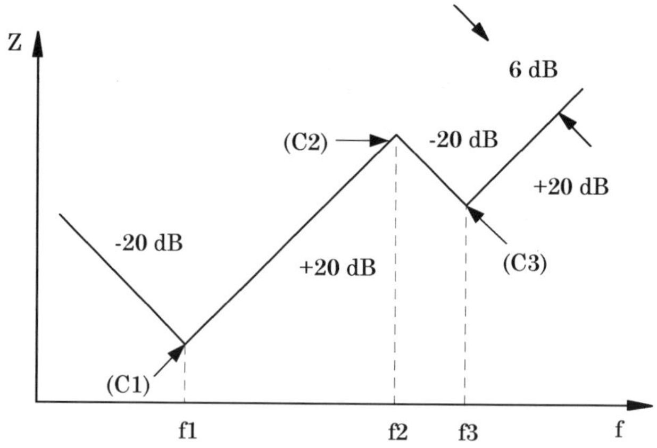

Figure 4.14 Bode plot of two capacitor in parallel.

If parallel decoupling must be used, although highly discouraged, the capacitance values should differ by two orders of magnitude or 100x. The total capacitance of parallel capacitors is basically only that of the larger value device per Thevenin analysis, with the smaller value device providing minimal benefit for energy charge storage.

To optimize effects of parallel bypassing and allow use of only one capacitive value, reduction in trace and loop inductance is required. Loop inductance includes both package and lead interconnects to the PDN. The shorter the interconnect length (not internal inductance), the greater the performance with regard to proving energy charge. However, if multiple values of capacitors are provided to establish a large spectral bandwidth for decoupling, enhanced performance will be achieve by using only one value of capacitor.

When parallel decoupling must be provided, do not forget that a third capacitor exists in the network–the power and return plane pair.

4.8 Power and Return Planes Providing Internal Decoupling Capacitance

A benefit of using multilayer assemblies is the ability to have an optimal power distribution network with significant energy storage capability. This performance is by virtue of having lower-impedance interconnects between planes and components than component-to-trace-to-plane configuration. Lower impedance interconnects also allows for less voltage drop to occur for components susceptible to switching noise. If any imbalance exists within the power distribution network (power and/or return plane pair), common-mode RF energy will be developed and propagated by any means possible throughout the entire assembly.

The relationship of two planes at opposite potentials creates a decoupling capacitive structure. This capacitor generally provides adequate decoupling for most low-speed (slower edge rate) designs; however additional signal or plane layers add cost to a printed circuit board assembly. If components have signal edge transitions (*tr* or *tf*) slower than 2 ns, use of high-performance, high self-resonant frequency discrete decoupling capacitors may not be required. Bulk capacitors are still needed however to maintain proper voltage levels due to current consumption by all components sharing the same voltage rail.

Depending on the physical distance spacing between planes, the dielectric value of the material, and placement within a board stackup, various values of capacitance can exist. Computational analysis will reveal actual capacitance value.

4.8.1 Calculating Power and Return Plane Capacitance

It is easy to calculate total capacitance of parallel plate structure using (4.9) based on the configuration of Fig. 4.15.

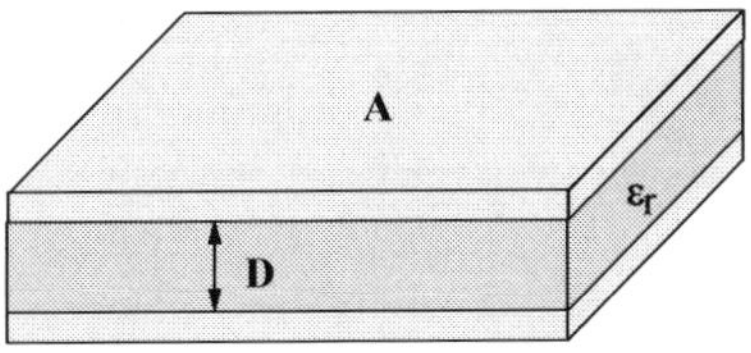

Figure 4.15 Physical parameters to calculate plane capacitance.

$$C_{pp} = k\frac{\varepsilon_r A}{D} \quad (4.9)$$

where:
C_{pp} = capacitance of parallel plates (pF)
ε_r = dielectric constant of the board material
A = common area between the parallel plates (square inches or cm)
D = distance spacing between the plates (inches or cm)
k = conversion constant (0.2249 for inches, 0.884 for cm).

This value of plane capacitance can also be described by (4.10). Actual capacitance is generally less than calculated owing to parasitics that cannot be anticipated or included, such as vias with anti-pads that reduce total amount of available copper to store transfer charge. Introducing relative permittivity of the dielectric material, ε_r, and ε_o, [permittivity of free space or vacuum is approximately equal to 1], we can obtain total capacitance of the parallel-plate pair.

$$C = k\frac{\varepsilon_o \varepsilon_r A}{d} = k\frac{\varepsilon A}{d} \quad (4.10)$$

where:
C = capacitance between the power and ground planes (pF)
ε_r = relative permittivity of the medium between the plates, typically ≈ 4.3
(Varies for linear material, usually between 2 and 10)
ε_o = permittivity of free space (usually equal to 1)
A = area of the parallel plates (m^2)
d = separation of the plates (m)
k = conversion constant (0.2249 for inches, 0.884 for cm).

Planes function efficiently as an internal decoupling capacitor due to their ability to store a large amount of charge. Optimum performance begins to become noticeable when distance spacing between the plates is less than 0.25 mm (0.010 in), with 0.125 mm (0.005 in.) preferred for higher-speed applications [4].

Other factors to consider when using planes for energy storage is the self-resonant frequency of the overall printed circuit board assembly along with all discrete capacitors provided. There will be a crossover point (self-resonant frequency) where the inductance of all discrete capacitors, in the inductive phase, meets the capacitive phase. At this specific frequency there will be a sharp anti-resonant response. When evaluating *S11* plots of a printed circuit board related to input impedance, numerous changes in phase will be observed in the upper MHz region. This occurs due to a combination of the impedance of the planes and discrete capacitors interacting with each other including input capacitance of components.

When anti-resonance occurs, there will no longer be a wide spectral distribution of decoupling over a desired bandwidth. If a clock harmonic is at the same frequency at this sharp anti-resonant frequency, the board will radiate common-mode switching noise, or the printed circuit board may become an unintentional radiator along with possible non-compliance with EMI requirements. Should this occur, decoupling capacitors (with a different self-resonant frequency, i.e., a different value) will be required to shift the resonance of the printed circuit board's power and return planes to force the anti-resonant frequency to pop-up where there is no harmonic of the switching noise or clock signal.

The printed circuit board is not really the transmitter; rather, the highly repetitive circuits or clocks are the cause of RF energy present that radiates or couples to unintentional circuits. Because decoupling capacitors will not solve this type of problem owing to resonances, system-level containment measures will be required such as shielding.

One simple method to change the self-resonant frequency of the power and return plane pair is to change distance spacing between planes (*z*-axis) or their physical size (area in the *x*-/*y*-axis). Increasing or decreasing distance separation and/or relocation within the stackup, or making the planes a different physical size will change total capacitance value. Equations (4.9 and 4.10) illustrates this concept should one need to re-tune a power and return plane pair's self-resonant frequency that creates an anti-resonant frequency with discrete devices. One disadvantage of using this design technique is that the impedance of the signal routing layers, or transmission lines, referenced to these plane may also change impedance value when distance spacing between a routing layers changes, which in turn may become a performance concern.

A designer must make compromises during layout, especially if a large amount of energy storage is more important than maintaining impedance control for transmission lines such as 50 ohms, where a higher trace impedance may be adequate for most circuit designs because not everything has to be at 50 ohm!

Multilayer printed circuit boards generally have a self-resonant frequency between 200 and 400 MHz. One must use computational analysis to analyze the impedance of signal traces if the technique of changing separation distance of planes for enhanced decoupling occurs.

4.9 Vias and Their Effects in Solid Planes

One caveat in solving Eq. (4.9) is that the inductance caused by antipads (clearance holes for through-vias) can minimize the theoretical effectiveness of using plane capacitance for decoupling.

Use of vias in planes will decrease total capacitance based on the number of via anti-pads and the amount of real estate that has been etched out from the planes. A capacitor works by virtue of energy storage that is contained between two metallic structures. With less metal (copper) current density distribution is decreased. As a result, a smaller area exists to support the number of electrons that create current density distribution. Figure

4.16 illustrates the value of capacitance between parallel power planes in two configurations: a solid power plane with 30% of the area removed by vias and clearance pads (anti-pads).

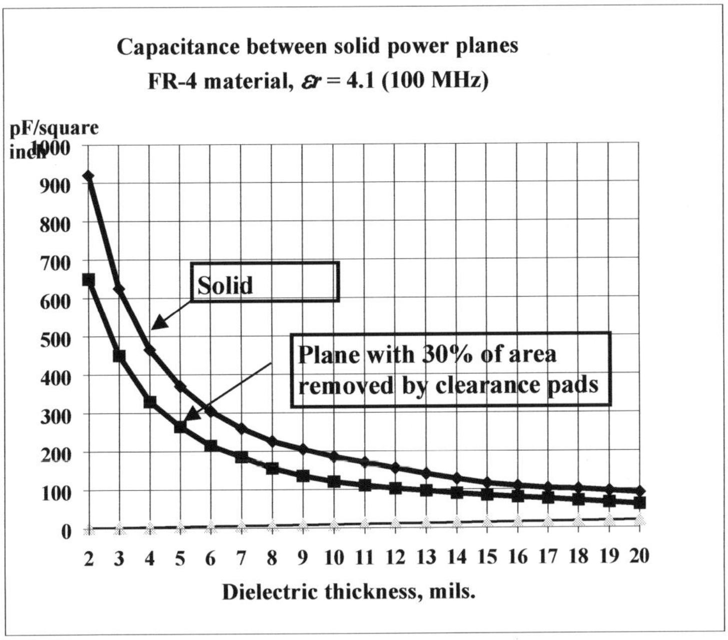

Figure 4.16 Effect of vias in power/return planes causing change in capacitance value.

4.9. 1 Combined Effects of Plane Capacitance with Discrete Capacitors

The effects internal planes were not considered in Fig. 4.13. Only two discrete capacitor values were evaluated to understand how an anti-resonant signal can occur when used in parallel. Effects are however illustrated in Fig. 4.17 when planes are taken into consideration in addition to the two parallel capacitor combination. Power and return planes have very little inductance and practically no *ESL* and *ESR*, which explains why planes minimize RF noise energy generally in the higher frequency range.

Because discrete decoupling capacitors are used in multilayer printed circuit boards, we must consider their benefit when lower frequency, slow edge rate components are utilized generally in the frequency range below 25 MHz. Research into effects of the power and return planes along with discrete capacitors reveals interesting results [3].

In Fig. 4.17, lower illustration, the impedance of the "bare board" closely approximates the ideal decoupling impedance that would result if only pure capacitance, free of interconnect inductance and resistance exists. This ideal impedance is given by equation: $Z = 1/(2\pi f C_o)$.

Discrete capacitors have low impedance at series resonance, f_s, and infinite at parallel resonance, f_p, or the point where a second capacitor interacts with the first one. These resonant frequencies can easily be calculated by (4.11), where n is the number of discrete capacitors, C_d is the discrete capacitor value, and C_o total capacitance of the power and return plane pair, conditioned by the source impedance of the power supply.

$$fs = \frac{1}{2\pi\sqrt{LC}} \qquad fp = fs\sqrt{1+\frac{nCd}{Co}} \tag{4.11}$$

For frequencies below series resonance, discrete decoupling capacitors behave as an ideal capacitor with an impedance of $Z=1/(2\pi f C)$. For frequencies near resonance, the impedance of a loaded printed circuit board is actually less than that of the ideal assembly. However at frequencies above fs, the decoupling capacitors begin to exhibit inductive behavior as a result of associated interconnect inductance. The frequency f_a, at which the magnitude of the board impedance is the same with or without discrete devices (where the unloaded printed circuit board intersects that of the loaded, non-ideal printed circuit board), is calculated by (4.12) where f_s is the series resonance frequency (zero on a Bode plot) and f_p is infinite impedance at parallel resonance (pole).

$$f_a = f_s\sqrt{1+(nC_d / 2C_o)} \tag{4.12}$$

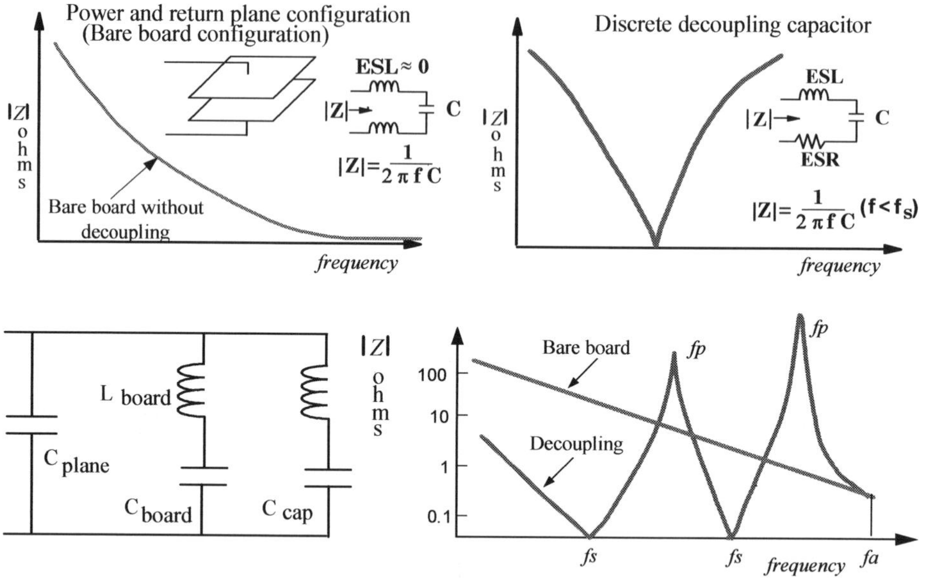

Figure 4.17 Decoupling effects of combined power/return planes with capacitors.

Resonances correspond to poles on a Bode plot. Series points (f_s) are *nulls* or *zeros*, while capacitors in parallel (f_p) have an anti-resonant frequency called *poles*. When multiple capacitors are utilized, poles and zeros will alternate. There will be exactly one parallel resonance (f_p) between each pair of capacitors or two zeros (f_s). In other words, if there are two different value capacitors, two zeros will exist and only one pole. If three different value capacitors are implemented, three zeros and two poles will be present.

For frequencies above fa or the frequency where the impedance of the power and return plane pair becomes smaller than the self-resonant frequency of additional ("*n*") discrete capacitors, no additional benefit occurs, as long as the switching frequencies of components are within the bandwidth or decoupling range of the power and return plane pair. This is because the board impedance remains below the target impedance already established or desired. At frequencies near a parallel resonant pole, the magnitude of the board impedance is extremely high and decoupling performance is far worse. This analysis clearly indicates that minimizing series inductance (trace and via interconnect to the planes) is crucial to achieving ideal capacitor behavior over the widest possible frequency range. Lowering interconnect inductance also increases both the series and parallel-resonant frequency, thereby extending the performance range of ideal capacitor behavior [6].

With multiple power/return plane pairs, for example +5 V/return and +3.3 V/return, both with different spacing between their respective planes, it is possible to have multiple decoupling capacitors provided which is highly desired for functional reasons. With proper layer stackup, both higher and lower frequency decoupling can be achieved using buried capacitance without use of discrete devices, with exceptions detailed in Section 4.22.

The inductance between two physical locations within a plane is extremely low, somewhere in the pH range. Inductance within a via is approximately 1 nH. Trace inductance is typically 2.5 to 10 nH "per cm" [3].

4.10 Effects of ESR and ESL in Decoupling Applications

One primary concern that affects functional operation of a decoupling capacitor and its ability to provide charge quickly to the PDN is equivalent series inductance and resistance [7]. To minimize interconnect inductance, a capacitor must be connected directly to planes at the mounting pad "without" any breakout trace to a via located some distance away. The value of *ESR* is usually fixed and is based on the internal resistance within the dielectric material between the plates, not the interconnect to the PDN.

4.10.1 Effects on Performance-Changes in *ESL* Values

How severe is the effects of *ESL* on minimizing ripple or noise? Figure 4.18 clearly shows this. Using a capacitor value of 100 nF (0.1μF), an *ESR* of 0.1Ω and a slew rate of 2A/ns, operational results in *ESL* are illustrated. With constant capacitance and changes in *ESL*, or increase in interconnect inductance, there is an increase in ripple voltage. Under certain situations, ripple voltage may exceed the functional specification of most digital components with regard to voltage margin.

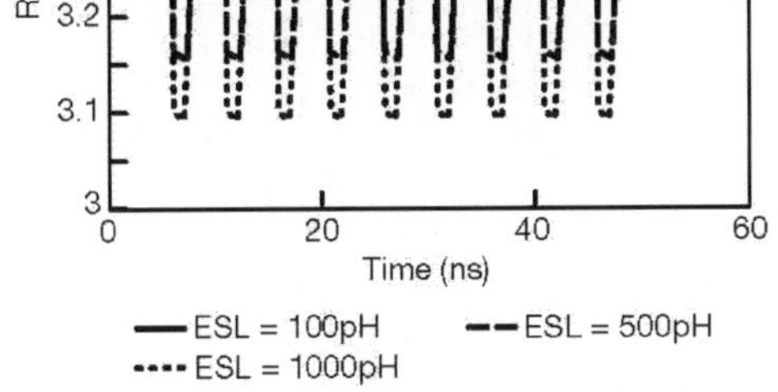

Figure 4.18 Effect of *ESL* on decoupling performance with a 100 nF capacitor. (Figure provided courtesy AVX Corporation [7])

4.10.2 Effects on Performance-Changes in *ESR* Values

The effect of different values of *ESR* within a PDN is now examined. The capacitor value is 0.1μF. Figure 4.19 illustrates that the greater the *ESR* in the network, an increase in ripple voltage occurs. This change is directly related to the *RC* time constant internal to the capacitor body. Switch-mode power supply designers know that resistance in a circuit causes functional disruptions and that it is important to keep the value of resistance (*ESR*) as low as possible to minimize the generation of noise in the PDN.

Another area of interest involves the slew rate and frequency of component current consumption. Figure 4.19 shows voltage ripple using a 100 nF (0.1μF) capacitor with an *ESR* of 1 Ω. The current sink for this simulation is 25, 50 and 100 MHz, with rise times of 0.5, 0.25 and 0.125 A/ns, respectively [7]. As mentioned earlier, the RC time constant affects performance, with lower frequency clock speeds producing less noise.

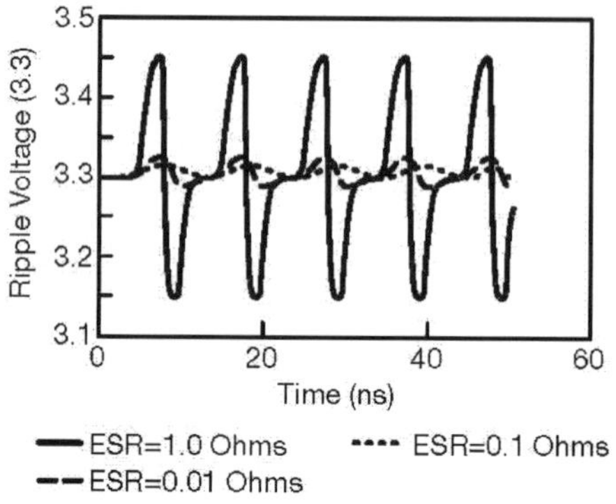

Figure 4.19 Effect of *ESR* causing a ripple on the voltage rail with a 100 nF capacitor. (Figure provided courtesy AVX Corporation [7])

4.11 Planes as RF Return Path for Transmission Lines

Multilayer printed circuit boards generally contain one or more pair of voltage and return planes. These planes function as both a low-inductance interconnect and high energy storage decoupling capacitor. In other words, these planes prevent development of RF currents generated from components switching logic states and permit devices to function within specific operational parameters.

Another critical function for planes has little to do with power distribution. This involves performing as a low impedance RF return path for signals routed on an adjacent layer, discussed in Chapter 2.

Multiple connections from return plane(s) to chassis ground, or 0V reference, minimize voltage gradients between circuits. These gradients are a major source of common-mode RF fields created due to impedance mismatches. This is in addition to bypassing undesired common-mode RF energy to chassis ground or away from switching logic devices. In many cases, multiple ground stitch connections to chassis are not practical. In such situations care must be taken to analyze and determine where RF loop currents could occur.

Planes located physically next to a signal routing layer provides enhanced magnetic field flux cancellation in addition to decoupling RF currents created from power fluctuations owing to components injecting switching noise into the PDN. An image plane or RF return path is a solid plane at either voltage or return potential physically adjacent to a routing layer. The reason why potential is not important is that components and the power delivery system are DC biased. AC noise due to components switching logic states can cause functional disruption in the frequency domain thus, it is a requirement that a low-impedance path exist for noise to be removed generally through the use of bypassing RF energy between planes at different potentials when a capacitor is used in decoupling mode, establishing a low impedance RF transmission line. For this to work as desired, all routing (signal) layers must be physically adjacent to a plane regardless of potential. Chapter 2 provides detailed discussion of image planes and the need for a low-impedance RF return.

4.12 Multi-Pole Decoupling Methodology

An alternative method of providing for large bandwidth power distribution delivery is using the multi-pole decoupling concept developed at Sun Microsystems [8]. This differs from the single value capacitor concept known as the "Big-V" due to its impedance characteristic when plotted on a graph. The primary difference is the use of non-standard value capacitors, and not just those having a "0" or "1" printed on the case package.

In order to properly implement multi-pole decoupling, it is critical to know the target impedance of the circuit board's power distribution network when operating under maximum conditions, a numeric value nearly impossible to know since many vendors do not publish total current consumption in their data sheet when all pins switch under maximum capacitive load. From here, one typically plots the "Big-V" curve using almost all values of capacitance possible at the same time on a log-log chart as was shown in Fig. 4.11 and 4.12 and then repeated in Fig. 4.20, top graph. In this top graph for simplicity reasons, the "Big-V" curve is calculated for capacitors from 100 μF to 10 nF in decades of 10, which are the most commonly used discrete values for decoupling applications generally all below 100 MHz. Note that these discrete capacitors provide minimal value for decoupling higher frequency circuits above 100 MHz.

In Fig. 4.20 bottom graph, we plot the impedance of these same series of capacitors at their same self-resonant frequency, but this time using additional capacitors of the same capacitive value per plot. In order to achieve low impedance using smaller value capacitors, many more are required in parallel that larger value ones. It was easily seen in Fig. 4.11 and 4.12 that smaller capacitance value will have higher impedance at their self-resonant frequency. The purpose of this analysis at Sun Microsystems was to determine total number of capacitors required on a printed circuit board to achieve 1-mΩ target impedance across the frequency spectrum [8]. Table 4.6 provides numeric analysis on how the lower plot of Fig. 4.20 was created. This chart is the basis for creating a multi-pole decoupling topology.

Multi-pole decoupling is a design technique for achieving a large spectral bandwidth of decoupling from the DC power supply switching frequency to about 80 MHz. Of interest is that in order to achieve low impedance at a particular frequency, many discrete capacitors are required, with only a few for larger value (i.e., 100 μF) to many at smaller value (10 nF). Capacitor values small than 10 nF provide minimal benefit for decoupling due to their limited energy storage capability and the need to use dozens, if not hundreds to achieve the same energy charge level as a larger size capacitor on power hungry assemblies containing many FPGAs, each having potentially over 1,000 pins per device and multiple voltage requirements.

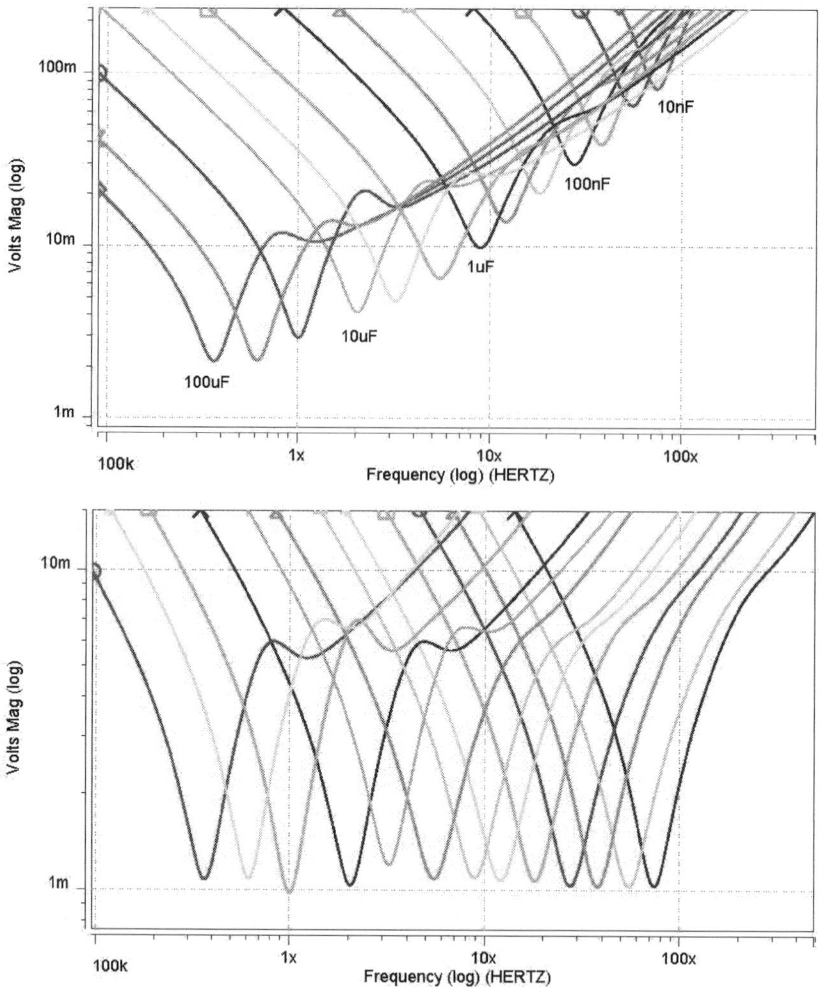

Figure 4.20 Multiple capacitor plots to achieve low impedance for a large spectral bandwidth of operation.
Source: [8, 9]

Using basic circuit analysis, the greater the number of capacitors placed in parallel, both *ESL* and *ESR* will become smaller in direct portion to the number provided (Thevenin equivalent value). For the lower plot of Fig. 4.20, there are two capacitors in parallel required to achieve 1-mOhm impedance with the value of 100 μF. To achieve this same low impedance at a higher frequency, it will take 60 capacitors with a value of 10 nF. If the goal for a high quality PDN is to provide low impedance across a large spectrum of frequencies, many capacitors in parallel will be required when using smaller capacitive values which add cost both in material and routing expenses to achieve the same amount of energy charge. Using many capacitors to achieve the same target impedance for the power distribution network could exceed available real estate on a printed circuit board layout leaving no room for active components.

Table 4.6 Various capacitor values and number to achieve 1 mΩ target impedance. Ref: [8]

Cap Value	Case Size	Dielectric Material	Measured Value	ESR (mΩ)	L internal (nH)	L mount (nH)	SRF (MHz)	Q	ESR divided by 1	# to Meet 1 mΩ Target
100 uF	1812	X5R	80.3 µF	1.8	2.112	0.600	0.341	0.7	2	2
47 uF	1210	X5R	42.1 µF	1.9	1.487	0.600	0.537	1.1	2	3
22 uF	1210	X5R	17.7 µF	2.5	1.300	0.600	0.867	1.3	3	7
10 uF	0805	X5R	7.26 µF	3.6	0.773	0.600	1.60	1.6	4	9
4.7 uF	0805	X5R	4.12 µF	4.2	0.544	0.600	2.32	2.1	4	5
2.2 uF	0805	X5R	1.98 µF	6.1	0.413	0.600	3.55	2.2	6	8
1.0 uF	0603	X5R	0.79 µF	9.1	0.391	0.600	5.69	2.3	9	12
470 nF	0603	X5R	404 nF	13	0.419	0.600	7.85	2.3	13	16
220 nF	0603	X7R	172 nF	19	0.438	0.600	11.9	2.3	19	28
100 nF	0603	X7R	75 nF	29	0.443	0.600	18.0	2.3	29	30
47 nF	0603	X7R	39 nF	38	0.451	0.600	24.7	2.4	38	40
22 nF	0603	X7R	17 nF	64	0.492	0.600	36.6	2.1	64	53
10 nF	0603	X7R	8.9 nF	80	0.518	0.600	50.4	2.4	80	60
						Total			**27**	**273**

Note: Measured values for capacitance and inductance are shown together with the series resonant frequency, Q and 600 pH mounting inductance.

The concept of multi-pole decoupling is excellent for achieving a large bandwidth of performance below 80 MHz. The question one generally has is "How do I implement this unique methodology"? Figure 4.21 clearly provides the answer. The Big-V curves for five different capacitor values are shown. The large flat line that increases in value is the target impedance for optimal performance. For the broadband curves under the target impedance, these are achieved by adding *ESR* to the capacitor or its mounting process. In a prior discussion, a relatively high *ESR* is not desired under most conditions since this *ESR* may increase impedance of the capacitor to above the desired level. Adding a small amount of *ESR* to the capacitor lowers the *Q* associated with the capacitor's self-resonant frequency (*Q*=*ESL*/*ESR*). There are commercial capacitor vendors who provide discrete devices with increased *ESR* for this reason, but at an additional per unit cost. Under this scenerio, having a larger value of *ESR* is highly desired for the multi-pole decoupling methodology. By adding a small amount of series resistance (or purchasing a capacitor with higher *ESR*), the *Q* of the circuit decreases, or bandwidth of operation increases across a broader spectral range, as well as minimizing the anti-resonant frequency that only occurs when two capacitors are in parallel with a capacitive difference in value.

Essentially, the narrow bandwidth of the Big-V below the target impedance is converted to a larger parabolic curve covering a greater bandwidth, or more frequencies under the desired target impedance. It is critical that all smaller value capacitor (i.e., those in the low nF range having a higher self-resonant frequency) enters the target impedance value before the larger value capacitor (with a lower self-resonant frequency) becomes inductive. By having the smaller value capacitors enter the target range with a greater spectral bandwidth ensures no anti-resonant frequencies are also established. As long as the impedance of all capacitors are below the target impedance desired, all capacitors functions as a decoupling element. In Fig. 4.21, the straight line is the spectral bandwidth of discrete capacitors remaining in the capacitive region below the target impedance without any anti-resonance present.

Example: If the target impedance of a printed circuit board is 100 mΩ and the zero of the Big-V is 1 mΩ, the component works as desired for decoupling, plus or minus a few MHz. If any series resistance (*ESR*) is added to the impedance calculation, then *Q* [*Q*=*ESL*/*ESR*, Eq. (4.2)] decreases and the capacitor now covers a larger bandwidth, at the expense of the impedance increasing to that of total *ESR*, which must remain below the target impedance at all times.

	C [uF]	ESR [ohm]	L [nH]
R-L	-	0.01	2
C1	33	0.006	1
C2	3.6	0.006	1
C3	1.3	0.006	1
C4	0.75	0.006	0.8
C5	0.56	0.006	0.5

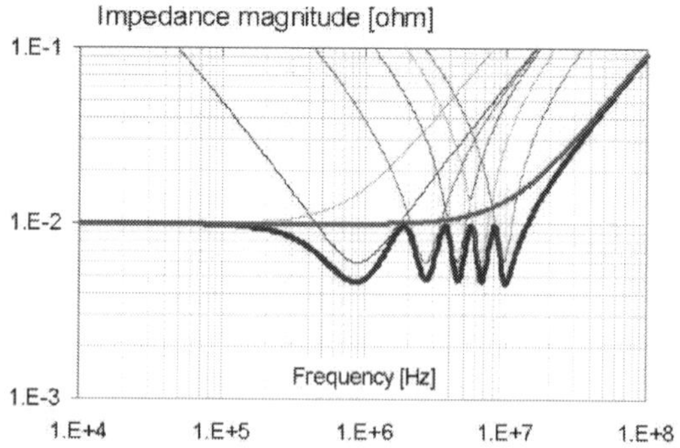

Figure 4.21 Multi-pole decoupling concept with various capacitor values. Source: [8, 9]

Computational analysis is mandatory for multi-pole implementation to determine where the anti-resonant frequency occurs. Depending on the phase of the capacitor, all devices must remain below the target impedance for a desired bandwidth of operation (Fig. 4.21).

4.13 Effects of Proper Decoupling Implementation

If proper implementation of decoupling capacitors is performed, significant improvement in minimizing switching noise within a PDN occurs, shown in Figs. 4.22 and 4.23. The spikes in Fig. 4.22 represent digital transitions from a component due to lack of energy delivery. With proper energy storage delivery (Fig. 4.23) switching noise is essentially removed.

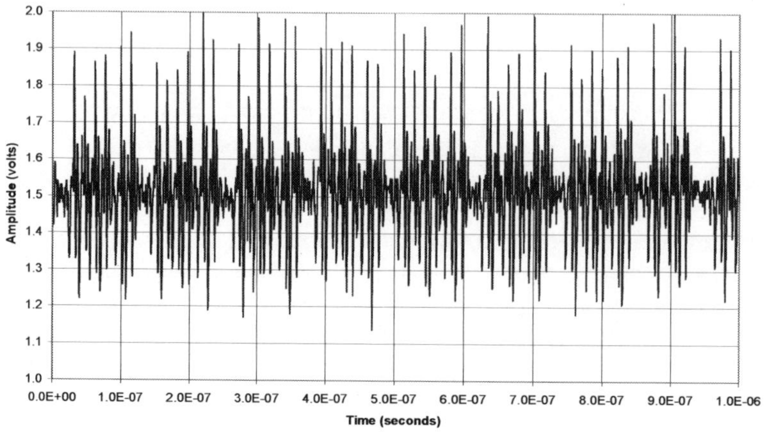

Figure 4.22 Power noise measured across power supply (1.5V) with capacitors removed. Source: Dr. Bruce Archambeault

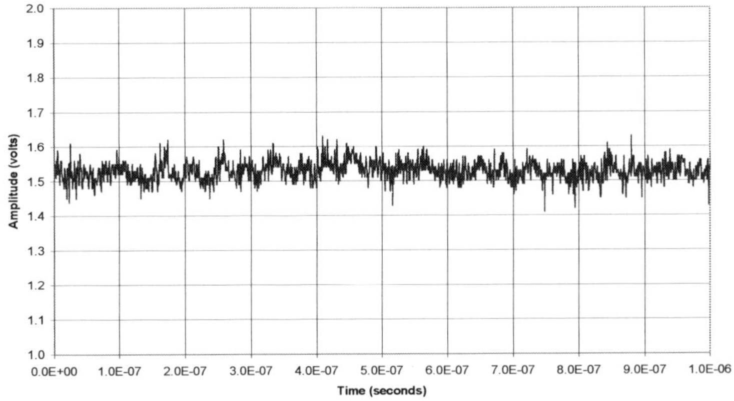

Figure 4.23 Power noise measured across power supply (1.5V) with capacitors in place. Source: Dr. Bruce Archambeault

4.14 Simplified Description of the Capacitor Brigade

Generally, engineers only consider use of decoupling and bulk capacitors as a minimal requirement to ensure a stable PDN. Although true, one must understand how various types of capacitors function within a printed circuit board. Capacitors can only provide charge to a localized area with a radius of performance limited by interconnect inductance and dielectric losses.

Fig. 4.24 shows a capacitor brigade configuration as it applies to digital components and the ripple effect from capacitance established within a PDN. The key item to note is the speed of operation of each capacitive element to charge the next element in the brigade. These need not be too fast in transferring charge but should contain sufficient energy storage potential to maintain the voltage level for active components located within its radius sphere of operation.

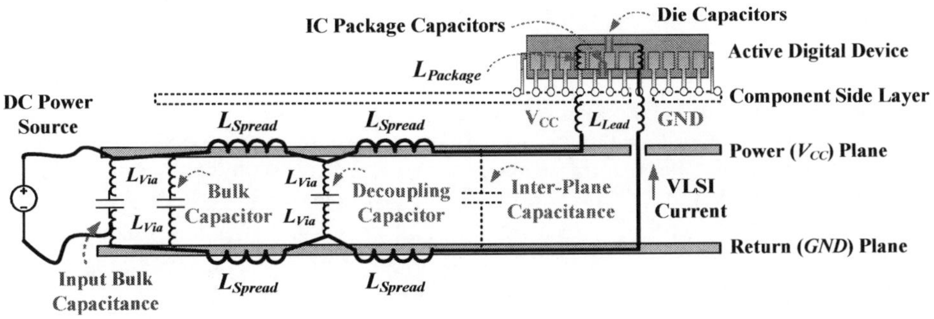

Figure 4.24 Simplified illustration of the capacitor brigade. Artwork courtesy: Elya Joffe

To describe the capacitor brigade:

- The first surge of current comes from on-die capacitance between gates.
- The on-die capacitance is recharged from the silicon wafer's internal power and return planes.
- The internal on-die power and return planes are recharged by internal package capacitance.

- The package capacitance is then recharged directly from the PDN's power and return planes.
- The PDN at the component's pins is thus recharged by a localized decoupling capacitor.
- The localized decoupling capacitor is then recharged by plane capacitance.
- Localized plane capacitance must now be recharged by bulk capacitors located a further distance away.
- The bulk capacitors must be recharged by the planes within its general vicinity.
- The planes in this portion of the PDN is now recharged by the power supply located even further away from the components.
- The power supply is then recharged by a steady-state voltage source, either DC or AC..

4.15 Radius of Operation-Effectiveness of Maintaining Voltage Levels

The current provided by a decoupling capacitor charges planes in a radial fashion to many components at the same time, like rocks thrown into water creating a ripple effect. All components within the wavefront will observe this ripple of charge regardless of physical location relative to actual placement of the capacitor on the printed circuit board. Decoupling capacitors "only" recharge planes, when planes exist, unlike single and double sided printed circuit board assemblies. Both analog and digital components consume current "from" the planes. Capacitors do not provide charge to components directly, except in single and double-sided assemblies where trace routing is mandatory between capacitor and component.

With "many" discrete capacitors provided on a printed circuit board, the wavefront at a particular location will appear to be a constant DC voltage source. The reason why this constant DC source appears to exist is that many propagating waves are present simultaneously spreading out their charge in a radial manner from all devices located close to each other. These many propagating wave will phase add or subtract at a specific x-/y-axis location. If dozens or hundreds of capacitor are charging and discharging asynchronously, each with a radially propagating wave, the voltage level *will appear to be pure DC* which is what we want. Numerous propagating waves phasing together to create a virtual DC voltage rail is the primary reason why an incorrect value of decoupling capacitor (such as 100 nF providing charge to GHz components which is a poor choice with a limited bandwidth of operation) works as well as it does, but "only" if there are many on the printed circuit board to create this virtual DC voltage level effect.

To minimize the number of discrete capacitors, each with two vias (sometimes making routing difficult on signal layers), having power and return planes closely spaced provides enhanced functionality in the higher frequency range than a discrete capacitor, which start to become less effective above 250 MHz due to loop inductance.

An illustration of the ripple effect from one capacitor is shown in Fig. 4.25.

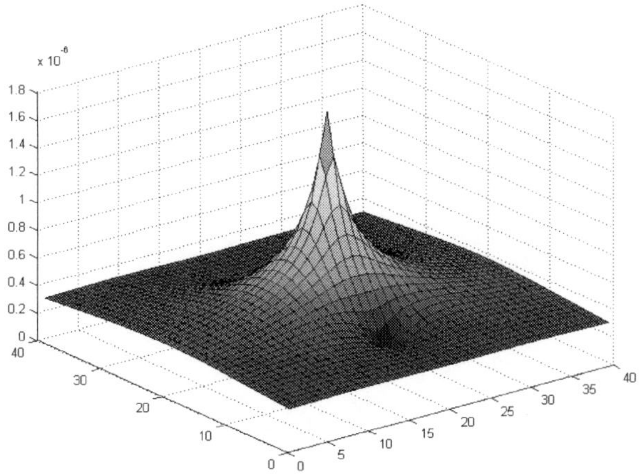

Figure 4.25 Power and return plane capacitive displacement current at 500 MHz through a via located 11.4 mm (450 mils) away the capacitor.
Source: Dr. Bruce Archambeault

4.16 Equivalent Circuit Model of a Printed Circuit Board

Before determining where to locate decoupling capacitors, a physical representation of a conceptual printed circuit board is shown with active and parasitic elements in Fig. 4.26.

In this figure, there are current loops between power and return caused by trace inductance, internal component wire bonds, lead frames, socket pins, component interconnect leads and the terminals of a decoupling capacitor. The key to effective decoupling is to minimize R_2, L_2, R'_2, L'_2, R_3, L_3, R'_3, L'_3, R_4, L_4, R'_4, and L'_4. Placement of power and return pins adjacent to each other within components helps reduce R_4, L_4, R'_4, and L'_4.

Impedance of a PDN must be minimized to ensure optimal delivery of current per Ohms law. The easiest way to minimize resistive and inductive components (impedance) is to provide solid planes spaced closely together. Spacing planes close together removes inductive elements that exist such as those found on single and double-sided assemblies. With less lead interconnect inductance from the silicon die to the printed circuit board planes (with regard to R_4, L_4, R'_4, and L'_4), the overall impedance of the decoupling loop is reduced, especially if BGA and flip chip mounting technology is chosen including using embedded active elements.

Figure 4.26 [1] makes it clear that EMI is a function of loop geometry and frequency from all sections of the assembly that includes the input power section (Loop current ***I1***), the processor section (Loop current ***I2***) and connection to output circuitry (Loop current ***I3***). Current loop **I4** is easily minimized using BGA and flip chip technology which minimizes L_4 and L'_4 and is the portion that contributes to the most amount of inductance in the entire assembly.

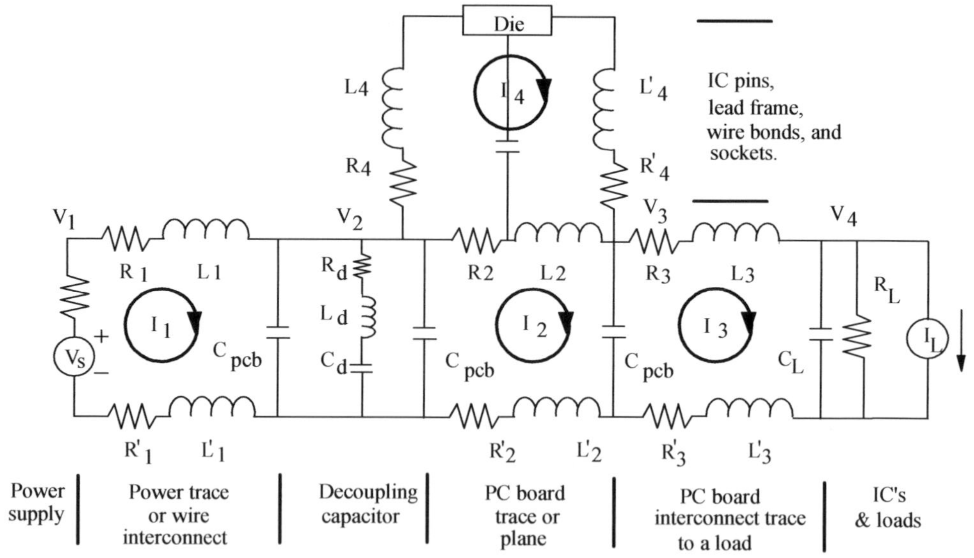

Electrical representation of a PCB.

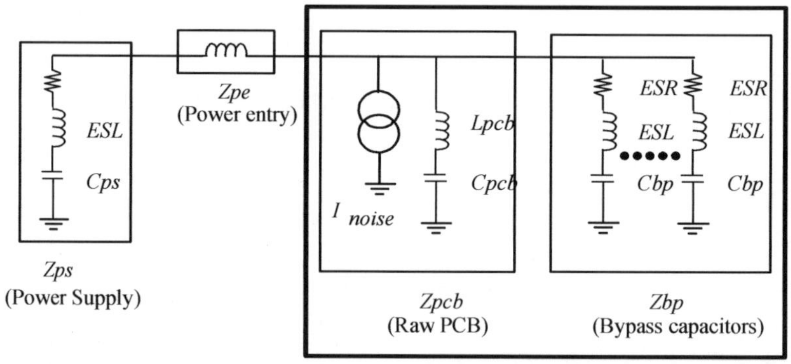

Schematic representation of a PCB.

Figure 4.26 Equivalent representation of a PCB showing all parasitic elements.

Installing a local decoupling capacitor, *Cd* in Fig. 4.26, lowers the impedance of current loop *I2*. This provides significant benefit related to delivery of energy charge as well as lowering the equivalent plane impedance at that physical location. This capacitor also provides a benefit of shunting unwanted AC switching noise that may be present on a power plane to the return (ground) plane, assuming the return plane is connected to chassis ground.

It is mandatory that decoupling loop impedance be much lower than the rest of the power distribution system. Not having low loop impedance will cause higher frequency RF switching noise to remain almost entirely within a localized area along with potential functional disruption to components observing voltage swings beyond operational margins due to lack of a stable power distribution network.

If the impedance of a localized current loop is smaller than the rest of the system, some fraction of high-frequency RF energy will transfer or couple to the larger loop formed by the power distribution network between the power supply and components. With this situation, RF switching currents also injected in the larger loop structure from an external immunity event can cause harmful interference. This situation is illustrated in Fig. 4.27 showing typical loop inductance, which when added together can be very large.

From *E=-L(dI/dt)*, any large amount inductance in the loop will cause common-mode RF noise to be created and propagated [10].

To summarize:

The most important item of concern when using discrete decoupling capacitors is to minimize lead and trace interconnect inductance and to locate the capacitor as close as possible to component pins if using single or double sided printed circuit boards.

Locating a decoupling capacitor physically adjacent to a device is not critical if both capacitor and component mounting pads go directly to planes and "not" utilize any physical trace between the two elements, since the inductance in the trace between capacitor and components creates undesired common-mode current.

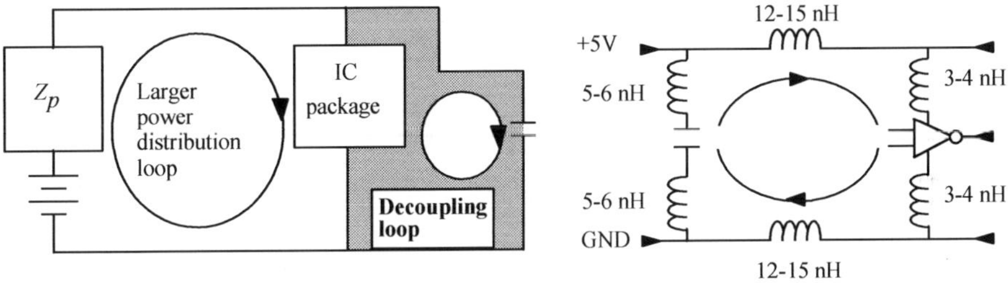

Figure 4.27 Power distribution model for loop area impedance calculation.

4.16.1 Transmission line (trace) inductance

When utilizing discrete capacitors for the purpose of decoupling, the method that provides optimal performance must be chosen based on knowledge of electromagnetic theory and not rules-of-thumb, application dependent, which means we must have the lowest amount of loop inductance in all three axis (includes via inductance). For certain layout configurations such as single or double sided assemblies, it is a functional requirement to run a trace from the capacitor pads to device and also to the power and return distribution network. Various combinations of this layout technique are presented in Fig. 4.28 for assemblies containing both power and return planes [6, 10].

Numerical analysis (Table 4.7) shows impedance of trace inductance between capacitor and a digital component. As trace length increases, inductance becomes very large. With large inductance per *E=-L(dI/dt)*, significant common-mode RF current € is created across this very small transmission line.

If the length of a routed trace from pad to via is 12.7 mm on one side, typical distance on most printed circuit board layouts, total inductance for both terminals would be 25.4 mm. If we insert a via next to a pad without any trace, total loop inductance becomes smaller by nearly one-half.

To help visualize effects within Table 4.7, we continue to examine Fig. 4.28. For the poor placement configuration, running a trace from capacitor to component and then connecting the composite configuration to power and return planes ensures maximizes loop inductance. It does not matter where the capacitor is physically located in the loop network since the capacitor charges the distribution system almost instantaneously, or much faster than component current consumption requirements. The loop area of the current path between the circuit through a via, plus the trace distance through the capacitor back through another trace and via to the other plane is physically much larger than desired. In summary, if we must place a capacitor next to a component, using four vias (two from the capacitor and two from the component in parallel) with no trace routed between, this provides magnitudes less inductance than even a short trace (Table 4.7).

POOR PLACEMENT CONFIGURATION

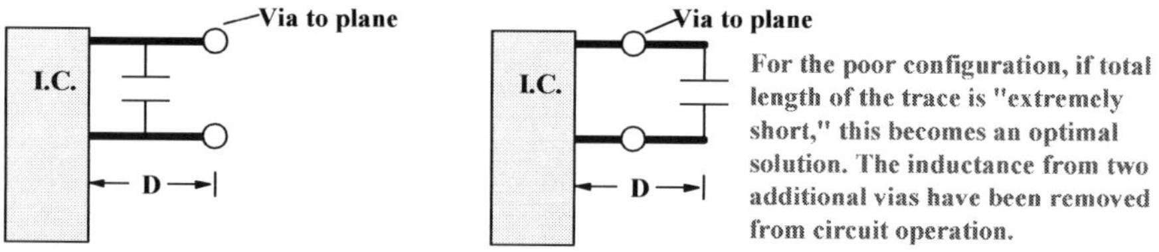
Via to plane
I.C.
D
Via to plane
I.C.
D
For the poor configuration, if total length of the trace is "extremely short," this becomes an optimal solution. The inductance from two additional vias have been removed from circuit operation.

Capacitor and component share common via, except different locations. Length of traces are the same. Inductance of both circuits are identical, since the total circuit consists of two traces and two vias.

BETTER CONFIGURATION — **ENHANCED CONFIGURATION**

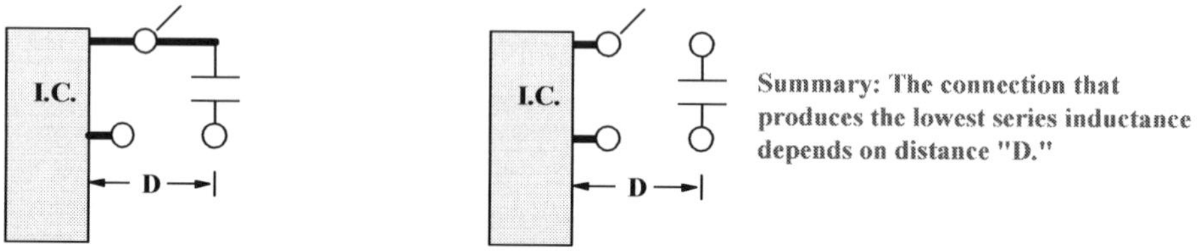
Via to plane
I.C.
D
Via to plane
I.C.
D
Summary: The connection that produces the lowest series inductance depends on distance "D."

Capacitor and component share only one via. Total inductance of the circuit includes one trace (some inductance) plus two vias. Useful when located very close to the component.

Inductance in the loop area is minimized. Impedance of planes is magnitudes less than the interconnect trace. Recommended for extremely dense layout designs.

Figure 4.28 Various mounting methodologies.

Table 4.7 Inductance of a microstrip trace (per pin pad, not including via inductance).

Trace length	Trace width	Distance to reference plane	Impedance
2.54 mm (0.100 in.)	0.127 mm (0.005 in.)	0.127 mm (0.005 in.)	880 pH
2.54 mm (0.100 in.)	0.200 mm (0.008 in.)	0.005 mm (0.005 in.)	733 pH
2.54 mm (0.100 in.)	0.254 mm (0.010 in.)	0.005 mm (0.005 in.)	662 pH
2.54 mm (0.100 in.)	0.127 mm (0.005 in.)	0.254 mm (0.010 in.)	1.2 nH
2.54 mm (0.100 in.)	0.200 mm (0.008 in.)	0.010 mm (0.010 in.)	1.0 nH
2.54 mm (0.100 in.)	0.010 mm (0.010 in.)	0.010 mm (0.010 in.)	945 pH
12.6 mm (0.500 in.)	0.127mm (0.005 in.)	0.005 mm (0.005 in.)	4.4 nH
12.6 mm (0.500 in.)	0.200 mm (0.008 in.)	0.005 mm (0.005 in.)	3.7 nH
12.6 mm (0.500 in.)	0.010 mm (0.010 in.)	0.005 mm (0.005 in.)	3.3 nH
12.6 mm (0.500 in.)	0.127 mm (0.005 in.)	0.010 mm (0.010 in.)	5.9 nH
12.6 mm (0.500 in.)	0.200 mm (0.008 in.)	0.010 mm (0.010 in.)	5.1 nH
12.6 mm (0.500 in.)	0.010 mm (0.010 in.)	0.010 mm (0.010 in.)	4.7 nH
25.4 mm (1.00 in.)	0.127 mm (0.005 in.)	0.005 mm (0.005 in.)	8.8 nH
25.4 mm (1.00 in.)	0.200 mm (0.008 in.)	0.005 mm (0.005 in.)	7.3 nH
25.4 mm (1.00 in.)	0.010 mm (0.010 in.)	0.005 mm (0.005 in.)	6.6 nH
25.4 mm (1.00 in.)	0.127 mm (0.005 in.)	0.010 mm (0.010 in.)	11.8 nH
25.4 mm (1.00 in.)	0.200 mm (0.008 in.)	0.010 mm (0.010 in.)	10.2 nH
25.4 mm (1.00 in.)	0.010 mm (0.010 in.)	0.010 mm (0.010 in.)	9.4 nH

4.17 Conflicting Rules for Printed Circuit Board Decoupling

There are many rules-of-thumb related to the use of decoupling capacitors, some of which are listed in Fig. 4.29 and examined further in this Section to determine if they are valid or not and under which specific conditions of use. The rules in Fig. 4.29 are not exhaustive and there are many more. For each rule, engineering analysis must be performed to determine which rule works and provides maximum value to ensure optimal signal integrity, which in turn minimizes creation of common-mode RF energy or undesired EMI.

With rules-of-thumb listed in Fig. 4.29, which ones are valid? The answer is "*it depends!*" It depends on many parameters, applications and other considerations generally not known or recognized by design engineers. All rules are valid if implemented correctly (under certain conditions of use) although they may conflict with other similar rules. In reality, rules help in understanding options to choose from during layout analysis, but may not be the right choice in the long run.

Use small-valued capacitors for high-frequency decoupling

Use 0.01 mF for local decoupling

Use one capacitor per power/return pin pair

Locate capacitors near the power pins of active devices

Use capacitors with a low ESR!

Avoid capacitors with a low ESR

Locate capacitors near the ground pins of active devices

Locate of the decoupling capacitors is not relevant

Use 0.001 mF for local decoupling

Never put traces on decoupling capacitors

Use the largest valued capacitors you can find in a given package size

Local decouping capacitors should have a range of values from 100 pf to 1 mF

Number of capacitors is not critical

Figure 4.29 Conflicting rules-of-thumb related to decoupling capacitor placement. (Courtesy: Dr. Todd Hubing – IEEE EMC Society Distinguished Lecturer Presentations)

A question commonly asked by printed circuit board designers, as well as digital engineers, is where to physically place decoupling capacitors? Little distinction is made between application of use that includes decoupling, bypassing and bulk applications, all of which have a specific requirement for optimal functional use. In addition, rules-of-thumb are generally implemented without questioning the validity of the rule and whether the rule applies or not; in many cases, these rules are valid under certain conditions and invalid under others.

Computational analysis is required for accurate implementation of decoupling capacitors on a printed circuit board. One should not rely strictly on rules-of-thumb, especially on higher speed designs.

When it comes to rules of thumb for power bus decoupling, the following questions generally arise, and are now examined per major topic items, extracted from Fig. 4.29.

These major topic items are:

1. Where do we locate a decoupling capacitor-near the power pin, return pin or it does not matter?
2. Should we install decoupling capacitors where there are no components physically located?
3. Should we use a capacitor with high *ESR* or low *ESR*?
4. Is there a relationship between capacitance value and packaging dimensions?
5. Should we use one capacitor per power/return pair, or share capacitors between components?

4.17.1 Where do we locate a decoupling capacitor-near the power pin, return pin or it does not matter?

The answer where to locate a decoupling capacitor is "*it depends.*" It depends on total loop area in all three axis (x, y and z). The z-axis is the interconnect path from top to bottom of the board in the vertical polarity directly through a via. Every transmission line including vias contains finite inductance (Table 4.7). The key to success is to minimize total loop inductance, *including the z-axis* as vias can be highly inductive.

In Fig. 4.30, we see different loop areas established between components and capacitor. These are all similar visually but have significant different levels of performance. The least amount of loop inductance is what we want to achieve, which in this figure is the bottom left illustration.

1. **<u>Next to the power pin or return pin</u>**.

This is a concern only when traces are routed between capacitor and component using a physical trace. The inductance of the trace can be substantial. If a trace must be used then we must know the DC potential of the plane next to the component. If layer 2 is the return plane, placing the capacitor next to the power pin provides for smaller loop inductance due to physical geometry. If layer 2 is a power plane, place the capacitor next to the return pin. So the answer to the question is "*It depends.*"

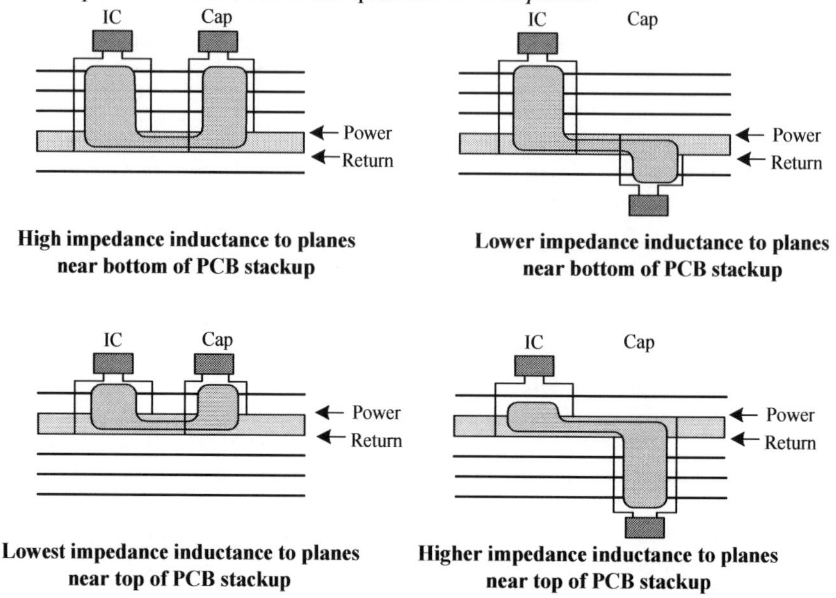

Figure 4.30 Loop area with decoupling capacitor placed adjacent to power/return pin.

2. **Locating the capacitor anywhere in the vicinity of the device.**

This is one of the more most important items to remember when using rules of thumb and is valid, most of the time. If we via directly to planes for both capacitor and component without any interconnect trace (similar to a via-in-pad concept) we have the smallest amount of loop inductance possible if both power and return planes are on layers 2 and 3 of a multilayer stackup. By using this technique to connect both component and capacitor to the planes, we can locate the decoupling capacitor almost anywhere, even a distance away *within reason.*

Why does this technique work on locating a decoupling capacitor up to several inches or cm's away and still have optimal performance? The answer is overly simple. This simple explanation deals with velocity of propagation of charge between capacitor and component. The electric charge from the capacitor propagates between planes in prepreg or core at a velocity that is directly related to the dielectric constant of the material. A capacitor located, for example several inches away from a component, can provide charge to that device in approximately 120 ps if the dielectric permits this speed of propagation to occur. If the component has a 1 ns switching rate, the charge from the capacitor will be available to the pins of the device as needed, well before the next edge transition or nearly 8 times faster than required. Per the capacitor brigade discussion (Section 4.14), capacitors charge planes not components if everything connect to the power distribution planes directly by vias without use of an interconnect trace. Component consumes current from planes. As long as planes have sufficient energy charge everything works great. If the decoupling capacitor is placed adjacent to the component, which is recommended by many vendor application notes, charge can be delivered in a few picoseconds, magnitudes faster than required providing the device switches at 1 ns and there is no interconnect trace between device and via. Few components require charge this fast. Placing a capacitor next to the device is thus required only if there is *a routed trace with inductance which prevents transference of charge quickly*. Application Notes by vendors do "*not*" take into account velocity of propagation, which is an outdated rule-of-thumb.

For example, if we have a 2000 pin FPGA with hundreds of power and return pins, it is impossible to place one capacitor per pin directly at the device per common practice or by vendor application notes recommendations. Decoupling capacitors are often forced to be located a distance away but will still work due to the fast velocity of propagation of electrons within the planes. However, if there is any significant trace/loop inductance from either component or capacitor, the capacitor must then be located as close as possible for optimal performance, since loop inductance will minimize current flow in addition to creating a voltage gradient across the transmission line interconnect, which in turn creates unwanted common-mode EMI based on the magnitude of loss present.

4.17.2 Should we install decoupling capacitors where there are no components physical located?

What is generally not realized by board designers are that decoupling capacitors are required not only next to or near components requiring a localized power supply, but also throughout an entire assembly even where there are no components physically located. This is especially true for analog circuits and slow speed digital devices that do not have any significant amount of current demand during digital logic state switching.

Why is it critical to put decoupling capacitors in regions where there are no components? We are not providing a decoupling function but that of lowering plane impedance to prevent a bounce condition from causing functional disruption of circuits physically located a distance away.

Research [11, 12] on switching noise created by large power consuming components bounces both power and return planes with every edge transition. Note use of the word "return." Engineers are generally concerned about "ground bounce." In reality there is no such thing as ground in a system, discussed in Chapter 5. We utilize 0V or return planes but commonly call them as ground planes. How often do we worry about or concern ourselves with power bounce, it's always ground bounce? What if the system is battery operated and both planes bounce at the same time? Where is ground in this assembly, or where is ground if an AC powered device uses only two lines for power and no third ground (safety) wire? Is the negative terminal of a battery at ground potential or 0V reference?

Any bounce on planes at RF frequencies will propagate from the switching element in a radial manner for a considerable distance affecting other components with this noise connected to the same power and return planes. It is generally assumed that localized decoupling provides clean DC voltage levels everywhere, yet we may not know if voltage ripple is within acceptable margins for digital and especially analog components. Digital devices can tolerate several millivolts of ripple yet analog are very sensitive to almost any fluctuations in power and return. With planes containing switching noise at low levels, some components a distance away may not be provided with high power quality and thus cease to function in an optimal manner.

Another concern is the overall impedance of the plane pair, which is not one constant value everywhere. Plane impedance varies based on frequency of operation, number of components utilized along with other second order elements. As much as we try to have plane impedance in the milliohm range, many times this cannot be achieved throughout the entire assembly. Decoupling capacitors bring the impedance of planes to a low value (Section 4.6). Planes in the upper frequency range can have impedance in the hundreds of ohms even if there are no components provided in that region of the board. Should analog devices sharing a common return plane be physically located in this area with high impedance, large plane bounce may occur per Ohms law causing functional disruption.

In [11, 12] it was observed that without low plane impedance at a particular *x-/y-* axis location on a printed circuit board, plane bounce of up to 120 VDC was observed with a 1-volt, 1-amp source using computational analysis. Per Ohms law, if the impedance of the plane is high, such as 120 ohms at that specific location (perform an S11 analysis on the assembly) and with 1-amp of switching noise, plane bounce of 120 volts could occur. Remember, this situation only exist "*under theoretical conditions if no decoupling capacitors are present.*"

Where did this high plane bounce value come from? This value highlights critical aspects of power distribution network design if done poorly without taking into consideration application and use of decoupling and bypass capacitors to lower plane impedance. If we utilize *any* capacitor at this specific *x-/y-* axis location, the impedance of the plane will become very low, down to the *ESR* value of the capacitor but within a limited bandwidth of operation. For our simulated example, there were no decoupling capacitors incorporate to examine what happens under *theoretical worst case operating conditions to understand a physical concept related to electromagnetic compatibility.*

Small value capacitors generally in the pF range must be scattered all over the printed circuit board layout, which lowers plane impedance for higher frequency signals with the intent of minimizing plane bounce in areas of the assembly where there are no components. If switching noise in this specific area finds a via and reflects back to another section of the printed circuit board, functional disruption may occur due to phasing effects in the cavity between the power and return planes. The lower the impedance of the capacitor package and interconnect methodology, the lower the plane bounce that affects both voltage delivery and return.

Discrete capacitors function only up to a few hundred MHz when used in decoupling mode. If switching elements are in the upper MHz to GHz range, discrete capacitors provide no value for recharging, however capacitors in the low pF range are excellent for lowering plane impedance and shunting switching noise on the power plane into the 0V system or chassis plane. Although these capacitors have a very high self-resonant frequency, they also have minimal charge storage capability along with higher *ESR*.

Placement of 1 nF (1,000 pF) capacitors (capacitors with a high self-resonant frequency) on a 2.54 cm (1-in.) grid assists in shunting RF switching noise that may be present due to high impedance at a specific location, especially if there are no bypass or decoupling capacitors provided in this area, or components for that matter. What is achieved by placing small value capacitors with a high self-resonant frequency is *not* to provide charge but to act as bypassing that allows switching noise present on a power plane to divert to the return plane that is hopefully connected to a chassis ground and ultimately earth ground. Depending on application, if the printed circuit board is connected to a metal chassis in multiple locations, all power plane noise is now shunted to a virtual RF ground, or chassis ground [11, 12].

A lumped model analysis of the printed circuit board shows that small value capacitors not only provides for shunting of RF switching noise in a power or return system, but also gives a small amount of charge for decoupling use. To have any real benefit from using these small value capacitors, many will be required to achieve the total amount of electric charge that switching components may require. Depending on the resonant frequencies of a printed circuit board, capacitors placed in this grid may be as small as 30–40 pF [6, 13].

> *In addition to providing low plane impedance to minimize plane bounce, a capacitor that is located across the power and return planes (decoupling mode) now acts as a bypass capacitor that shunts RF switching noise present on the power plane to chassis ground through the return plane(s). If for no other reason, these "bypass" capacitors will provide more benefit to a design in minimizing development of common-mode EMI because standing waves on planes or in the cavity between power and return (actually a transmission line acting as a dipole antenna) are now shunted to 0V (a.k.a., ground), making the driven element of the dipole antenna an inefficient radiator!*

4.17.3 Should we use a capacitor with high ESR or low ESR?

As presented in Section 4.12, *Multi-pole Decoupling*, broadband benefits occur over a large spectral bandwidth of frequencies if all decoupling capacitors are below a desired target impedance, which means their *ESR* must be very low. Depending on application of use and *Q* of the circuit, we either have very low impedance over a very narrow range of frequencies, along with high anti-resonant frequency, or a lower *Q* using capacitors with higher *ESR* that is still below the target impedance but provides a greater spectral bandwidth of operation.

Depending on application, one must know what the target impedance of the design must be for optimal decoupling performance and over what range of frequencies. Capacitors with high *ESR* are more expensive yet provide enhanced operation under certain conditions.

The decision is up to the designer on which decoupling method provides greatest benefit based on intended use, therefore the answer to the question of which type of capacitor to use with regard to *ESR* is "*It depends*."

4.17.4 Is there a relationship between capacitance value and packaging dimensions?

This is another area of concern a designer must take into consideration when choosing a capacitor for a specific application. The rule of thumb to "use the largest value capacitor in a fixed package size possible" applies only under certain conditions. If the capacitor value is large, its self-resonant frequency may be well below a desired bandwidth of operation because large paths through lengthy surfaces set up greater inductance. We must choose a capacitor to be resonant at a particular frequency or range of frequencies and have still have low enough impedance over a desired bandwidth.

Another concern when using higher power consuming devices, as mentioned many times in this Chapter, is interconnect inductance and resistance. We must minimize total inductance to the greatest extent possible. Inductance can be reduced by not having a long interconnect between the mounting pad of the capacitor and its interconnect via. Using any trace length for this implementation is the equivalent of using long leads commonly found in radial or axial based capacitors configurations.

The other item of concern is reducing *ESL* and *ESR*. Discrete capacitors have internal *ESR* based on the dielectric composition of the material that makes up the capacitor body, which could be substantial in addition to a small amount of package inductance. In order to minimize both internal *ESL* and *ESR*, a capacitor with reverse aspect ratio is sometimes used. A reverse aspect ratio capacitor is shown in Fig. 4.31.

The resistive elements within multilayer ceramic capacitors that contribute to total *ESR* are electrode plate resistance, termination material resistance, interface resistance at the electrode termination junction and dielectric resistance. Shorter and wider plates decrease plate resistance contribution. The wider plates will also increase the area of interface contact, thus reducing this resistive contribution and thus lowers total impedance of the capacitor for enhanced decoupling [7].

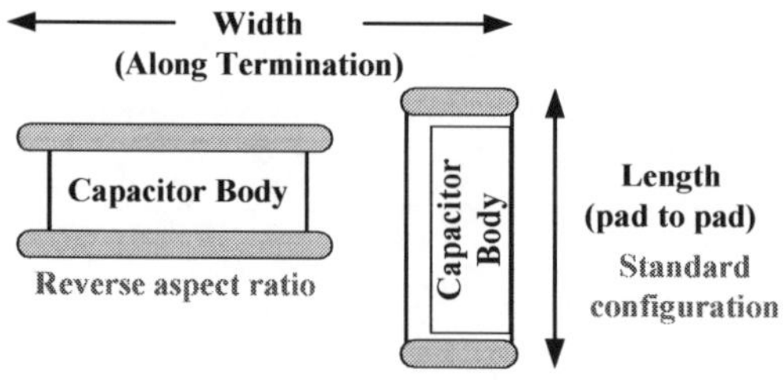

Figure 4.31 Capacitor bodies with reverse aspect ratio.

4.17.5 Should we use one capacitor per power/return pair or share capacitors between components?

This rule of thumb is controversial because of theory versus reality. This is in addition to component vendor application notes generally written by technical writers and not design engineers. Almost every application note says to put one decoupling capacitor per power and return pin pair. *This is incorrect information based on outdated knowledge.* If there are 10 pairs of pins, we technically need 10 capacitors using this rule. If there are 100 pin pairs such as those found in FPGAs or large-scale BGA, we need 100 capacitors. Is this realistic? Just because an application note says to put one capacitor per power pin as a general rule-of-thumb, does this mean the application note is correct? Was any engineering analysis performed by the component designer or is this a handed down rule-of-thumb because it is known to work but may be overkill if actually implemented. Is one capacitor per power pin even required for large scale devices?

As detailed in Section 4.15, *Radius of Operation*, when capacitors discharges their electrons, or recharge a power and return plane pair it does so in a radial manner. This means the plane is fully recharged within a specific radius located around the device. This is true only if there are no traces routed between capacitor and component, only a via directly to the planes which minimizes loop inductance.

If there are two power/return pin pairs adjacent to each other, both will be energized from the same planes that are recharged by the same decoupling capacitor. As long as the decoupling capacitor provides sufficient charge to keep the planes from bouncing, these planes are able to provide enough current to multiple power/return pins simultaneously. If the component consumes 1 mA of current per pin and the capacitor provides 10 mA of charge, then theoretically this one capacitor can provide enough charge to support 10 pin pairs. This means 9 less capacitors to load the power supply in addition to permitting more area to route traces without having to traverse a field of extra vias.

Another way to view this requirement is when there are two digital components physical next to each other, with power and return pins near close by. Most application notes say to use one capacitor per pair, however in this application one capacitor is sufficient to maintain charge for two devices with functional degradation. This is illustrated in Fig. 4.32 with an application note from a major semiconductor manufacturer validating this design requirement (Fig. 4.33).

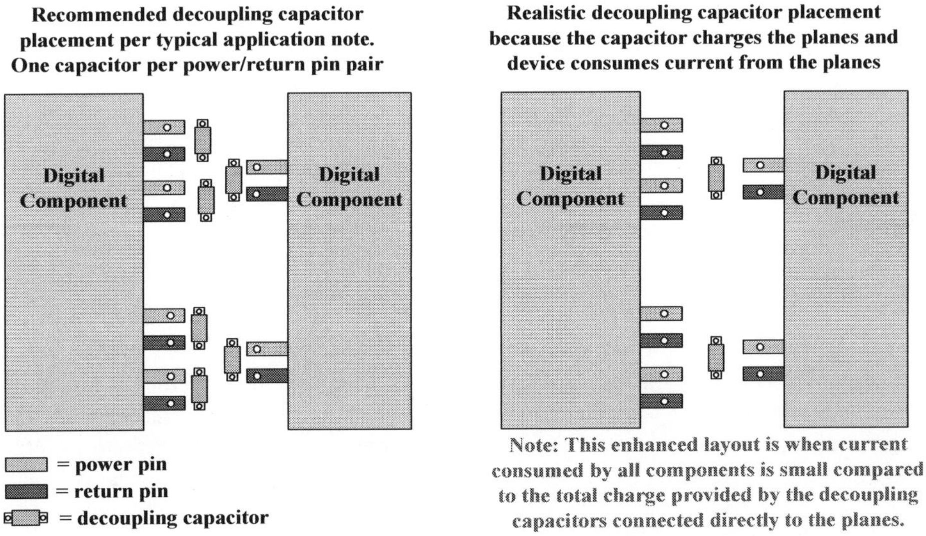

Figure 4.32 Sharing decoupling capacitors among pins and components.

The question at hand is "Should we use one capacitor per power/return pair"? The answer is, again, "*It depends.*" It depends if the capacitor can provide sufficient charge to keep the voltage level stable without bounce (both power and return), and how much current digital components consume. To experience this situation personally, remove 50% of the decoupling capacitors on a printed circuit board and measure performance, both signal integrity and EMI. We always over populate a printed circuit board with decoupling capacitors due to application notes that are overly conservative. In addition, if we via the capacitor directly to the planes, location is not important as long as the charge reaches the device before the next switching transition.

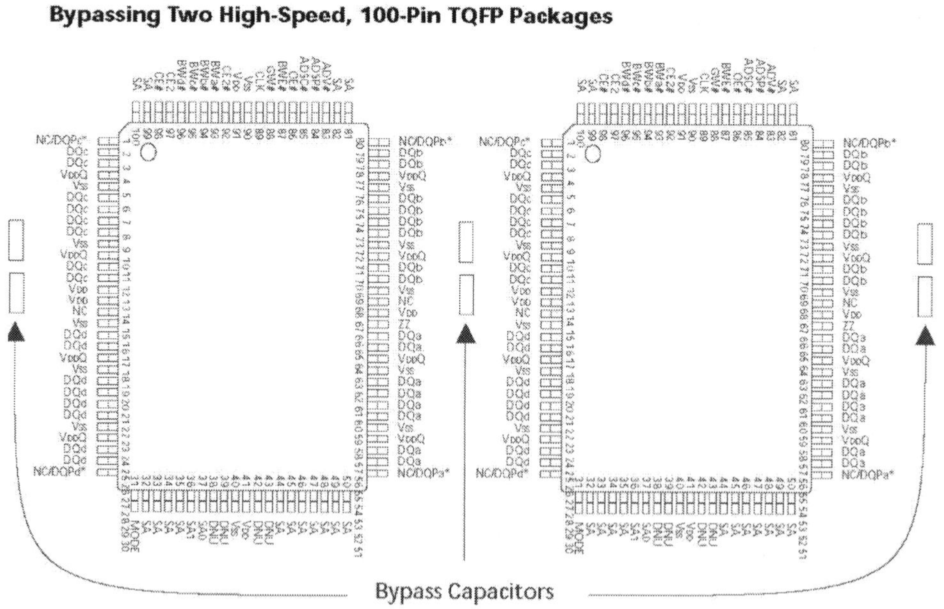

Figure 4.33 Simplified decoupling capacitor placement for large scale components. One decoupling capacitor per power/return pin pair not required (Note: "Bypass" should be called "Decoupling" for correct application of this word) Figure courtesy: Micron Technology Inc. [16]

Although many application notes provide the standard rule-of-thumb for one capacitor per power pin, Fig. 4.33 illustrates this rule as inappropriate. Notice how many power and return pins are adjacent to each other in the package pin-out, all supplied by one decoupling capacitor. Of course, all designs must be evaluated individually to determine if this design technique is appropriate.

4.18 Inductance of Mounting Pads for Components and Capacitors

Multiple vias and short trace lengths reduces overall inductance of the interconnect between capacitor and component depending on implementation and manufacturing requirements. Increasing the size of the SMT pad and using multiple vias to the reference planes provides significant improvement, especially in higher-frequency applications. If the via(s) is located internal to the mounting pad (via in pad), additional benefit is achieved however, if there is any significant amount of current flow in the via this enhanced design technique is called a fuse and must never be implemented for decoupling requirements (Fig. 4.34). Use of microvias are best suited to signal traces only.

Using numerical analysis, calculated inductance are on the order of:

- Pair of surface traces 10–15 nH/inch (25.4-38.1 nH/cm)
- Pair of vias for decoupling capacitor 0.40–1 nH (200–500 pH each)
- Plane inductance 0.1 nH

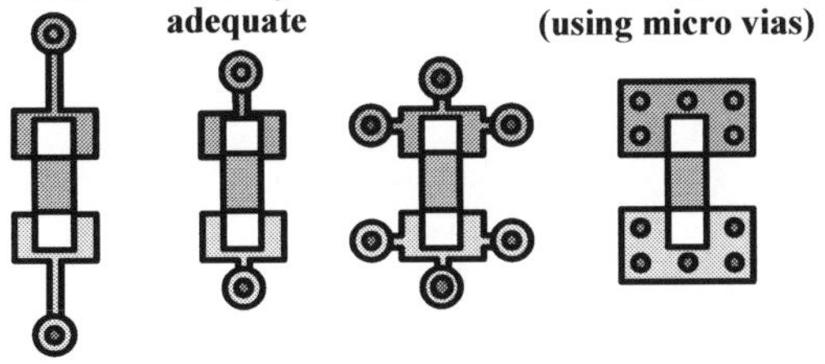

Figure 4.34 Comparison of connection inductance for SMT components.

These inductance numbers illustrate that any trace and via inductance is significantly greater than plane inductance, and that any inductance in a power distribution network causes functionality problems besides that of planes.

Maximizing the physical width of the trace routed from the capacitor body to plane(s) also minimizes total loop inductance. The wider the trace width the lower the inductance, a perfect rule-of-thumb when determining if a trace should be larger or smaller in physical size with relation to transmission line inductance. The same goes for connecting component pads to planes. For each additional via, inductance will be reduced similar in concept to resistors placed in parallel (Thevenin equivalent circuit). If multiple vias can be utilized this is a preferred way to do a layout however, this is an expensive process for drilling additional through-holes in addition to blocking routing channels on internal layers, and which is generally not desired by manufacturing or board layout designers.

Figure 4.35 illustrates various patterns to minimize trace inductance and enhance performance of decoupling capacitors for multilayer designs. Use of very small SMT components may not allow placement of vias between the mounting pads; thus use of vias on the inside edge becomes mandatory.

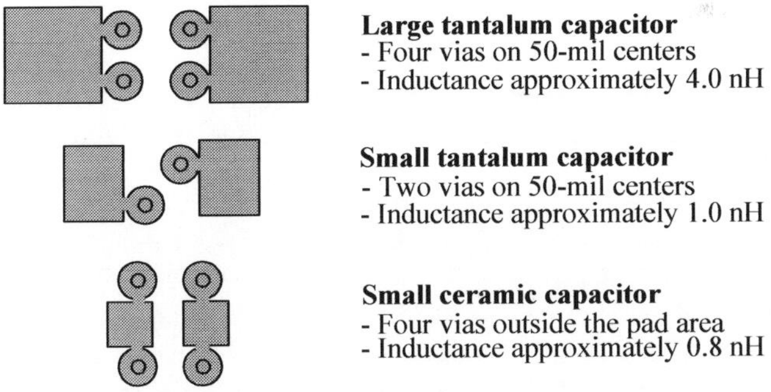

Figure 4.35 Comparison of connection inductance for various SMT components.

<u>Ways of decreasing loop inductance for decoupling capacitors</u>

1. Ensure power and return planes are located physically next to the surface layer(s).
2. Use smallest package size as possible (minimize package inductance).
3. Use very short connections between capacitor pads and vias to the planes if an interconnect trace must be used.
4. Use multiple capacitors in parallel (reduces *ESL* and *ESR–Thevenin equivalent circuit*).
5. Position interconnect vias physically between the capacitor pads (see below).

To further expand on a process of minimizing loop area inductance, Fig. 4.36 is provided based on the illustration of Fig. 4.35. By placing the vias between the mounting pads, if physically possible, one can significantly reduce the loop area for the component.

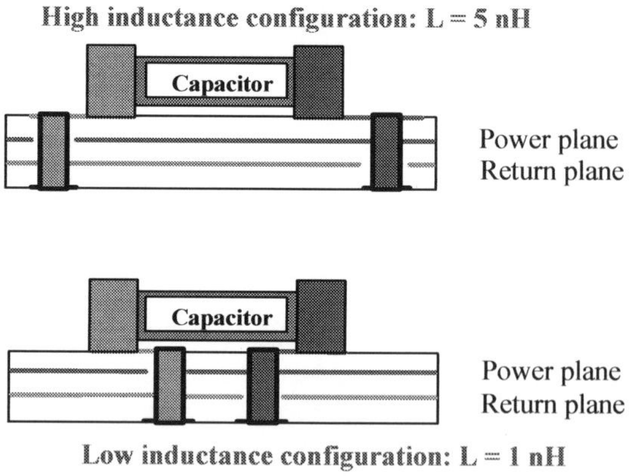

Figure 4.36 Placement patterns for optimal performance–multilayer implementation.

For multi-voltage systems, such as +5V/+3.3V, planes should be located next to the top and bottom layers of the printed circuit board, decoupled from the side closest to the power/return planes(s) to minimize total loop inductance. In addition to the function of decoupling, capacitors also provides a transmission line return path for RF current to travel between planes at different potentials. Optimal placement of capacitors in critical locations, or physically next to an existing layer jump with a signal containing a large spectral content, provides significant benefit for printed circuit boards with only one power and return plane to ensure a coaxial based transmission line exists. The important item to remember is not the specific numeric capacitance value of the capacitor (i.e., 1 nF or 100 nF), but the *magnitude* of impedance at a particular frequency which must be extremely low for optimal performance over a desired bandwidth.

Figure 4.37 illustrates another layout technique on how to locate decoupling capacitors to minimize inductance. Regardless of whether the printed circuit board is simple or complex, almost all products require a trace to be present between a component lead and capacitor, or interconnect via (via-to-pad is ideal). These interconnect, also called a pin-escape, breakout and similar terminology exist by virtue of the fact that component pins are spaced tightly together with small packaging densities. A trace must be routed from the component to a via located nearby for connection to a signal, power or return plane. It is not possible, manufacturing wise, to have physically large vias embedded in a component's mounting pad. Solder may flow into the via preventing the component from having a secure bond connection in addition to other manufacturing concerns such as solder bridging along with higher impedance bonding connection.

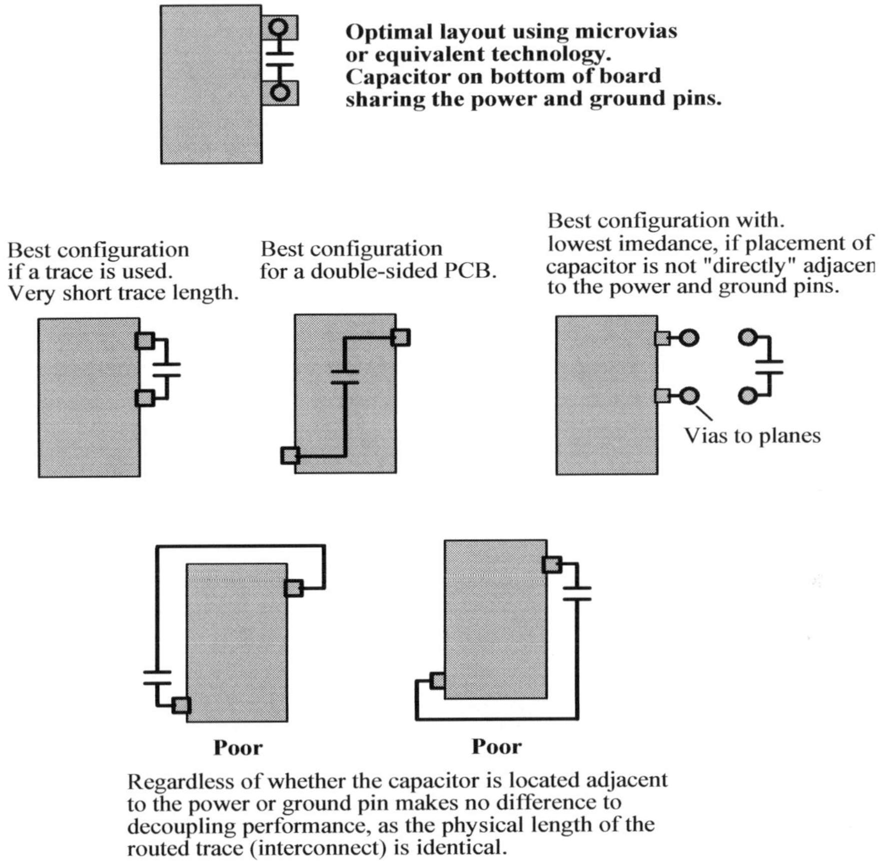
Optimal layout using microvias
or equivalent technology.
Capacitor on bottom of board
sharing the power and ground pins.
Best configuration
if a trace is used.
Very short trace length.
Best configuration
for a double-sided PCB.
Best configuration with.
lowest imedance, if placement of
capacitor is not "directly" adjacen
to the power and ground pins.
Vias to planes
Poor
Poor
Regardless of whether the capacitor is located adjacent
to the power or ground pin makes no difference to
decoupling performance, as the physical length of the
routed trace (interconnect) is identical.

Figure 4.37 Placement recommendation showing loop inductance when using a trace.

4.19 Bypass and Decoupling Value Calculation

Bulk capacitors such as electrolytics, tantalum and high-frequency monolithics are all required on a printed circuit board, application and frequency dependent. Monolithic or ceramic capacitors must have low impedance and a self-resonant frequency higher than clock harmonics or switching frequencies requiring suppression. Typically, one selects a capacitor with a self-resonant frequency in the range of 10–30 MHz for circuits with edge rates of 2 ns or less. Many printed circuit boards happen to be self-resonant in the 200–400 MHz range. Proper selection of decoupling capacitors, along with knowing the self-resonant frequency of the printed circuit board structure (power and return planes acting as one large decoupling capacitor) will provide enhanced EMI suppression above the highest functional frequency of discrete devices that may be provided. Table 4.4, provided earlier in this Chapter, shows the self-resonant frequencies of discrete devices. Surface-mount capacitors have a higher self-resonant frequency by approximately two orders of magnitude (or 100x) as a result of less lead inductance (from the plates of the capacitor to the interconnect via that includes the breakout trace from the device). Electrolytic capacitors are ineffective for high-frequency decoupling and are best suited for power supply subsystems or power line filtering and to minimize power supply dropout.

It is common practice to select a decoupling capacitor for a particular application, usually the first harmonic of a clock or processor. A capacitor is sometimes selected for the third or fifth harmonic, since this is where a majority of RF current is usually observed as EMI. Use of 100 nF capacitors may be too inductive at frequencies above 50 MHz and may also not be able to provide charge to the planes quickly due to lead and loop inductance that minimizes propagation of current flow.

The 100 nF (0.1 μF) capacitor is the most common value used in printed circuit boards for decoupling. When asking a designer why they choose this value, they are unable to give a reason besides "It's a common value that we have used in the past." Engineering analysis is rarely performed to determine if this value is even proper. Although a manufacturer's data sheet indicates use of common value (e.g., 100 nF), what assurance does one have that this value is the correct one in the first place? Was this value calculated by component designers or was it just left to historical usage? Should board designers use a rule-of-thumb since this is standard practice, or was the value identified in data sheets provided by a technical writer without justification because this was the value they always used in past application notes? Did the component manufacturer recommend a value under minimal or maximum capacitive switching load conditions along with other parametric values required for optimal performance?

For historical purposes, around 1965 the United States Air Force discovered interference problems with transmitters on planes. If they installed a decoupling capacitor to this airborne electronic system, the aircraft would be able to communicate using a new type of radio. Use of 0.1 μF (100 nF) electrolytic capacitors was selected with long wire leads (large ESL and ESR values). The operating frequency of the equipment was at 200 kHz. Consequently, if 0.1 μF is acceptable for 200 kHz systems, then it must be acceptable for 200 MHz or 2 GHz products!

The key to optimal performance is to calculate a value for functionality reasons. A capacitor should not be used based on historical usage without understanding how and why a particular capacitive value was decided upon. This does not mean we should abandon use of 100 nF capacitors as they still work extremely well for decoupling purposes, as being the one value to use, but to consider other values and dielectric properties as appropriate.

When performing component placement on a printed circuit board, one should make physical placement provisions (install mounting pads) for high-frequency RF decoupling in case we need to provide this discrete element, although use may not be needed until functional operation validates this need. Verify that all bypass and decoupling capacitors chosen are selected based on intended application of use. This is especially true for clock generation circuits or those with periodic signals. The self-resonant frequency of the planes must take into account all significant clock signals requiring suppression, which is generally considered the fifth harmonic of the original frequency. Capacitive reactance (self-resonant reactance in ohms) is calculated by (4.13).

$$X_c = \frac{1}{2\pi f C} \quad (4.13)$$

where:
X_c = capacitance reactance (ohms)
f = resonant frequency (Hertz)
C = capacitance value (Farads).

The minimum capacitance required for optimal performance is determined by the maximum amount of voltage drop allowable across the capacitor as a result of a transient current surge. This voltage drop is exacerbated when all components operates under maximum capacitive load. An appropriate value for decoupling applications can be easily calculated by (4.14).

$$C = \frac{I\Delta t}{\Delta V} \quad \textit{(Farads)} \tag{4.14}$$

where:
C = capacitance value (Farads)
I = total amount of current required to maintain functional operation (Amps)
Δt = duration of the transient event (seconds)
ΔV = allowable voltage drop (Volts).

For example, a typical digital component has an input transient surge of 20 mA for 10 ns. The voltage drop must be less than 100 mV to ensure proper logic transitions. The optimal value for a decoupling capacitor to prevent plane bounce is easily calculated by (4.14) and illustrated below.

$$C = \frac{(20\ mA)(10\ ns)}{(100\ mV)} = 2\ nF\ or\ 0.002\ uF$$

A problem with using standard equations lies with not knowing actual lead and loop inductance that minimizes transfer of current charge. Voltage spikes occur across inductance in addition to restriction of current flow. For any magnitude of ripple noise, the maximum amount of series inductance permitted is described by (4.15).

$$L = \frac{V\Delta t}{\Delta I} \quad \textit{(Henries)} \tag{4.15}$$

where:
L = inductance of the loop circuit (Henries)
V = maximum noise spike (Volts)
Δt = duration of the transient event (seconds)
ΔI = transient current in the decoupling loop (Amps).

For this same component, with a transient surge of 20 mA and an edge rate (rise/fall time) of 2 ns, the inductive values that restrict a noise spike to 100 mV peak is:

$$L = \frac{(100\ mV)\ (2\ ns)}{(20\ mA)} = 10\ nH$$

This means that total lead and series inductance cannot exceed 10 nH. This is a challenging task when traces are required to be routed between component and capacitor. One should not forget to include the inductance associated with the bond wires internal to a component package, if not a Ball Grid Array or flip chip configuration.

4.20 Capacitive Effects on Signal Traces (Wave Shaping)

Capacitors are sometimes used to wave shape differential or single-ended mode signals, generally in analog and I/O circuits. These capacitors are rarely used in clocking networks. The capacitor, *C*, alters the signal edge (slew rate) by rounding or slowing the time period that the signal edge transition takes to change from logic state 0 to logic state 1 or vice versa, shown in Fig. 4.38.

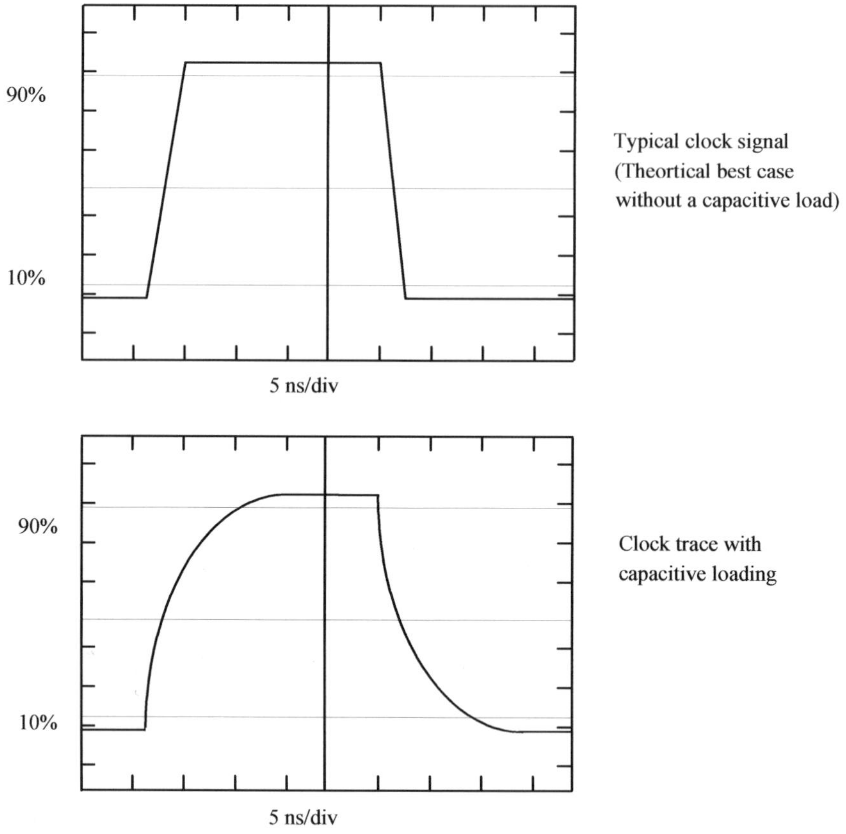

Figure 4.38 Capacitive effects on clock signals.

Figure 4.39 illustrate a change in the slew rate (clock edge) of a signal under capacitive load. Although the transition points remain unchanged, the time period or edge rate, t_r, is different. This elongation or slowing of the signal edge is a result of the capacitor charging and discharging. The change in transition time is described by the equations in Fig. 4.39. The Thevenin circuit is shown without load. The source voltage, V_b, and series impedance, R_s, are internal to the driver or clock generation circuit. The capacitive effect on the trace is a result of this capacitor being located in the circuit.

When Fourier analysis is performed on this signal edge transition (conversion from time to frequency domain), a significant reduction of RF energy occurs along with a decrease in spectral RF distribution. Hence, we see improved EMI compliance. Care is required during the design stage to ensure slower edge rates will not adversely affect operational performance, or signal integrity and is best determined through use of computational analysis.

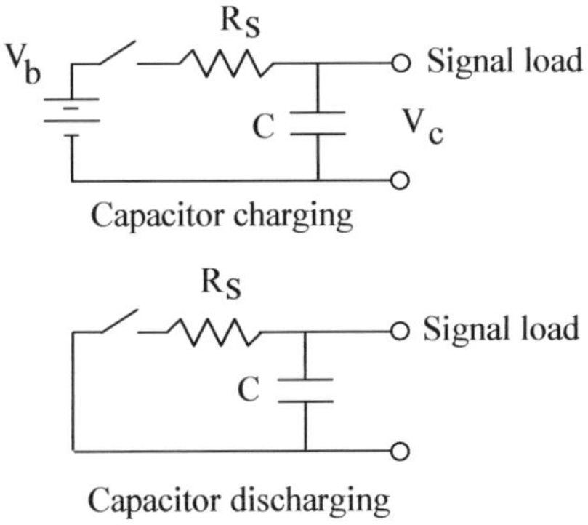

Charging

$$V_c(t) = V_b\left(1 - e^{-t/RC}\right)$$

$$I(t) = \left(\frac{V_b}{R}\right)e^{-t/RC}$$

Discharging

$$V_c(t) = V_b e^{-t/RC}$$

$$I(t) = \left(\frac{-V_b}{R}\right)e^{-t/RC}$$

Figure 4.39 Capacitor equations, charging and discharging.

The value required to alter the shape of a signal can be calculated in two ways. Although capacitance is generally selected for optimal performance at a particular resonant frequency, use and implementation depends on installation, lead and trace length inductance and other parasitic parameters that may change the resonant frequency of the device. The installed value of capacitive reactance is the primary item of interest. Calculating capacitance will be approximate and is generally accurate enough for implementation.

Before selecting a filter capacitor value to wave shape a signal, the Thevenin impedance of the network must be known. The impedance should be equal to two resistors placed in parallel. For example, using the Thevenin equivalent circuit Eq. (4.16), assume Z_S = 150 Ω and Z_L = 2.0 kΩ where Z_S is source impedance and Z_L is load.

$$Z_t = \frac{Z_s * Z_L}{Z_s + Z_L} = \frac{150 * 2000}{2150} = 140\Omega \qquad (4.16)$$

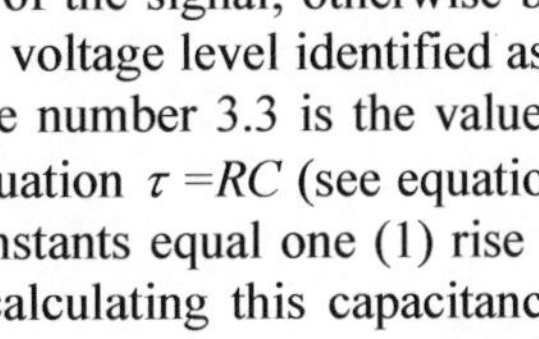

Capacitors must be chosen so that the edge time ($t_r = 3.3R * C$) equals an acceptable rise or fall time transition for proper functionality of the signal; otherwise baseline shift may occur. Baseline shift refers to the steady-state voltage level identified as logic LOW or logic HIGH for a particular device family. The number 3.3 is the value of the time constant for a capacitor to charge, based on the equation $\tau = RC$ (see equations earlier in this paragraph). Approximately three (3) time constants equal one (1) rise time. Since we are interested in only one time constant for calculating this capacitance value, the value of the time constant period, k = $1/3t_r$, which becomes $3.3t_r$ when incorporated within the equation (inverse of $1/3t_r$).

METHOD 1

Equation (4.17) is used to determine the maximum capacitance value for wave shaping, based on knowing the edge rate of the clock signal.

$$t_r = k\, R_t C_{max} = 3.3 * R_t * C_{max}$$

$$C_{max} = \frac{0.3\, t_r}{R_t} \tag{4.17}$$

where:

t_r = edge rate of the signal (the faster of either the rising or falling edge) (seconds)

R_t = total resistance within the network (ohms)

C_{max} = maximum capacitance value to be used (Farads)

k = one time constant.

Note: C will be in nanofarads if t_r is in nanoseconds
C in picofarads if t_r is in picoseconds

For example, if the edge rate is 5 ns and the impedance of the circuit is 140 Ω, the maximum value of C is easily calculated by (4.18).

$$C_{max} = \frac{0.3 * 5}{140} = 0.01\ nF\ \ or\ 10\ pF \tag{4.18}$$

A 60 MHz clock with a period of 8.33 ns on and 8.33 ns off, R = 33 Ω (typical for an un-terminated TTL part for example) with an acceptable $t_r = t_f = 2$ ns (25% of the on or off value). Therefore,

$$\left(C = \frac{0.3 * t_r}{R_t}\right) \qquad C = \frac{0.3\left(2 * 10^{-9}\right)}{33} = 18\ pF \tag{4.19}$$

METHOD 2

- For wave shaping, determine the highest frequency to be filtered, f_{max}.
- For differential pair traces, determine maximum tolerable value for each capacitor. To minimize signal distortion, use (4.20).

$$C_{max} = \frac{100}{f_{max} * R_t} \tag{4.20}$$

where: C is in nanofarads, f in MHz, and R in ohms.

To filter a 20 MHz signal with R_L = 140 Ω, capacitance value with low source impedance, Z_c, would be

$$C_{min} = \frac{100}{20 * 140} = 0.036\ nF\ \ or\ 36\ pF \tag{4.21}$$

When using capacitors to alter the characteristics of a signal transition, remember the following:

- If degradation of the edge rate is acceptable, including consideration for the timing margin generally up to three times the value of calculated C_{max}, increase the capacitance value to the next highest standard value.
- Select capacitor with proper voltage rating and dielectric material for intended use.
- Select capacitor with a tight tolerance level. A tolerance level of +80/-0% is acceptable for power supply filtering but may be inappropriate for high speed signals.
- Install the capacitor with minimal lead and trace inductance.
- Verify functionality of the circuit with the capacitor installed. Too large a value capacitor can cause excessive signal degradation.

4.21 Bulk Capacitor Application

Bulk capacitors ensure a sufficient amount of DC power (voltage and current) is available at all times, especially when digital components consume a significant amount of energy when operating under maximum capacitive load. Maximum capacitive load refers to the total amount of drive current for both core and I/O partitions of the component in addition to charging current for driving signal lines. With a large number of signal loads, a significant amount of current must pass through the device sourced by the power distribution network.

Components switching digital logic states cause fluctuations on the power and return planes, or inject RF switching current into the PDN if there is poor power distribution (called SSN-simultaneous switching noise in the time domain or plane bounce in the frequency domain). When planes have a lack of current charge, a bounce condition may occur that could exceed power level margins of components, thus functional operation is not ensured. Bulk capacitors provide energy storage to planes to minimize bounce at a slower rate of operation. Once planes are charged by a very large localized storage reservoir, the planes in the localized area then re-charges decoupling capacitors nearby, which in turn provides a stable PDN for components. Components consume current from localized planes. Refer to the Section 4.14 (Capacitor Brigade) for greater details.

Bulk capacitors (i.e., tantalum or electrolytic dielectric) are required in addition to higher self-resonant frequency decoupling to provide stable DC power and to minimize RF modulation in the power distribution network. One bulk capacitor should be located for every power consuming circuits, and also in the power supply network section generally where the voltage enters the planes from the switching regulator, however the rule of thumb of one capacitor per power and return pin pair is "invalid" for this specific application of use.

A signal integrity or EMI event may occur due to lack of bulk capacitance spread out throughout the assembly. This because most of the time we utilize too many decoupling capacitors hoping that these 100 nF capacitors (most commonly used value) provide sufficient charge to the planes, but if they are unable to be fully recharged, their ability to provide for a stable PDN becomes limited. If we remove some of these 100 nF capacitors, since most of the time we install more capacitors than we need, less stress is placed on the power supply in its efforts to recharge these devices. Most power supply assemblies utilize switching MOSFETs that generate a large amount of switching noise required to supply a large amount of current to the power distribution network. With less stress on the MOSFETs, less radiated EMI is observed from this circuitry area of the power supply. Removing up to 50% of standard value decoupling capacitors may be the solution to an EMI problem instead of redesigning circuits and a board layout due to less loading on the power supply switching components.

In addition to locating bulk capacitors in power consuming circuits, the following locations need to be considered when using decoupling capacitors. These capacitors can also be used, if properly chosen, to perform the job of bypassing undesired common-mode RF energy present on the power rail to 0V reference or chassis ground. Thus we get two benefits from one device; bypassing and decoupling.

- Power entry connectors.
- Power terminals on interconnects for daughter cards, backplane connectors, peripheral device controllers and secondary circuits.
- Near all high power-consuming digital components.
- The furthest location from the input power connector to minimize plane bounce due to IR drop in the network.
- High-density component placement remote from the DC power section.
- Adjacent to clock generation circuits, buffers and drivers.

When using bulk capacitors, the voltage rating must be chosen such that the nominal level equals no more than 50% of the capacitor's actual rating provided by the manufacturer to prevent self-destruction due to thermal heating should a voltage surge occurs. For example, with plane power at 5 Volts, one should use a capacitor with a minimum of a 10 Volt rating however, it is common practice to use a 6.5 V device.

Memory arrays require additional bulk capacitance owing to the extra current required for functional operation, especially during a refresh cycle.

Selection of bulk capacitors must not be based on values that were used with slower speed devices of the past. Today's technology requires faster power deliver. Bulk capacitors must be able to provide a sufficiently greater amount charge quickly to support a PDN in addition to decoupling capacitors that are also required to be recharged. Considerations of resonance, placement on the printed circuit board, lead-length inductance, existence of power and return planes, and the like must all be included when selecting a capacitor either for bulk or decoupling application.

For selecting a bulk capacitor value, the following procedures are provided to help determine optimal selection [13].

EXAMPLE 1

1. Calculate maximum current consumption *(ΔI)* anticipated by all devices using the same voltage level. Assume all gates switch simultaneously consuming additional amounts of current and include the effect of power surges by logic crossover (cross-conduction currents).
2. Know ahead of time maximum amount of power supply noise permitted *(ΔV)* by devices for functionality purposes. Factor in a safety margin.
3. Calculate maximum common-path impedance acceptable to circuits from results determined in #1, and #2 above.

$$Z_{cm} = \Delta V \,/\, \Delta I \tag{4.22}$$

4. If solid planes are used, allocate impedance Z_{cm} to the connection between the power and return planes.

5. Calculate impedance of the interconnect cable if provided, Z_{cable} $(R + j2\pi fL_{cable})$ from the power supply to the printed circuit board. Add this value to Z_{cm} to determine the frequency below which the power supply wiring is adequate $(Z_{total} = Z_{cm} + Z_{cable})$.

$$f = \frac{Z_{total}}{2\,\pi\, Z_{cable}} \tag{4.23}$$

6. If the switching frequency is below calculated f of (4.23), the power supply wiring is fine. Above f, bulk capacitors, C_{bulk}, are required. Calculate the value of the bulk capacitor for an impedance value Z_{total} at frequency f (4.24).

$$C_{bulk} = \frac{1}{2\,\pi\, f Z_{total}} \tag{4.24}$$

EXAMPLE 2

A printed circuit board has 200 CMOS gates (G), each switching 5 pF (C) loads within a 2-ns time period. Power supply inductance is 80 nH.

$$\Delta I = GC\frac{\Delta V}{\Delta t} = 200(5\ pF)\frac{5V}{2\ ns} = 2.5\text{ A (worst case peak surge)}$$

$$\Delta V = 0.200\text{ V (from noise margin budget)}$$

$$Z_{total} = \frac{\Delta V}{\Delta I} = \frac{0.20}{2.5} = 0.08\ \Omega$$

$$L_{cable} = 80\ nH$$

$$f_{ps} = \frac{Z_{total}}{2\,\pi\, Z_{cable}} = \frac{0.08\ \Omega}{2\ \pi\, 80\ nH} = 159\ kHz$$

$$C = \frac{1}{2\,\pi\, f_{ps}\, Z_{total}} = 12.5\ \mu\text{F}$$

Capacitors commonly found on printed circuit boards for bulk applications are generally in the range of 4.7–100 μF.

4.22 Buried Capacitance

Buried capacitance refers to having an integrated power and return plane structure with very small distance spacing between each other along with a high dielectric constant material separating the plates. The advantage of a buried capacitance layer is the ability to provide an extremely large amount of current to switching components with minimal inductance in the decoupling loop. There will always be a minuscule amount of inductance from vias to the planes, and if the planes are on layers 2 and 3 of a multilayer assembly, via inductance is extremely small especially if the via has been back-drilled from the bottom of a large stackup assignment to enhance signal integrity on very high-speed designs.

Back-drilling is a process where after a multi-layer assembly is manufactured, prior to component placement, a through-hole via is drilled out from the underside to remove any stub length from the layer the via is connected to down to the bottom of the board. If the via carrying a signal trace traverses the board from top-to-bottom where the signal must be routed on the bottom layer, back-drilling cannot be performed. This process is generally required for high frequency data signals. Back-drilling also remove stub effects in a transmission line for very high-speed signals and is illustrated in Fig. 4.40.

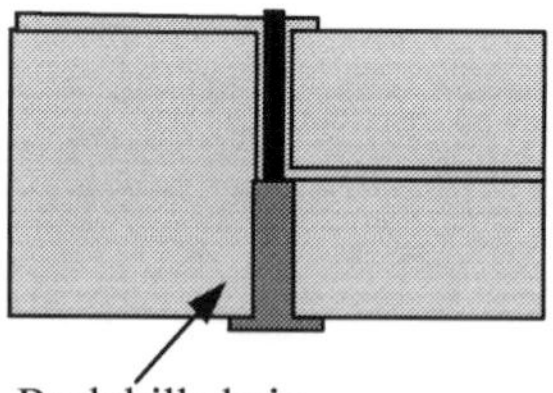

Figure 4.40 Backdrilled via implementation.

Discrete capacitors are effective up to approximately 200 MHz. Above this frequency discrete capacitors cease to provide any significant value for recharging planes due to loop inductance. To ensure functionality, we must use thin laminates between the power-return layers. The question many ask when using buried capacitance "Can we really eliminate many discrete decoupling capacitors from the board?" The answer is yes, but not because of increased capacitance inherent due to the thin laminate, but rather due to lower inductance above the series self-resonant frequency of both the discrete and plane capacitance combination that permits fast transfer of current charge.

Although buried capacitance may eliminate the need for some discrete devices along with additional cost, use of this technology may far exceed that of the capacitors being removed.

A detailed description of how a buried capacitive layer is manufactured is provided in Fig. 4.41 and consist of two planes (voltage and return) separated by a very thin core (typically less than 2 mils or 20 μm). Buried capacitance involves a special manufacturing process where material is imaged onto a substrate using power and return plane artwork or film. The lamination process is an integral part of the assembly.

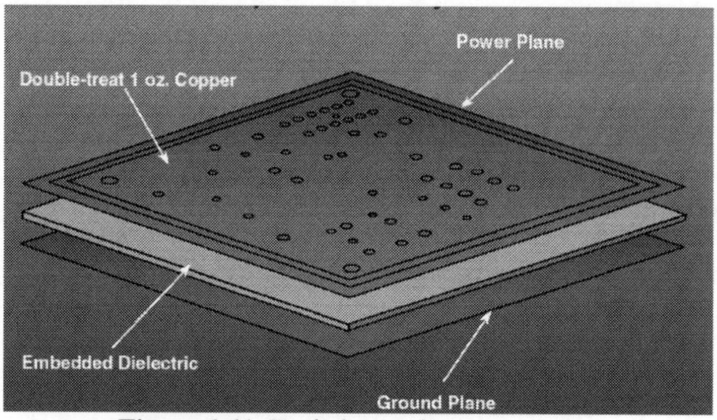

Figure 4.41 Buried capacitance structure.
Source: *National Center for Manufacturing Sciences*
"An Overview *of the NCMS Embedded Capacitance Project"*

Figure 4.42 illustrates implementation of buried capacitance. Using buried capacitive layers means the overall thickness of the stackup will becomes slightly thinner. If total thickness must be identical to pre-buried capacitance implementation, additional thickness in core or prepreg content can be utilized to make up the difference.

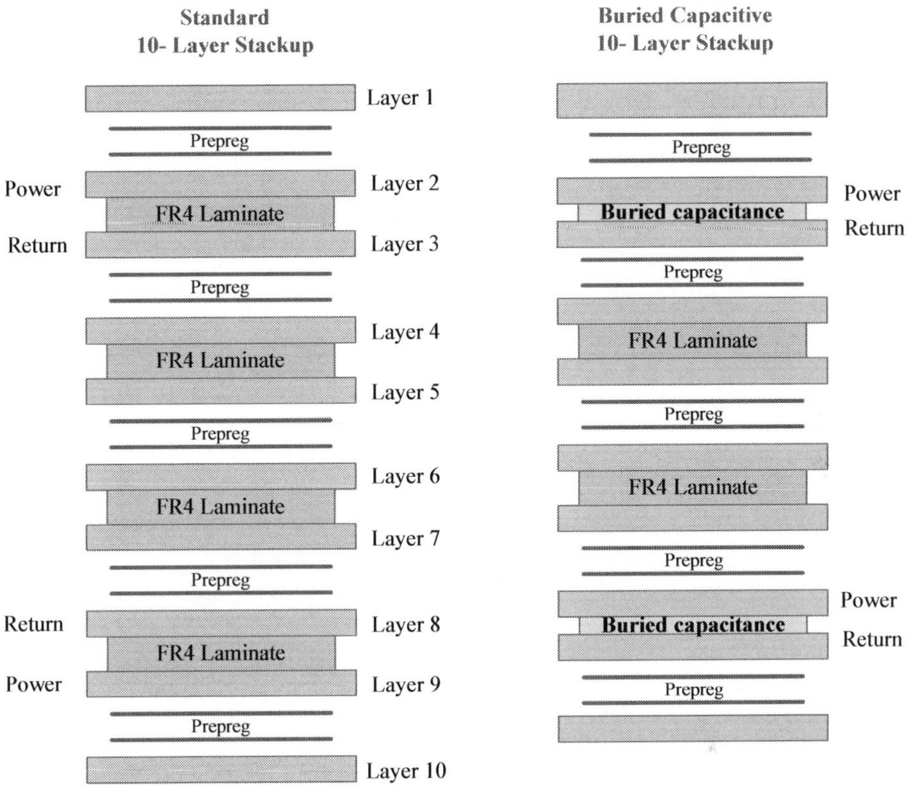

Figure 4.42 Implementation of buried capacitance in a ten layer stackup.

<u>Reasons why embedded capacitance is preferred over discrete components for decoupling application</u>

- **Permits increases component usage (microstrip layer only)**
  1. Frees up valuable real estate occupied by discrete capacitors.
  2. Potential for reducing the number of layers (and cost).
- **Lower interconnect inductance**
  1. Improves electrical performance (signal integrity).
  2. Reduces power bus noise and development of common-mode EMI.
- **Quality and reliability improvement**
  1. Reduces number of discrete components that would normally be used.
  2. Reduces number of solder joints and mechanical handling.

<u>Benefits of Implementing Buried Capacitance</u>

1. Cost: Assembly cost can be reduced by using less discrete devices along with drilling fewer vias.
   Note–overall cost of the material may however exceed removal of discrete components.

2. Quality and reliability: The reliability of the printed circuit board should increase due to a reduction in the total number of components, plated holes, and solder joints.
3. Increased density: By eliminating discrete capacitors, more room is available for other components and easier transmission line routing.
4. Design efficiency: Reduces design time because fewer components are required in schematic capture, board layout and component placement. It also becomes easier to route traces stripline since there are fewer vias to route around.
5. Easy to implement in existing designs: Simply copy existing artwork to replace existing power and return plane pairs with an embedded design along with changing overall board stackup dimensions to accommodate use of thinner dielectrics (maintain overall board thickness). Note-this may change the impedance of transmission lines which must then be re-analyzed to determine if signal integrity performance is degraded.

Plane capacitance is required for higher-speed digital circuits

1. Provides a greater amount of distributive capacitive charge quickly to all components simultaneously.
2. Less inductance is present in the decoupling loop which provides for enhanced performance.
3. Does not have a limited range of operation or a specific self-resonant frequency.
4. Enhanced decoupling performance above the functional range of discrete devices, especially in the higher frequency range (i.e., GHz).

To better understand the concept of buried capacitance, the power and return planes should be considered as a high capacitive source of energy storage that operates at higher frequencies with very little inductance. The plates are an equal-potential surface with no voltage gradient except for a small DC voltage drop. The capacitance of this structure is calculated simply as area divided by thickness times the permittivity of the dielectric material, Eq. (4.9). For a 10 in. (25.4 cm) square board, with 1-mil (0.0254 mm) spacing between layers, FR-4 dielectric ($\varepsilon_r = 4.1$), we have capacitance value of 9.2 nF (9200 pF). Table 4.8 shows comparative values of capacitance between power and return for the same physical size board relative to distance spacing between planes using Eq. 4.9 shown below

$$C_{PP} = k\,\frac{\varepsilon_r A}{D} \qquad \text{(4.9 repeated)}$$

where:

C_{pp} = capacitance of parallel plates (pF)
ε_r = dielectric constant of the board material
A = common area between the parallel plates (square inches or cm)
D = distance spacing between the plates (inches or cm)
k = conversion constant (0.2249 for inches, 0.884 for cm).

Vias will cause a reduction in total capacitance when we remove copper from planes. The magnitude of this concern is minimal for most product designs and can be ignored for all but extremely complicated and very high-speed systems requiring large amount of very fast switching current.

Decoupling capacitance (buried capacitance) is increased because the distance spacing between the planes *(d)* in the denominator is decreased significantly. The power and return planes are the means of distributing power to components quickly because of significantly lower lead inductance. Reducing dielectric thickness is highly effective for higher frequency decoupling applications. In addition, all resonances are decreased due to higher skin effect losses when planes are located physically closer to each other.

Table 4.8 Parallel plate capacitance - 10 in. (25.4 cm) square assembly, ε_r=4.1.

Distance spacing between planes	Total capacitance between planes	Total capacitance between planes per in^2	Total capacitance between planes per cm^2
1 mils [0.025 mm]	9.2 nF	920 pF	2.34 nF
2 mils [0.051 mm]	4.6 nF	460 pF	1.17 nF
4 mils [0.102 mm]	2.3 nF	230 pF	584 pF
5 mils [0.127 mm]	1.8 nF	180 pF	457 pF
10 mils [0.254 mm]	921 pF	92 pF	233 pF
15 mils [0.381 mm]	614 pF	61 pF	155 pF
20 mils [0.508 mm]	460 pF	46 pF	116 pF
40 mils [1.02 mm]	230 pF	23 pF	58 pF

The use of discrete decoupling capacitors may not be necessary when buried capacitance is used. With fewer discrete devices, less inrush surge current is required to recharge these capacitors, which is beneficial to minimizing board-induced noise voltage, plane bounce and development of common-mode energy in most applications. If the power supply has fewer components to source switching current to, enhanced benefit may be observed during conducted emission testing.

Application Guidelines for Buried Capacitance

Buried capacitance can be effectively used if the design meets requirements of (4.25), and is more effective with fast rise times, low current demands and recommended for frequencies above 30 MHz:

$$R_t * I_{tr} / A \leq 12.7\ (\text{cm}^2)\ [\text{or } 5\ (\text{inches}^2)] \qquad (4.25)$$

where:

R_t = rise time of the primary digital clock pulses nanoseconds to picoseconds
I_{tr} = peak current surge (ma)
A = area of the plane (in^2 or cm^2).

4.23 Summary-Guidelines for Power Distribution Networks

1. **Determine board impedance requirements.**
   To maximize performance, plane impedance must be below a pre-determined level. Calculate this value by estimating total current required by active devices and dividing that value into the maximum noise margin that can be tolerated.
2. **Establish total amount of capacitance required.**
   More capacitance results in lower power bus impedance, but only within a specific bandwidth of operation that the capacitor(s) is operating at which must be below this target impedance. When using embedded capacitance, the value should be greater

than or equal to the amount of discrete components already provided if it is to replace these devices. Embedded capacitive planes usually replaces some discrete capacitors (i.e., those with a value of 0.01 μF/10 nF or smaller).

3. **Ensure resonances are damped.**
   When board dimensions exceed ½ wavelength, power bus resonances may occur causing other EMI and signal integrity problems. Damp these resonances using resistive elements, or space the power and return planes at a distance less than 0.010 mils (0.254 mm) apart. One can also sprinkle low capacitive value components throughout the board to help damp resonances.
4. **Estimate total board impedance.**
   This can only be achieved using sophisticated, complex numerical modeling software. Planes must be modeled as *finite* radial transmission lines.

Power Bus Decoupling Strategy: With closely spaced planes (< 0.254 mm or 10 mils)

1. Size bulk capacitance to meet board requirements.
2. Size local decoupling capacitors to meet board requirements.
3. Mount local decoupling capacitors in the most convenient location without lead inductance.
4. **Don't put traces on capacitor pads or breakouts.**
5. Too much capacitance is OK.
6. Too much inductance is not OK.
7. The location of an individual decoupling capacitor is not critical if the interconnect between capacitor and plane is direct through use of vias to planes and *"not"* a routed trace.
8. The value of the local decoupling capacitor is not critical, but it must be greater than total inter-plane capacitance.
9. Inductance of the connection is the most important parameter when using decoupling capacitors, and must be as small as possible.
10. **Local decoupling capacitors are generally ineffective above a couple hundred MHz.**
11. Local decoupling capacitors cannot supply significant charge in the first few nanoseconds of a logic transition from digital components, relying on inter-package and on-die capacitance for immediate switching needs.

Power Bus Decoupling Strategy: With widely spaced planes (> 0.75 mm or 30 mils)

Inductance of planes can no longer be neglected or taken for granted. In particular, the mutual inductance between the vias of active devices and decoupling capacitor loop is critical for optimal performance. Mutual inductance will tend to cause the majority of the current to be drawn from the nearest decoupling capacitor and not from the planes, although this is dependent on the spectral bandwidth of the current (Fig. 4.43).

1. Local decoupling capacitors should be located as close to the active device as possible (near pin attached to most distant plane).
2. The value of the local decoupling capacitors should be 10 nF (0.01 μF) or greater.
3. **The inductance of the connection is the most important parameter of concern.**
4. Local decoupling capacitors can be effective up to 1 GHz or higher if they are connected properly with minimal loop inductance.

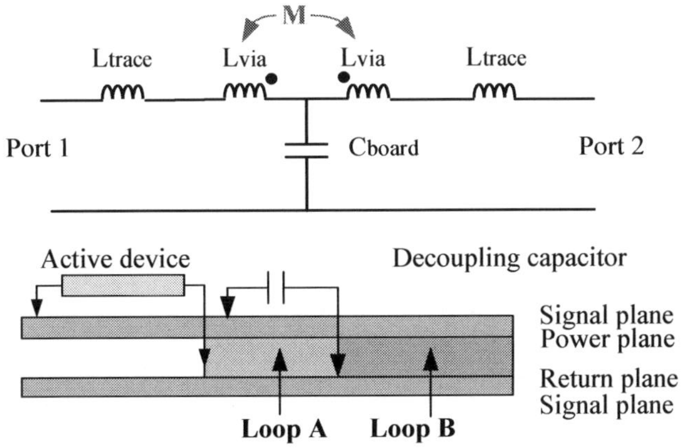

Figure 4.43 Mutual inductive coupling due to loop area between vias.

REFERENCES

1. Montrose, M. I. 1999. *EMC and the Printed Circuit Board – Design, Theory and Layout Made Simple*. Hoboken, NJ: John Wiley & Sons/IEEE Press.
2. Paul, C. R. 1992. "Effectiveness of Multiple Decoupling Capacitors." *IEEE Transactions on Electromagnetic Compatibility*, Vol. 34(2), pp. 130–133.
3. T. P. Van Doren, J. Drewniak, & T. H. Hubing. 1992. "Printed Circuit Board Response to the Addition of Decoupling Capacitors," Tech. Rep. TR92-4-007, University of Missouri, Rolla EMC Lab.
4. Radu, S., R. E. DuBroff, T. H. Hubing, & T. P. Van Doren. 1998. "Designing Power Bus Decoupling for CMOS Devices." *Proceedings of the IEEE International Symposium on Electromagnetic Compatibility*, pp. 175-379.
5. Montrose, M. 1996. "Analysis on the Effectiveness of Image Planes within a Printed Circuit Board." *Proceedings of the IEEE International Symposium on Electromagnetic Compatibility*, pp. 326–331.
6. Drewniak, J. L., T. H. Hubing, T. P. Van Doren, and D. M. Hockanson. 1995. "Power Bus Decoupling on Multilayer Printed Circuit Boards." *IEEE Transactions on Electromagnetic Compatibility,* 37 (2), pp. 155–166.
7. Cain, J. "The Effects of ESR and ESL in Digital Decoupling Applications," AVX Corp.
8. Novak, I., et. al. "Comparison of Power Distribution Network Design Methods: Bypass Capacitor Selection Based on Time Domain and Frequency Domain Performances." DesignCon 2006.
9. Weir, S. "Bypass Filter Design Considerations for Modern Digital Systems, A Comparative Evaluation of the Big "V," Multi-pole, and Many Pole Bypass Strategies." DesignCon 2006.
10. Montrose, M. I. 1999. "Analysis on Trace Area Loop Radiated Emissions from Decoupling Capacitor Placement on Printed Circuit Boards." *Proceedings of the IEEE International Symposium on Electromagnetic Compatibility*, pp. 423–428.
11. Montrose, M. I. 2007. "Power and Ground Bounce Effects on Component Performance Based on Printed Circuit Board Edge Termination Methodologies." *Proceedings of the IEEE International Symposium on Electromagnetic Compatibility*.

12. Montrose, M. I. 2012. "Analysis on Decoupling Capacitor Placements Associated with Power and Return Plane Bounce." *Proceedings of the Asia-Pacific International Symposium on Electromagnetic Compatibility*, pp. 425-428.
13. Johnson, H. W., & M. Graham. 1993. *High Speed Digital Design*. Englewood Cliffs, NJ: Prentice Hall.
14. Montrose, M. 2000. *Printed Circuit Board Design Techniques for EMC Compliance-A Handbook for Designers*. Hoboken, NJ: Wiley/IEEE Press.
15. Van Doren, T., J. Drewniak, T. Hubing. 1992. "Printed Circuit Board Response to the Addition of Decoupling Capacitors," Tech. Rep. #TR92-4-007, UMR EMC Lab.
16. Micron Technology Inc. Application Note TN-00-06. Bypass Capacitor Selection for High-Speed Designs.
17. Xu, M., T. Hubing, J. Chen, T. Van Doren, J. Drewniak, & R. DuBroff. "Power-Bus Decoupling With Embedded Capacitance in Printed Circuit Board Design." *IEEE Transactions on Electromagnetic Compatibility*, Vol. 45, No.1, February 2003, pp. 22-30.
18. Ricchuti, V., "Power-Supply Decoupling on Fully Populated High-Speed Digital PCBs." *IEEE Transactions on Electromagnetic Compatibility*, Vol. 43, No. 4, November 2001, pp. 671-676.
19. Erdin, I. 2003. Delta-I Noise Suppression Techniques in Printed Circuit Boards, for Clock Frequencies Over 50 MHz," *IEEE Symposium on Electromagnetic Compatibility*, Boston, MA, pp. 1132-1134.
20. T. Zeef and T. Hubing, "Reducing power bus impedance at resonance with lossy components." *IEEE Transactions on Electromagnetic Compatibility*, Vol-37, No. 2, May 1995. pp. 307-310.

5 Referencing Made Simple (a.k.a. Grounding)

5.1 An Overview on Referencing (a.k.a. Grounding)

Referencing is required for every electrical device regardless of use or application. The reason why is simple. All components, both analog and digital, switch logic states using either a sine wave (analog) or infinitely fast AC slew rate signal (a.k.a.– digital waveform in appearance using a rounded edge). Components are referenced to specific voltage levels that are required to transition logic states. This means that voltage levels are based on the difference from 0V and a desired reference. This reference is commonly called "ground," a poor choice considering that there is no such thing as ground in electrical products, examined extensively in this Chapter.

For digital signals such as 3.3V, this voltage level is 3.3V above a 0V reference point. Our concern as designers is to ensure our reference is stable and does not deviate or fluctuate from a baseline value. Any deviation from baseline can create signal integrity problems as well as development of common-mode currents should the magnitude of the voltage deviation be beyond voltage levels required for optimal functionality. Ripple or voltage variation on planes is sometimes referred to a "ground bounce" when in fact it should be identified as plane "bounce" since it could be either at voltage or return potential.

Design engineers generally use the word ground, such as ground plane, ground trace, safety ground, chassis ground, or other variants of the word ground. In reality there is no such thing as ground in electrical systems as a word by itself.

The world "ground" is often confused with the process of providing an RF current reference return path for signals on transmission lines. RF return current does not require a power or ground reference to have its (RF) current return to the source. RF current propagates through free space or any metallic conductor without a potential associated with it.

Referencing, as a term, is often associated with printed circuit board component operation since components may be (application dependent) isolated (separated) or never connected to earth ground. If a device operates off battery power where is earth ground?

Within electrical engineering the word *"reference,"* which is a primary aspect of product design, is often not well understood as a concept by engineers. Referencing is not always easy to understand or comprehend intuitively, and does not usually allow for straightforward modeling or analysis since many uncontrolled factors affect performance, especially parasitics that exist. Every circuit is ultimately referenced to a common point and cannot be left to chance; it must be designed into the system from the very beginning. One cannot assume a stable 0V (signal return image) reference is present. Desired performance is not easily achieved if no thought is given to a specific design and application ahead of time. With regard to chassis grounding, what if the unit is battery operated or uses a two-wire AC mains cable without third-wire ground? Where is chassis or earth ground in this system?

Creating an optimal reference, or grounding methodology, minimizes unwanted noise pickup. Proper implementation of printed circuit board referencing and bonding cable shields to a 0V reference (or chassis ground, application dependent) will minimize common-mode RF development and enhance signal integrity. One advantage of a well-designed reference system is protection against unwanted interference reception and radiated emissions for basically near-zero cost in material usage.

Herein, we will be using the terms Referencing and Grounding interchangeability, or in areas that helps make "*Referencing and Grounding Made Simple,*" although technically use of the word *ground* may not be correct for the context in which it is being discussed.

5.2 Definitions

The word "grounding" is vague and means different things to different people. For logic designers, "ground" refers to a reference level for logic circuits and components. For system and mechanical engineers "ground" is the metal housing or chassis. For electricians "ground" refers to the third wire safety connection mandated by their respective National Electric Codes. For airline pilots "ground" refers to where they keep the plane when not in the air; for farmers "ground" is where they grow food, and for the homeless "ground" is where they sleep at night! So what is the correct definition of ground? It depends on who you are and what you do or who you ask.

When someone says "I have a ground plane in the printed circuit board," are they referring to analog ground, digital ground, signal ground, chassis ground, thermal ground, common ground, signal ground, multi-point ground, etc.? The following words used in this chapter are defined as follows:

- ***Bonding.*** Making a low-impedance electrical connection between two or more metal surfaces.
- ***Circuit.*** Multiple devices with source impedance, load impedance and interconnect between. For digital devices, multiple sources and loads may be part of one circuit where all elements are referenced to the same point or use a common signal return conductor. For EMC applications, circuits usually originate in one location (source driver) and terminates elsewhere (load).
- ***Circuit Referencing.*** The process of providing a common 0V reference for circuits to communicate with. Circuit referencing is critical for functional performance. This reference is not intended to carry functional current such as power or power return, only RF propagating fields.
- ***Earthing.*** The connection of a safety ground wire to earth ground at the service entrance of a building, or a ground rod driven deep into soil.
- ***Equipotential Reference Plane.*** A metallic plane used as a common connection for power and signal referencing. This plane may not be at equipotential levels at RF frequencies owing to its electrically large size and related self-impedance.
- ***Reference (Ground) Loop.*** A circuit that includes a conducting element (plane, trace or wire) assumed to be at 0V potential (ground). At least one reference loop must be present. Although a reference loop (as a signal image return) is required for functionality purposes, the severity of the problem for RF currents flowing through the loop depends on unwanted signals that may co-mingle and which may cause system malfunction.
- ***Grounding.*** A generic term with as many definitions as there are engineers. This word must be preceded by an adjective to make sense, technically (i.e. grounding what?).
- ***Grounding Methodology.*** A chosen method for directing return currents in a pre-defined manner appropriate for an intended application.
- ***Ground Stitch Location.*** The process of making a solid connection from a printed circuit board to a metallic structure for the purposes of providing system wide referencing regardless of methodology utilized.

- ***Holy Ground.*** (*See* single-point ground). Sometimes referred to as the actual location for 0V referencing, but whose use is highly discouraged for obvious reasons.
- ***Hybrid Ground.*** A methodology that combines both single-point and multi-point grounding, depending on functionality of the circuit and frequencies present.
- ***Multi-point Ground.*** A method of referencing different circuits together to a common equipotential or reference point. Connection may be made by any means possible in as many locations as required. In reality, the term multi-point grounding is invalid in electrical engineering and technically should be referred to as "multiple connections to a single-point reference" (where a common and low impedance planar structure is the reference point).
- ***Referencing.*** The process of making an electrical connection or bond between two circuits that allows 0V potentials to be identical.
- ***RF Ground.*** Providing a reference for the transfer of RF current, generally a metallic enclosure or plane. Sourcing RF current to a 0V potential (ground) will allow products to comply with both signal integrity and EMC requirements.
- ***Safety Ground.*** The process of providing a return path to earth ground to prevent hazard of electric shock through proper connection and routing of a permanent, continuous, low-impedance, adequate fault capacity conductor.
- ***Shield Ground.*** The process of providing a 0V reference or electromagnetic shield for cables to a metallic chassis housing.
- ***Single-point Ground.*** A method of referencing many circuits together at a single location using a physically long conductor. All signals will thus be referenced to the same location. The location may be a very low impedance structure.

Again, when using the word "ground" it is mandatory to use an adjective to define what type of reference system we are referring to, such as signal ground, chassis ground, safety ground, digital ground and analog ground, for example.

5.3 Defining Various Types of Grounding Systems

When discussing any type of ground structure in system design, there are many variants, some of which are listed below. This not all-inclusive but represents the majority of grounding systems one may encounter.

Power/safety ground

- An intended (neutral) or designed in safety ground connection used with 50/60/400 Hz systems to prevent electric shock should an abnormal fault condition occur. If the amount of leakage current in the transmission path is more than a few micro- to milliamps, injury or even death could result.
- Usually identified by a green or green with yellow stripe wire, but could be braided assemblies protected with heat shrink tubing or other means.
- Implementation or use of a safety ground must be made permanent, which ensures that a user can never defeat this low-impedance bond connection under any condition.

Lightning ground

- A controlled path for lightning to reach Earth (or flow through a structure) through a very large rod or metallic structure located deep within earth ground.
- Lightning is generally a 1 MHz event and up to 100 kAmps per millisecond event.

- Lightning ground requires a high quality "low impedance" connection to the grounding rod in the soil.
- Buildings generally provide lightning rods on all corners of the building which will divert energy to the outside portion of the building, thus creating a Faraday cage structure to keep high-energy transients from entering the building.

RF ground

- Usually referred to as a virtual ground, or a metallic structure that captures propagating RF fields through the process of skin effect within a metallic chassis; sources a propagating field into chassis ground or 0V-potential.
- An RF ground works for RF fields from "anything greater than DC to daylight" and μA to Amps.
- Making a virtual connection requires a minimum ground impedance bond from enclosure to chassis ground, if one is provided, else the connection made to the 0V potential of the battery generally called the ground (-) terminal.
- Provides an RF return path for propagating fields to return to its source, generally through parasitic capacitance.
- Covers the frequency spectrum of interest.
- Requires minimum impedance to direct and capture maximum current/flux flow.

ESD ground

- Provides a controlled path for ESD currents that may enter a printed circuit board and sends this energy to chassis ground or primary 0V reference.
- Is generally implemented as a guard band on a printed circuit board, or large copper fill on both top and bottom layers that has a dedicated connection to chassis ground or 0V reference.
- ESD events have generally a 0.7-3.0 ns rise time (and may be faster than 0.2 ns), 100-300 MHz and 10-50 Amps.

Circuit/signal reference (*not ground!*)

- Provides a return path for intended signal flow and for AC/DC power return; mA to Amps.
- Minimum low impedance path for maximum deliver of both voltage and current.
- Generally implemented as a plane or grids within a printed circuit board.
- Digital and analog circuits require use of a reference, not chassis ground however, both structures can be connected to create a single-point reference to a chassis plane if the application permits.

5.4 Common Grounding Symbols

When creating a schematic, one generally uses a symbol to represent ground when in reality there is no such thing as a real ground on a printed circuit board; it is 0V reference. This symbol may represent digital and/or analog partitions and is used to help designers identify functional areas of the printed circuit board and maintain isolation, but should this reference be identified by one symbol or two? Are they in reality at the same reference or different? These two planes must be connected somewhere, either through a trace (0 ohms) or ferrite bead (generally near 0 ohms at DC) so again, are they at the same potential or not, electromagnetically?

It is highly recommended to use only one symbol on any schematic, unless the application or design requires a unique, separate symbol for operational reasons (i.e., identify digital from analog circuits), or help identify functional areas such as chassis and ESD ground from 0V reference. However, all symbols in Fig. 5.1 must be joined together, somewhere. If there is more than one ground symbol on a schematic, then technically, there is job security for EMC engineers!

Figure 5.1 illustrates commonly used symbols to represent ground. Note, for digital and analog ground, the term return is substituted for ground in this Chapter.

Symbol	Function	Application
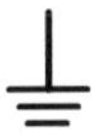	**AC safety ground**	**Protection against AC mains shock hazard**
	**Chassis ground**	**Connection to non-current carrying chassis**
	**Digital return**	**Path carrying the return of a digital signal**
	**Analog return**	**Path carrying the return of an analog signal**

Figure 5.1 Typical ground symbols commonly found on printed circuit board schematics.

5.5 Different Types of 0V-Referencing

The following types of grounds are commonly found in product design. Under most application of use they are identical.

Often a 0V-reference may serve multiple needs each with a different application. Although all reference systems can exist at the same time, eventually they must all come together at a single location. How this single point location occurs is discussed later in this Chapter.

- Signal Ground
- Common Ground
- Analog Ground
- Digital Ground
- Safety Ground
- Noisy Ground
- Quiet Ground
- Earth Ground
- Hardware Ground
- Single-point Ground
- Multi-point Ground
- Shield Ground

What about: *RF Ground?*

What kind of symbol do we use for RF Ground on a schematic?
(RF grounds are normally signal image returns)

5.6 Fundamental Grounding Concepts

There several fundamental grounding concepts that are going to be discussed in this Section. Understanding each concept makes it easy to design both printed circuit boards and system assemblies. There are however, misconceptions or ideas about grounding versus referencing. Choosing the right methodology is critical for system design and success.

5.6.1 Grounding Misconceptions

We present both myths and facts about the concept of grounding. As previously stated, the word "ground" is an invalid word in electrical engineering if not prefixed by an adjective to describe the type of reference or ground system that one is referring to.

Most engineers acknowledge that "ground" is some form of a current return path and that a good ground reduces circuit noise. This belief causes many to assume that we can sink noisy RF currents into the planet Earth, generally through a building's main grounding structure. This is valid if we are discussing safety grounding to prevent electrical shock hazard, not signal voltage referencing. Using this reasoning, spacecraft would function poorly if referenced to earth ground.

The following are common misconceptions regarding the issue of grounding, not referencing [7].

Myths

- Ground is always a current return path for signals (image returns).
- A "good ground" reduces system operational noise by sinking noise currents into the earth.
- One should connect to a "low noise ground reference" and not to building steel.

Facts – What is a true electrical ground?

- A high-quality, low impedance conductor to a 0V-reference.
- One that contains a large surface area located physically close to the system to minimize interconnect loop inductance.
- One that is "not" part of the signal current return path.

5.7 Primary Concerns Related to the Issue of Grounding and Referencing

There are two primary areas related to grounding and referencing:

1. Grounding for product safety (includes protection against effects of lightning and electrostatic discharge).
2. Signal referencing for components (AC signals or RF currents).

We begin discussion with safety ground followed by signal referencing.

5.7.1 Grounding for Product Safety

If connection is made by a low-impedance path to mother Earth, this is generally called safety grounding. A signal or 0V reference may or may not be connected to earth potential. Connection of different ground methodologies may be unsuitable for a particular application and may exacerbate EMC problems, if not present a risk of electric shock hazard.

Safety grounding minimizes or prevents a voltage difference between exposed conducting surfaces from becoming electrically live. The more conductors we make available to reduce the voltage difference to extremely low levels, the less chance of electric shock hazard.

The primary concern associated with safety ground is the protection of people and domestic animals from hazard of electric shock. When a product is at hazardous voltage potential levels, serious injury or death may occur.

If the system is powered by an AC voltage source above a certain amplitude level, defined below, exposed metal must then be bonded to a "green or green/yellow stripe" colored wire generally provided within the AC mains power cord. This requirement also applies to battery-operated devices if power conversion circuitry is built into any external module or physically located on a printed circuit board powered by AC mains voltage. If the unit operates from a DC voltage source, then only the external power supply or wall connected adapter assembly needs to comply. If a conflict occurs between EMC compliance and product safety, safety takes precedence. No exception exists; safety first.

Electric shock occurs when current passes through the body. Current on the order of a few milliamps can cause a reaction in persons of good health and may create biological harm due to involuntary reaction. Higher currents can have more damaging effects. Voltage below 42.4 VAC peak or 60 VDC are not generally regarded as dangerous under dry moisture environmental conditions. Electrical parts that may be touched or handled should be at earth potential, or properly insulated/isolated to prevent electric shock. [4]

Under normal conditions, any voltage [absolute value] greater than 42.4 VAC peak, or 60 VDC on a printed circuit board (or system) is considered hazardous and requires special attention by a product safety compliance engineer working in conjunction with the EMC and printed circuit board layout designer. Medical devices have more stringent levels.

When are hazardous voltages used on printed circuit boards? Telecommunications circuits operate at +/-48 VDC. Power supplies are sometimes incorporated in a printed circuit board connected to AC mains voltage and routed to secondary assemblies through cables and interconnects. Solenoid drives generally operate at 78 VDC, 100-120 VAC or 230 VAC. Process control equipment generally uses voltages far in excess above 42.4 VAC peak. These are only a few examples of circuits that use hazardous voltages.

In Fig. 5.2, stray impedance in the interconnect between voltage potential V_1 and chassis is identified as Z_1. The stray impedance between chassis and earth ground is shown as Z_2. The total potential of the chassis is the impedance of Z_1 and Z_2 acting as a voltage divider. The chassis potential relative to the voltage driven source is calculated by (5.1). This potential could reach hazardous levels, enough to cause a shock hazard to exist.

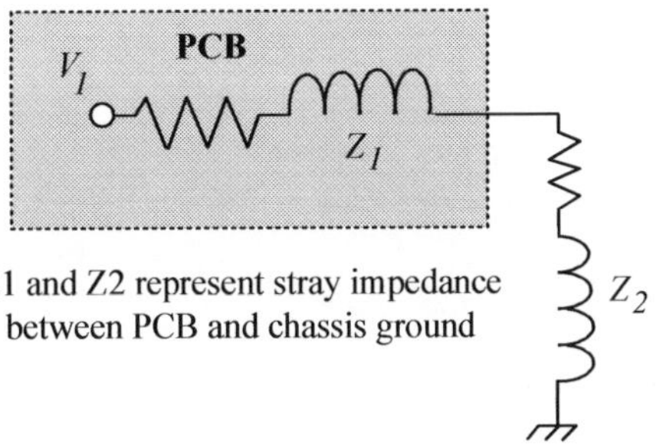

Figure 5.2 Stray impedance in a circuit from a voltage source to chassis ground.

$$V_{\text{chassis}} = \left(\frac{Z_2}{Z_1 + Z_2} \right) V_1 \tag{5.1}$$

---

[4] This description of electric shock hazard is extracted from the international product safety standard, IEC/EN 60950, *Specification for safety of information technology equipment, including electrical business equipment.*

It must never be assumed that as long as everything is connected to earth ground through an appropriate means (green/yellow wire, braid strap and the like), then all is well. This ground connection will have finite impedance at RF frequencies that increases as frequency increases, depending on the quality of the bond connection. In general, safety earth ground is not required for EMC compliance. A metal chassis may provide a good low-impedance bond connection to an RF reference or other structure. In many instances, this bond connection must be in parallel with safety earth ground for devices connected to an AC mains source.

If we observe common-mode emissions emanating from a power cord, a safety earth ground connection may be required, application dependent. A line filter installed at the mains power inlet removes harmful current, both safety and common-mode. Internal to the line filter in Fig. 5.3 are "Y" capacitors, which is a component located between a voltage potential (hot/neutral or hot/hot) and chassis ground to remove common-mode noise. The "X" capacitor is a component located across both power lines to balance out differential-mode signal propagation without any connection to ground. When using "Y" capacitors, hazardous leakage current can be diverted to a metal enclosures, if not sent to a hard-wired earth ground. This can only occur if the value of the "Y" capacitor is large. This is a primary reason why "Y" capacitors must be safety agency approved and are limited to very small capacitive values to minimize leakage current and the potential for shock hazard. The same for the "X" capacitor which also must be safety agency approved to always fail open, instead of shorting out and making a direct connection between the two AC mains voltages.

A line filter showing both types of capacitors is in Fig. 5.3.

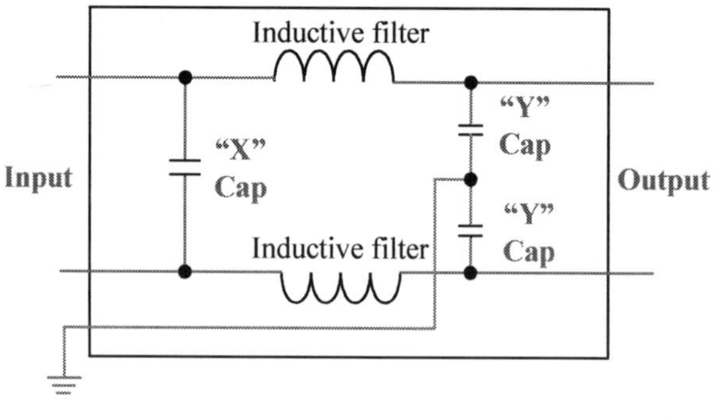

Figure 5.3 AC Line Filter Configuration.

At times, it is required that the safety earth ground path be removed during EMC testing. One means of achieving this requirement is by inserting an inductive coil in series with the earth return wire external to the box during testing [4]. This inductor ensures high frequency leakage currents remain within the system and not injected into the facility mains. Higher frequency RF currents are also thus prevented from radiating to the external environment by a secondary structure such as a Faraday shield or Gaussian structure (i.e., sheet metal enclosure).

Safety grounding minimizes or prevents a voltage difference between exposed conducting elements. The more conductors available in a bond connection, a significant reduction in voltage difference between enclosure and earth occurs due to lower inductance (Thevenin equivalent circuit of impedance in parallel). This reduces the voltage potential difference to extremely low levels. With lower impedance, there is also a smaller chance of electric shock.

"Signal voltage referencing to ground" is a totally different emphasis than "safety grounding," but has a causal relationship. For signal referencing, the voltage difference typically must be less than a few millivolts. With safety grounding, a few millivolts present may create a significant shock hazard due to inductance in the ground path described by [$E=-L(dI/dt)$]. It takes no more than a sustained few milliamps to cause death, and with a minor amount of voltage present across a fixed impedance (per Ohms law), this current level could be substantial.

Current requires a return path to complete a closed-loop circuit. We usually only consider AC or DC current in power circuits, not RF return since RF return is generally reserved for signal lines, not power and 0V rails. Although a return path is mandatory, it need not be at AC ground potential. Free space is not at AC ground potential. Analog ground is to be isolated from digital or chassis ground to prevent disruption to sensitive circuits and does not require reference to AC chassis ground. Not all currents within a system mandate a safety ground or a signal voltage reference, such as low-voltage battery-operated devices. No shock hazard can exist with low-voltage circuits.

To guarantee a system functions within a specific design requirement, signal or 0V may not be the same as AC/DC current return except under specific conditions of use, assuming the system is, for example an AC controller for pumps and motors. Regardless of application for both safety and signal referencing, we must reduce any ground voltage difference between the two circuits or avoid having a voltage potential difference at all.

Why is safety grounding discussed in a book about EMC and printed circuit boards? The reasons are obvious. Many printed circuit boards contain hazardous voltages. These include: power supply assemblies, telecommunication networks, relay driven instrumentation control units, power switching modules and similar products. User safety cannot be separated from EMC requirements. The field of regulatory compliance includes both product safety (meeting essential safety requirements based on National Electric Codes or governmental mandated legislation), as well as EMC limits for emissions and immunity protection.

Product safety standards mandate both creepage and clearance distances between power circuits to prevent electric shock if one is to apply the mark from a third-party certification agency or company such as Underwriters Laboratories (UL), any Nationally Recognized Test Laboratory (NRTL) in North America, Canadian Standards Association (CSA), the CE mark based on the European Low Voltage Product Safety Directive, and numerous other entities having legal jurisdiction worldwide related to safety approval.

Maintaining physical distance spacing between two elements at hazardous voltage potential requires compliance with spacing requirements published in safety standards. This distance spacing is called creepage and clearance.

Creepage and clearance spacing is of concern because transmission lines or circuits carrying high-voltages above safe levels may have an abnormal failure sometime in the future. Failures that could occur include primary-to-secondary, primary-to-ground or secondary-to-ground circuits. To prevent shock hazard due to an unexpected failure of insulation or component breakdown, traces or transmission lines must be routed with a specific amount of physical spacing (distance) between high-energy (voltage) and secondary or ground circuits. This requirement is especially critical in power supplies and related circuitry and is specified in every product safety standards.

When routing AC or high-voltage traces, one must use sufficient trace width and spacing to comply with legally mandated creepage and clearance requirements as well as current carrying capability and heat dissipation. The following definition of creepage and clearance below is identical between all international product safety standards. Illustrations of creepage and clearance measurement distances are shown in Fig. 5.4.

- Creepage is the shortest path between two conductive parts, or between a conductive part and the bounding surface of the equipment, measured along the surface of the insulation.
- Clearance is the shortest distance between two conductive parts, or between a conductive part and the bounding surface of the equipment, measured through air.
- Bounding surface is the outer surface of the electrical enclosure considered as though metal foil was pressed into contact with the accessible surface of insulation material.

When dealing with ground currents several fundamental concepts must be remembered.

- Whenever a current flows across finite impedance, a finite voltage drop occurs. Per Ohm's law, there can never be "zero volt potential" in the real world. The units may be in the pico or femto range (voltage or current). Still, a finite value will exist.
- Current must always return to its source. This return may consist of numerous paths with various amplitudes provided for each return current present, proportional to the finite impedance within each and every path or transmission line (Ampere's law). Unintended currents can and will travel within alternate return paths which may not be designed to handle hazardous fault currents such as people or domestic animals.

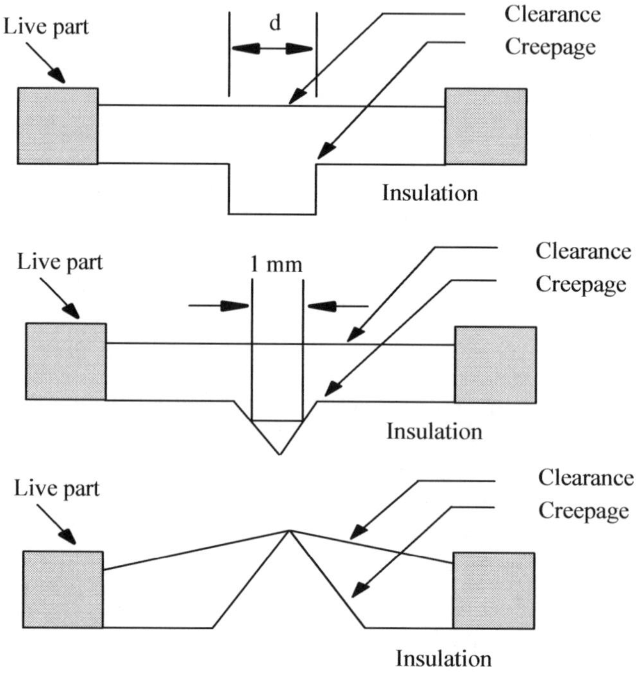

Figure 5.4 International creepage and clearance distance requirements between conductors.

To summarize, a voltage potential at a hazardous level must not exist anywhere in the system where people may inadvertently make contact. Under an abnormal fault condition or electrical short, a printed circuit board carrying high voltage levels energizing a metal chassis housing can assume full-voltage potential and create a shock hazard or death.

5.7.2 Signal Referencing for Components (AC Signals or RF Return Current)

The majority of design concerns related to EMC compliance lie in referencing circuits to each other. Both source and load must be at the same voltage reference for functionality reasons. In regards to a printed circuit board, components must be referenced to a common point (typically a plane) in order for logic circuits to determine a valid transition state. If the reference level between two circuits is not the same, functionality concerns occur such as noise margin erosion and threshold levels for logic switching. This is in addition to possible creation of a ground-noise voltage potential. Any ground-noise voltage will cause common-mode currents to be developed which is exactly what is not desired with regard to both radiated and conducted emissions and compliance.

When a transmission line (a.k.a. printed circuit board trace) travels between components, either the power or return plane will act as the RF return path. Note use of the words *return plane* instead of *ground plane*. RF return current does not care what the voltage potential is of the planes. If this is the case, why do we always call or refer to the RF return path as being coupled to a ground plane when it could be either power or return? This is shown in Fig. 5.5 with only the ground pins shown using a reference path for functionality reasons. If this was a battery operated device where is earth ground? Does DC current need earth ground to work, or only a return path for the flow of electrons to travel from the positive to negative terminals of a battery?

The word "*ground*" is usually used to describe an equipotential point that serves as a reference potential between components, which is incorrect use of this word. The correct word should be 0V potential. The word "ground" is not representative of actual application since digital ground may be completely different from analog ground, which may also be different from chassis ground, so which ground reference are we referring to? Ground is not the return path that RF current takes because this specific type of current is an electromagnetic field and is not at DC potential. The word "ground" totally ignores this fact!

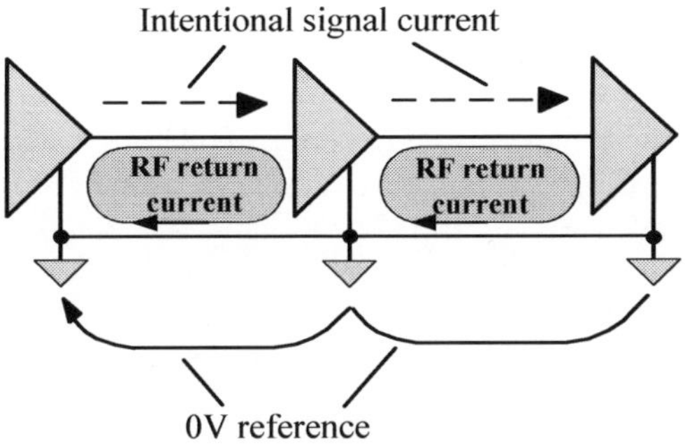

Figure 5.5 Referencing signal propagation for functional operation between components.

There may be less inductance between a noisy circuit and a ground or reference location than the connection to an equipotential point. RF current always takes the path of "least impedance." At lower frequencies, where $R >> \omega L$, the current will take the path of least resistance, as resistance dominates. At higher frequencies, where $R << \omega L$, inductance dominates due to $X_L = 2\pi fL$ in the equation [$Z = R + 2\pi fL$].

A better definition of signal return is a low-impedance transmission line for RF current to return to its source. This is for every application. RF current is the item of concern, not voltage. If a voltage difference exists between two components through a finite impedance, current will be established per Ohm's law, either AC or DC. Current in the RF return path determines the magnitude of magnetic coupling between circuits since we are dealing with electromagnetic field propagation within a dielectric in a closed loop path (Chapter 1). The physical size of the loop area determines the frequency of the radiated emissions.

The RF signal return path is determined by the type of product design, frequency of operation, logic devices used, I/O interconnects, analog and digital circuits and product safety requirements (electrical shock hazard).

A typical grounding method used to describe the signal-ground concept is shown in Fig. 5.6 where the load is connected to one ground reference and the source to another. Ground-noise voltage, V_n, is caused by inductive losses in the return path loop between the two difference references.

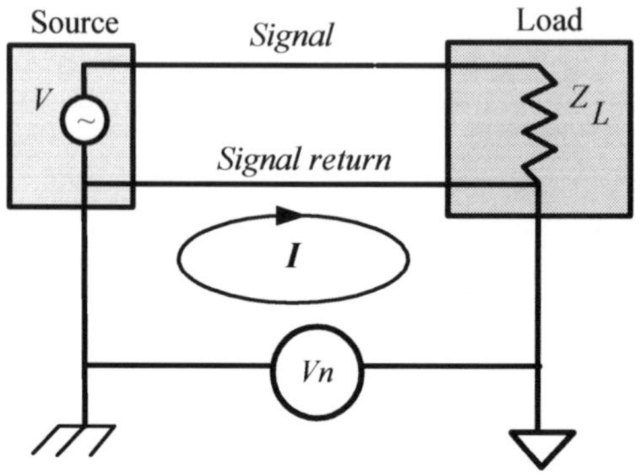

Figure 5.6 Typical grounding between circuits.

To illustrate this fundamental concept further, Fig. 5.7 shows two subsystems connected to a metallic plate, chassis or other metallic item which is generally called *ground* for lack of a better name, or understanding what this return path should be called. These subsystems may be analog, digital or another undefined source. If components are digital, then power (+V_{cc}) returns to its source referenced to the power plane (transmission line) through a pre-defined propagation path. DC current is constantly changing when devices switch logic states, both sourcing and sinking current. In analog circuits, the return current may contain low-frequency, high-frequency narrowband or broadband signals. Thus, analog signals generally have a dedicated return or "analog ground" that is different from digital logic devices due to their low-voltage margin of operation, under certain applications.

In Fig. 5.7a, the RF return current path of Subsystem #2 travels through the same return line as Subsystem #1. The two currents add at the power supply. Since return paths have finite impedance, resistive and/or inductive, a voltage potential will be developed between the two Subsystems. The ground point of Subsystem #1 is varying at a rate proportional to the signals switching within Subsystem #2. By virtue of coupling through a common impedance the power source now sees two separate voltage potentials simultaneously.

So far this has been a discussion about ground-noise voltage, which again is improper use of terminology. What about the voltage that is observed at the load? The voltage of the return point for Subsystem #2 is $[Zg_1I_1 + (Zg_1 + Zg_2)I_2]$. Subsystem #2 contains the signals of Subsystem #1 through Z_{g1} in addition to its own signal. This is called as common-impedance coupling.

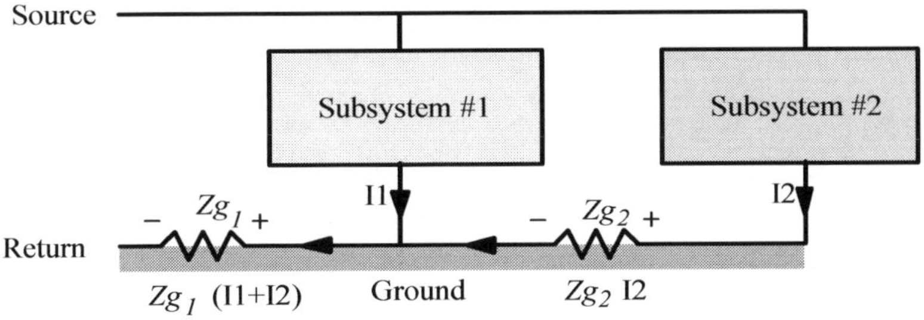

Figure 5.7a Common-impedance coupling in a ground or return structure.

Figure 5.7b illustrates a connection commonly found in data communication systems where a signal (i.e., RS-232) with a return is utilized and has a shield. We assume the source is a computer and load a peripheral located some distance away.

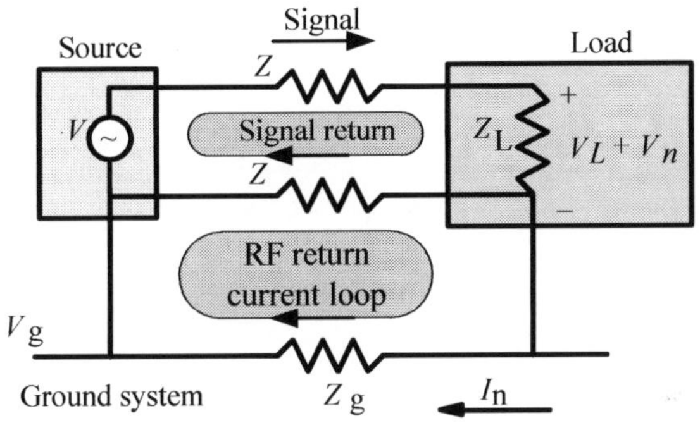

Figure 5.7b Conductive coupling of ground noise with external interconnect.

A reference (Z_g) is common to both devices and may be the third wire AC safety ground connection through the mains wiring in the building, or cable shield with no relationship or connection to DC signal return. This connection is required for functional safety and to prevent risk of electric shock. The power cord safety ground has high impedance at RF frequencies. In this situation, any RF energy in the transmission line from the external peripheral drives noise current I_n into the safety ground system. Assume load Z_L is greater in impedance than return wire impedance Z and AC ground wire impedance Zg. The ground noise voltage developed on the return wire adds to the signal voltage of the load. Total voltage, V_n that can cause harmful radiated or conducted interference is calculated by (5.2).

$$V_n = \frac{Z - Z_g}{Z + Z_g} * (V_g) \tag{5.2}$$

A misconception regarding ground impedance is the type of impedance present. Most engineers assume that ground is at DC potential or has low resistance. At higher frequencies of operation, such as 100 kHz and above, the primary impedance element is inductance, not *resistance* (Chapter 2). Additional losses due to *skin effect* starts to become noticeable in the upper frequency range generally 1 GHz and above. Skin effect losses are however negligible compared to the inductive losses at lower frequencies. Inductance is approximately 15 nH for a 0.508cm (0.020 inch) transmission line routed on a printed circuit board. Solving $X_L = 2\pi fL$, at 100 MHz, inductive reactance is 24 Ω/cm (9.43 Ω/inch). A #28AWG wire (radius of 6.3 mils) has an inductive reactance of 65.9×10^{-3} Ω/inch. As observed, there is a significant magnitude of difference between resistance and inductance components at 100 MHz. This is why resistance is not a concern at RF frequencies in transmission lines.

Designers must keep in mind the path that RF return current takes during printed circuit board layout. One cannot concern themselves only with functionality based on computational analysis such as SPICE. The design engineer and printed circuit board layout person must work together to determine all anticipated paths which RF return current could flow during component placement, including parasitics. The question to ask is "Where will all RF current flow irrespective of what we call ground, since the word ground does not exist on a printed circuit board?" Every conductor will have a voltage drop associated with it along with corresponding current generation per Ohms law. This current is usually at RF potential, since most systems contain digital components switching logic states bouncing both power and return planes alternatively, at nearly the same time.

During design, minimal cost is achieved when grounding or referencing is taken into consideration before implementation. A well-designed ground/reference system, not only on the printed circuit board but system wide, offers both improved emissions and immunity protection. A signal return system using long transmission lines that was not considered during design or re-utilized from a legacy product (because it once worked so why redesign) is a sign of system failure related to system functionality or EMC compliance.

Important areas of concern include the following [1, 2, 3]:

- Minimize or reduce current loops by careful layout of high-frequency components, namely observing RF return paths.
- Partition areas of the printed circuit board, or system, to keep high-bandwidth noise circuits from corrupting lower-frequency circuits. This can easily be achieved by splitting planes.
- Design the printed circuit board, or system, to keep interfering currents from affecting other circuits that may be connected to a common RF return path, both real and parasitic.
- Carefully select ground or 0V points to minimize loop currents, ground impedance and transfer impedance of the circuit.
- Consider current flow through the ground/reference system as it relates to noise being injected into or from a circuit.
- Connect very sensitive (low noise-margin) circuits to a stable 0V reference source.

5.8 Grounding Methodologies

When implementing a grounding or referencing system on a printed circuit board, several methodologies are commonly used. For each methodology a hybrid combination may exist. Different methodologies may be used at the same time, but only if the designer understands the concept of current flow and return paths, differentiating between RF return and safety or chassis ground.

Many types of ground terminology are found on printed circuit board schematics. These include digital, analog, safety, signal, noisy, quiet, earth, single-point, and multi-point, but not RF ground since RF ground is a virtual connection mandatory to allow active components in a metallic enclosure to pass EMC tests within a shielded enclosure, where the shield absorbs radiated RF fields and essentially sends these to [chassis or earth] ground in some manner, or absorbs and dissipates this energy in the metal (Chapter 6).

A ground ("referencing") methodology must be specified and designed into a product during the concept stage and not be left to chance or luck based on prior experience. Designing an optimal grounding/referencing system is also cost effective since additional revisions of a printed circuit board may not have to occur. During any printed circuit board design, a choice must be made between two basic concepts; single versus multi-point. Interactions with other grounding methods can exist if they are planned in advance for due diligence with respect to current flow and inductive loops. The choice of grounding/referencing methodology is dependent on product application.

Remember, if single-point grounding is to be used, be consistent in its application and ensure there is no possible secondary path that will create a large loop area for return current to propagate within. The same rule exists for multi-point. A multi-point ground should not be mixed with single-point unless the design allows for isolation or partitioning between planes and functional subsections.

The discussion that follows is divided into three main grounding methodologies. These methodologies are single-point, multi-point and hybrid, although other methods of grounding/referencing are available. In the discussion below, the word ground will be used in certain context but in reality, when discussing application or use on a printed circuit board, the term *"reference"* should be substituted. That is, unless, the printed circuit board makes a virtual or direct connection from DC return to chassis ground.

5.8.1 Single-Point Grounding Methodology

A single-point grounding methodology is one in which all returns are tied to a common reference point. This methodology prevents RF currents from propagating between dissimilar circuits usually at different voltage or reference levels. It also prevents sharing of common return paths for undesired RF currents which produces common-impedance coupling.

Single-point grounding is preferred when the speed of components, circuits, interconnects and the like is typically in the range of 100 kHz or less. This is because the effects of distributive transfer impedances are minimal and inductive reactance is less than the resistive element (Chapter 2). At higher frequencies, inductance in the return path will start to become noticeable including power/return planes and interconnects. This higher impedances can cause EMI if the length of the transmission line (a.k.a. trace) coincides with multiples of a quarter-wavelength based on the edge rate of periodic signals. With finite impedance in the RF current return path, a voltage drop is established along with creation of unwanted common-mode RF currents.

Owing to significant impedance at RF frequencies generally above 100 kHz, transmission lines and return conductors act as a loop antenna based on the physical size of the structure. A convoluted loop is still a loop, regardless of shape. At frequencies above 1 MHz, a single-point ground connection generally is not used due to inductive reactance that exceeds distributed resistance. However, exceptions do exist if the design engineer recognizes pitfalls and designs the system using highly specialized and advanced grounding techniques.

In Fig. 5.8, two methods are shown for single-point grounding: series and parallel. The series connection is in a daisy-chain configuration which allows common-impedance coupling to occur between each system when an interconnect is provided between boxes, not shown. There is finite inductance between each element and a common reference (ground), all of which are connected in turn to a single AC chassis ground point. Distributed or parasitic capacitance is also present between circuits to a single-point reference. This reference, shaded in the illustration, is at the same potential everywhere and always magnitudes less in impedance than any interconnect inductance. With both inductance and parasitic capacitance (due to physical existence of the wires), resonances may occur generally at switching frequencies. For this configuration, three different resonances could be present, which may create an EMI situation by developing common-mode current across the fixed impedance of each wire.

When using the top connection methodology (Fig. 5.8), the total amount of current injected into chassis (AC) ground is the summation of $I1+I2+I3$. Since there is finite impedance within Z_1, a voltage potential between the systems and chassis ground is created, calculated by (5.3) and (5.4) where $\omega=2\pi f$.

$$V_A = (I1 + I2 + I3)\ \omega L_1 \tag{5.3}$$

$$V_C = (I1 + I2 + I3)\omega L_1 + (I2 + I3)\omega L_2 + (I3)\omega L_3 \tag{5.4}$$

With this widely used configuration, a large amount of current across this finite impedance will produce a significant voltage drop between circuits. The voltage reference between circuits and chassis ground may be sufficiently large to cause the system to create EMI. During the design cycle, one must be aware of this potential problem from using the series connection (daisy-chain). This configuration should also not be implemented when widely different power levels are present since high-power consuming circuits produce large ground currents which will find another transmission line to couple creating crosstalk and RF noise. Large ground currents in turn affect low-level analog devices. If this method must be used, the most sensitive circuits and components must be physically located close to the input power connector and as far away from higher power consuming components and circuits that could cause a reference plane to bounce.

A more optimal method of using single-point grounding is the parallel configuration, shown in the bottom half of Fig. 5.8. This method, like series connection, also has a disadvantage in that each RF current return path may be at different impedance thus exacerbating ground-noise voltage in each path. If more than one printed circuit board is utilized or if various subassemblies are combined within an end-use system, the RF return of each interconnect may be physically long. This interconnect may also possess significant impedance that will negate the desired effect of having a low-impedance reference connection to chassis ground. Many products fail emissions testing when multiple printed circuit boards are tied together in a parallel fashion, believing that a "single" point connection will solve all RF return current problems.

Like series connection, distributive capacitance is also present from each circuit to ground. The designer must minimize the inductance from each circuit to chassis ground, which is difficult, because this type of implementation requires a long wire or a dedicated trace. As a result, resonances between each circuit should be approximately the same. This may or may not affect circuit operation.

Another problem associated with using the single-point grounding methodology with a physical piece of wire is radiated coupling. Coupling can occur between transmission lines (wires), a transmission line and a printed circuit board, or wire and chassis housing. In addition to RF radiated coupling, signal crosstalk may be established depending on physical distance spacing between source and RF current return paths.

Coupling occurs by both capacitive and inductive means. The amount of common impedance coupling present is dependent on the spectral content of the RF return signal. Higher frequency components will radiate more than lower frequency devices due to the short length of the wire which are efficient antennas at these frequencies.

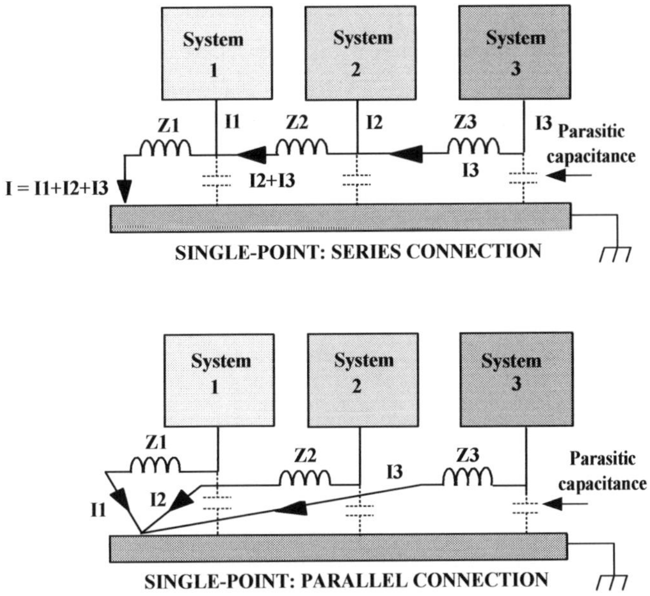

Figure 5.8 Single-point grounding methods.
Note: Inappropriate for high frequency operation, generally above 100 kHz-1 MHz

Single-point grounds are usually found in lower frequency and amplitude sensitive circuits, such as audio, analog instrumentation, 50/60/400 Hz and DC power systems, along with products packaged in plastic enclosures. Although single-point grounding is commonly used in lower frequency applications, it is occasionally found in extremely high-frequency circuits and systems with extreme care in design to minimize or prevent ground loops if subsystems must communicate with each other. Higher frequency application is permitted when a design team understands all problems that exist with inductance using different ground return structures.

It is nearly impossible to effectively implement single-point grounding in certain products because different subassemblies and peripherals must be bonded directly to a metal chassis, assuming a metal chassis is utilized. Distributed transfer impedance will always exist between the metal chassis and printed circuit board that inherently develops loop structures. Parallel single-point grounding places these RF loops in regions where they are least likely to cause problems (i.e., they can be controlled and directed rather than allowed to transfer energy inadvertently).

An example of poor implementation of single-point grounding is shown in Fig. 5.9. In this example, the A/D 0V reference is within both the digital and analog section under the assumption that the open connection point on the right side of the printed circuit board (bridge) will provide optimal single-point connection as long as the analog section is not bonded to any other reference point. Single ended signals are unfortunately routed across the gap in the area of the converter's moat or isolated area (left side of the converter).

If lower-frequency (kHz) noise is not a problem, the 0V reference connection should be placed as near to the A/D device as possible. Analog and digital power must be isolated from each other using an appropriate filter to keep digital noise from corrupting analog circuitry.

In the upper left corner of Fig. 5.9 there is a moat (split plane) and E_{cm}. This is where common-mode voltage is established with RF return current propagating back to the source through a high-impedance connection, or free space for those traces crossing in this area. Per Ohm's law, with RF current across a fixed impedance, RF voltage is established that will drive a dipole antenna structure. This antenna structure could be the planes of the printed circuit board since planes are in reality transmission lines. Board edge radiated emissions may also occur.

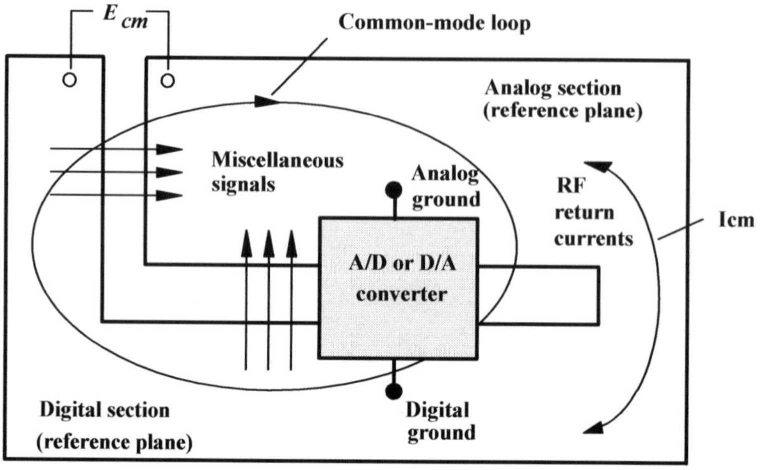

Figure 5.9 Bad implementation of single-point (connection) referencing.

In Fig. 5.9 various current and voltage sources are present. Almost all RF return current travels through the bridge. The bridge on the right-hand side is the low-impedance RF return path for all signals traveling to the analog section from digital, including those crossing the moat. Since a closed-loop path must exist for signal functionality, any RF energy crossing the *moat* must complete its return through the *bridge*. This RF current is identified as I_{cm}. For traces crossing the moat in the upper left corner, the return path is physically long and without flux cancellation, EMI is assured.

Since there is a moat present, a common-mode voltage potential will be established at the point furthest from the bridge. The impedance between the two power sources will be different based on the inductance of the power and return planes. With common-mode voltage and fixed impedance of the return path, common-mode RF current is developed which travels through both digital and analog sections. Once a loop is created with RF energy a magnetic field is created along with possible RF emissions.

Another bad implementation of single-point referencing in a mixed circuit system is shown in Fig. 5.10. The ground wires are non-current carrying conductors (RF return). The interconnect between circuits (Circuit #1 to Circuit #2, and Circuit #1 to Circuit #3) are identified as "GND" by the circuit designer on a schematic. These connections are, in reality, the signal return path for currents that travel between circuits, but which type of current is propagating, AC or DC since we have no idea what the word GND represents. These interconnects create a current loop, increases self-inductance of the transmission lines and develops a stray magnetic field between circuits and the 0V reference point. In addition, parasitic capacitance *C*, between Circuit #1, Circuit #2 and Circuit #3, associated with a common return is shown along with inductance *L*. This *LC* network may create a resonance that occurs at a frequency or harmonics of an oscillator, thus exacerbating a system wide problem.

To summarize, single-point grounding is not ideal when dealing with circuits operating above 100 kHz due to parasitic capacitance and inductance. Long wires have significant inductive reactance. If we consider inductance by itself, it is easily calculated by $E=-L(dI/dt)$. Common-mode current will be generated due to this high inductance. With equation $X_L=2\pi fL$, as frequency increases so does inductive reactance, which helps establish a voltage potential difference between locations. This voltage potential difference is also another cause of common-mode current generation.

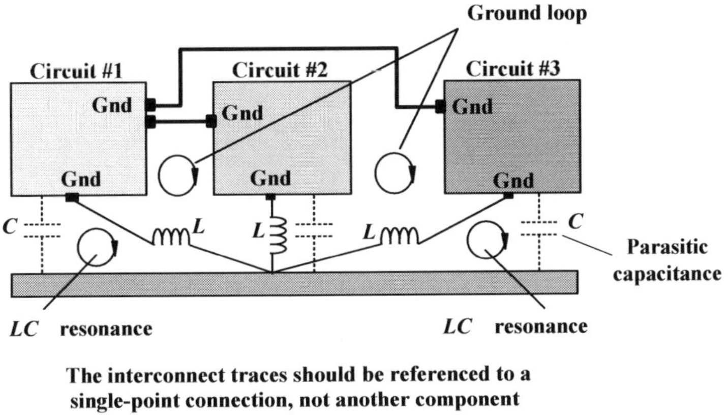

Figure 5.10 Poor implementation of single-point referencing.

A third poor layout example at the system level is the use of "single-point earthing/grounding, sometimes called "star-earthing/grounding" (Fig. 5.11). There are 0V splits in the return plane, assuming this design technique keeps devices' circulating return currents confined to certain circuit areas. Using this technique assumes we prevent crosstalk (e.g. digital noise affecting analog circuits) however, these planes still need to be connected together somewhere, shown with a ferrite bead. This technique works well below a few tens of kHz.

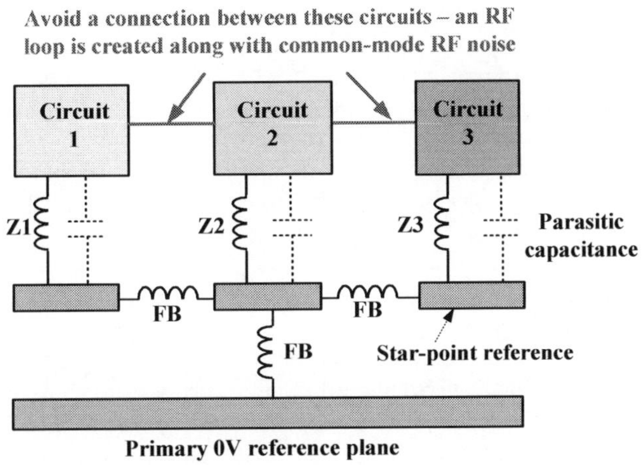

Figure 5.11 Routing single-point, star configuration – a poor design technique.

Splitting 0V planes, with regard to RF return current paths that comply with both Kirchhoff and Ampere's law, states that currents divide according to the admittances of the various alternative paths, including stray paths that must return in full to its respective source. This stray path could be free space or metallic interconnects (wires or traces).

Microprocessors and switch-mode converters commonly use this star-referencing layout technique to minimize use of having additional power and return planes, which is poor design practice for signal integrity, power integrity and EMC compliance.

The 0V reference plane within printed circuit boards should be a solid, continuous layer. Exceptions exist for specialized applications such as when high-bandwidth video components are located next to low level audio circuitry. For this application, a split reference plane provides significant benefit but mandates extra care to make sure no traces cross the split plane. Details why was provided in Chapter 2.

Common-mode RF current develops as a result of improper RF return of differential-mode signals and insufficient flux cancellation. If a split 0V (ground) plane is implemented for a specific design application, no traces can cross the split. Common-mode current is always developed across the split. The reason for creating a solid 0V reference in the first place is that it provides a low-impedance (high-admittance) return path for all RF currents and to minimize ground noise induced voltages that may develop if routed over a split.

5.8.2 Multiple Connections to Single-Point Reference (a.k.a. Multi-point Grounding)

Higher-frequency printed circuit boards generally require the 0V reference plane(s) be connected to a common reference point, usually a metal chassis or enclosure. The correct definition of multi-point grounding is "*multiple connections to a single point reference.*" This definition will become obvious with the discussion that follows.

Multiple connections to a signal point reference, or multi-point grounding, minimizes plane impedance present in the RF current return path since no wire or physically long interconnect is utilized. Loop inductance is also minimized significantly which is the primary reason why single-point grounding starts to become ineffective above several 10's of kHz. When a bond connection is made between a 0V reference plane and metal enclosure, in reality, the entire system is eventually at one reference potential shown with the *chassis ground* symbol in Fig. 5.12, lower right corner. Regardless of the number of circuits identified on a schematic using a symbol such as digital ground, analog ground, chassis ground, ESD ground, and the like, they eventually connect together somewhere which means in there is only "one" reference in the system, or true single-point referencing. Proper connection of different ground or reference potentials is thus critical for both signal integrity and EMC.

The moment we make a connection between printed circuit board and a single-point reference using this methodology, the loop structure for RF currents becomes very small physically and the planes an inefficient radiating dipole antenna.

Another primary advantage of making multiple connections to a single-point reference lies in disrupting the efficiency of the planes from behaving as a dipole antenna when RF common-mode current is present. Planes, as mentioned in earlier Chapters, are transmission lines with a driven element and return. This transmission line pair represents a dipole antenna. The only difference between transmission lines (a.k.a. traces) on a printed circuit board and planes is physical dimensions. Take a signal transmission line and make it infinitely wide, or make a power or return plane the width of a small trace. On single-sided printed circuit boards, we have only power and return traces and according to Ampere, these must also exist as a closed loop circuit with both source and load impedance present. The loop area of this power distribution network could be extremely large.

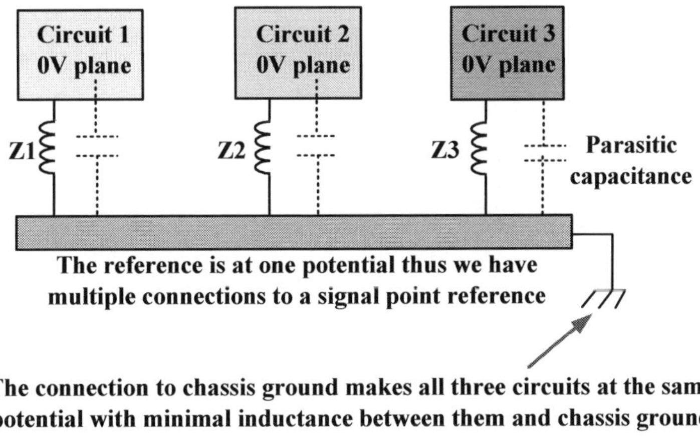

Figure 5.12 Multiple connections to a single point reference methodology. (a.k.a. multi-point grounding)

In Fig. 5.13, a dipole antenna is shown in two configurations. Between the two driven elements is a dielectric which could be core material (fiberglass), prepreg (glue) or air (Fig. 5.13a). RF energy propagates from the driven element to return with capacitance between the two wires acting as the dielectric (Refer to Dipole Antenna Model-Chapter 1). For a dipole to be an efficient radiating antenna, its length must be ½ wavelength or a multiple down to 1/20 of a wavelength. Take the 10th harmonic of a high speed signal (edge rate, not frequency, and do a Fourier transform to determine the highest frequency that could exist). We see that very small dimensions can allow a dipole antenna to become an efficient radiator at higher frequencies.

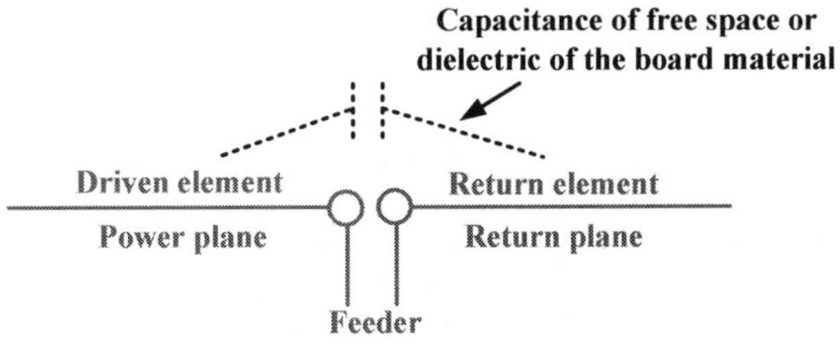

Figure 5.13a Dipole antenna representation.

By providing multiple connections to a single-point reference breaks up the ground element of the dipole antenna, thus making this an inefficient radiator (Fig. 5.13b). Multiple connections or shorting out this [ground leg] element minimizes RF energy present in the cavity between the power and return planes. Digital switching signals bounce (reflect) between components and vias provided within the cavity. These signals also reflect back from the physical edge of the printed circuit board into the middle portion of the assembly. Under this condition, we can have board edge radiated emissions as well as significant functional concerns. The propagating wave may phase add with other reflecting waves in the cavity causing a significant bounce condition to occur on either or both the power and return planes.

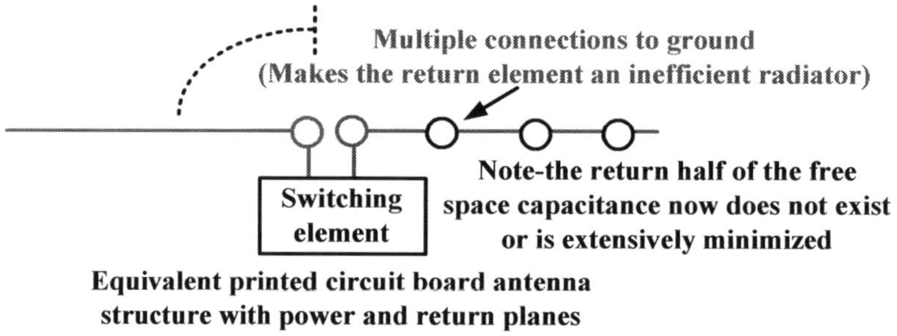
Multiple connections to ground
(Makes the return element an inefficient radiator)
Switching element
Note-the return half of the free space capacitance now does not exist or is extensively minimized
Equivalent printed circuit board antenna structure with power and return planes

Figure 5.13b Dipole representation of planes in a PCB with multi-ground points. (Inefficient radiating element with the ground terminal shorted out)

Low plane impedance is a result of lower inductive characteristics of solid planes, or by additional low-impedance stitch connection to the chassis reference point that reduces inductance (basic circuit analysis on parallel impedances). For a solid plane, inductance is the only element of concern, not capacitance. Capacitance does play a part in creating an internal decoupling capacitor or defining the velocity of propagation of a signal when used with a second plane at an opposite potential. We are ignoring capacitance for this discussion, at this time.

Inductance, L, is determined by knowledge of the impedance of the plane, for this example copper. Typical inductance for 10 x10 square inches (25.4 x 25.4 square cm) is shown in Table 5.1. The equations for solving an exact value for the impedance of planes is complex and beyond the scope of this book.

Table 5.1 Impedance of a 10x10 inch (25.4x25.4 cm) copper metal plane.

Frequency (MHz)	Skin Depth (cm)	Impedance (ohms/square)
1 MHz	6.6×10^{-3}	0.00026
10 MHz	2.1×10^{-3}	0.00082
100 MHz	6.6×10^{-4}	0.0026
1 GHz	2.1×10^{-4}	0.0082

Multi-point grounding minimizes ground impedance present in the power distribution system of the printed circuit board by shunting RF currents from circuits to chassis ground as was shown in Fig. 5.12 [1]. This lower impedance is primarily due to lower inductance from large, solid planes (or transmission lines) versus that of a transmission line (a.k.a. trace) or wire with higher impedance in its structure. In very high-frequency circuits, the power and return leads from components must also be kept as short as possible to minimize loop impedance, especially when a break-out trace is used from the component's pin to a via, which then connects to a plane. Any impedance in the interconnect loop causes a voltage potential to be developed across the interconnect bond. This in turn creates common-mode current.

If one can make interconnects to internal planes without running a physical trace on the top or bottom of the printed circuit board, similar to a BGA configuration or flip chip, loop inductance is significantly minimized.

When using lower-frequency components, multi-point connections should be avoided since RF currents from all circuits flow through a common ground or 0V reference. The common impedance of the reference plane can be reduced by using a different plating process on the surface of the material [7]. Increasing the thickness of the copper plane has no effect on minimizing plane impedance, since RF currents travel on the skin or surface of the material. Increasing plane thickness affects mainly the total current carrying capacity, heat dissipation and reference for transmission line impedance control.

A general rule of thumb is that for frequencies less than 100 kHz, single-point grounding may often be the best grounding methodology to use. Between 100 kHz and 1 MHz, single-point grounding may be used only if the longest length trace or bond-stitch connection is less than 1/20 of a wavelength, assuming slow edge rates and low-frequency spectra, *and* there are no disruptive influences between the connections.

Multi-point grounding also minimizes inductance between noise generation circuits and a 0V reference point. This minimization occurs because many parallel RF current return paths exist illustrated in Fig. 5.12. Even with many parallel connections to a 0V reference, ground loops may still be created between circuits. Ground loops are prone to magnetic field pickup of ESD energy if transmission lines are routed on the outer layers of the printed circuit board, or creation of radiated EMI from the inductance of the differential-mode loop creating common-mode current due to loop inductance, generally observed from via implementation.

In very high-frequency circuits, lengths of ground leads from components must also be kept as short as possible; inductance causes a voltage drop to occur which in turn creates common-mode currents. Trace lengths as long as 0.020 inch (0.005 mm) adds inductance to the transmission line of approximately 15–20 nH per inch [38-50 nH/cm], depending on trace width. This inductance may cause a resonance to occur when distributed capacitance between reference plane(s) and chassis ground forms a tuned resonant circuit. The capacitance value, *C*, (5.5) can be determined but only if we know the actual or true inductance of the transmission line.

$$Z = \frac{1}{2\pi f\sqrt{LC}} \tag{5.5}$$

where:
Z = impedance (Ohms)
f = resonant frequency (Hz)
L = inductance of the circuit (Henries)
C = capacitance of the circuit (Farads).

Equation (5.5) describes most aspects of frequency domain concerns. This equation, though simple in form, requires knowledge of how to calculate both *L* and *C* which by themselves are not easy to determine, use, and implement. Only with correct values will we be able to determine the real impedance of any transmission line or interconnect.

5.8.3 Hybrid Grounding

A hybrid ground system is a combination of both single- and multiple connections to a single point reference (multi-point grounding). This configuration is used when mixed frequencies and assemblies are present. Figure 5.14 shows two hybrid topologies. For the capacitive coupling version at lower frequencies, the single-point configuration is dominant whereas the multi-point configuration works better at higher frequencies. This is because the capacitor shunts high-frequency RF currents to ground after the single-point connection goes inductive (due to parasitic capacitance). The key to success is understanding what frequencies are present and desired direction of return current flow.

The inductive coupling version is used when multiple ground stitch locations must be connected to a chassis ground reference for safety reasons and lower-frequency connections. Note that parasitic capacitance still exist which could cause a resonance condition and undesired EMI by allowing eddy currents to propagate within each loop area. The stitch could be a physical device (e.g., ferrite bead), *L1-L3*, which prevents RF currents traveling from the reference plane up to the printed circuit board and circuits thus ensuring functionality, while allowing DC voltages to remain referenced to their respective 0V point.

Conversely, ferrite beads *L1-L3,* if physically provided, also keeps higher frequency RF currents created by the printed circuit board from propagating down to the reference plane which minimizes ground loops by forcing the return currents to travel through the lowest impedance path to ground through the single-point connection (wire), which should be at much lower impedance than the inductors or chokes. This type of ground methodology is used only under very specific conditions and applications.

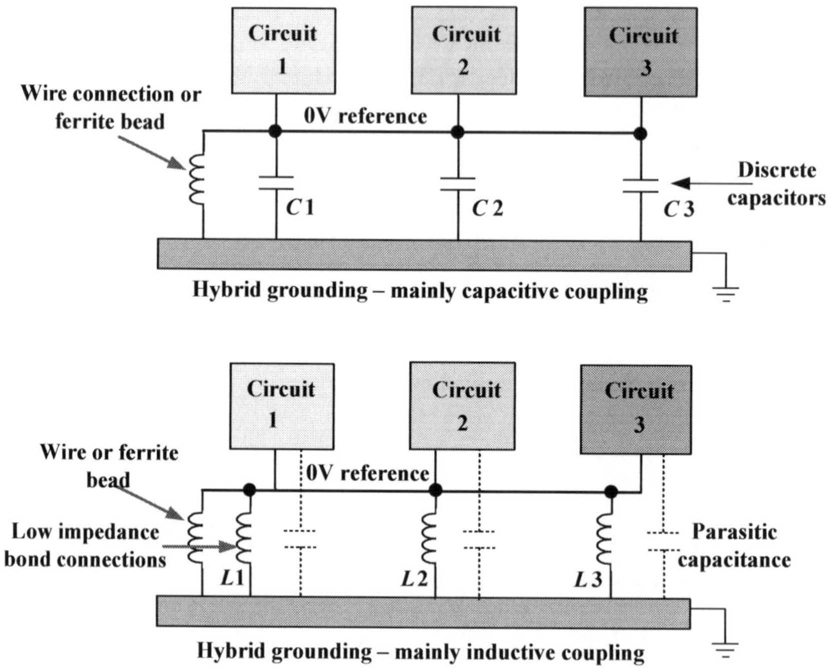

Figure 5.14 Hybrid grounding methodology.

Using capacitors or inductors in a ground topology allows us to steer or direct RF return currents in a manner that is optimal for signal functionality by minimizing loop areas. One must control printed circuit board layout by defining the path that all RF currents may take. Failure to recognize all current return paths may result in either emission or immunity problems.

Another example of a hybrid-ground system, Fig. 5.15, illustrates a cabinet rack with four separate assemblies. Each assembly contains various elements that include: (1) a communication system; (2) network interfaces; (3) control and display units; 4) modules for add-on adapter cards; and (5) power system. This configuration is common with many industrial products.

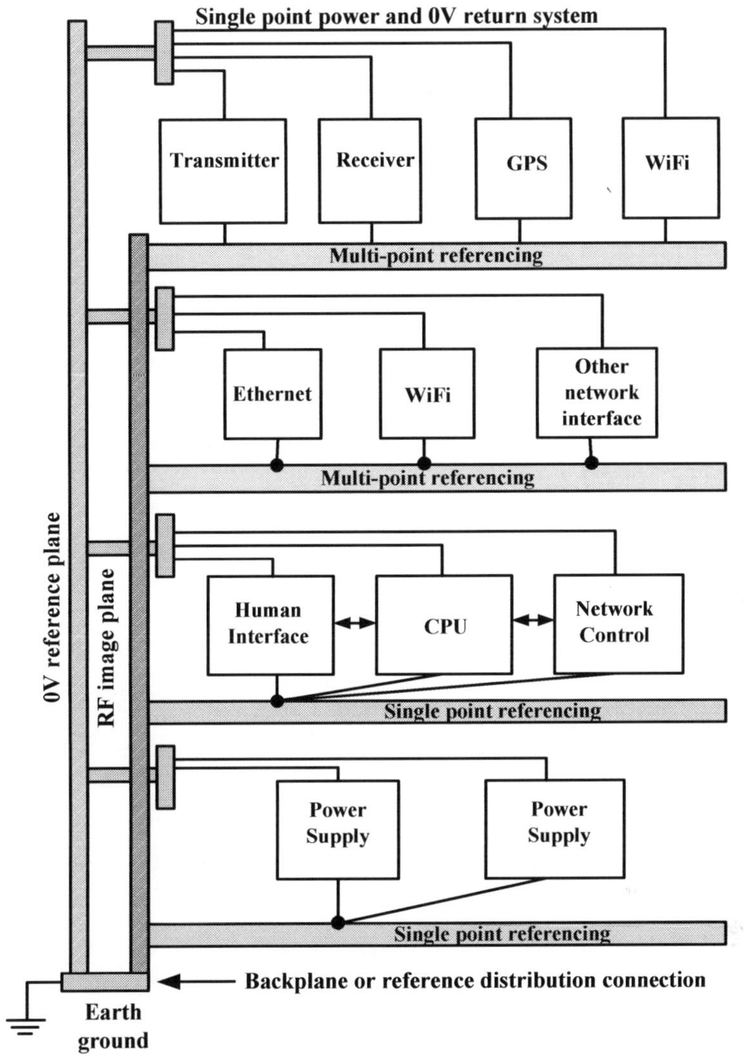

Figure 5.15 Typical cabinet configuration using hybrid grounding.

5.9 Controlling Common Impedance Coupling Between Transmission Lines

When two transmission lines share a common RF return path, common impedance coupling may occur. If the magnitude of coupling is above a certain threshold level, EMI concerns develop as well as possible functional disruption. To minimize or prevent common impedance coupling, three main methodologies exist.

- Lowering the common impedance to a minimum value.
- Avoid having a common-impedance path in the first place.
- Minimizing ground inductance.

5.9.1 Lowering the Common Impedance Path Inductance

A ground or reference system requires a metal transmission line: trace, wire, strap, chassis, planes and the like. All conductors have a frequency response dependent on material and geometry with DC resistance described by (5.6).

$$R = \rho l / A \text{ (ohms)} \tag{5.6}$$

where:
R = DC resistance (Ohms)
ρ = resistivity of the material (Ohms•mm²/m).
l = length of the conductor in the direction of current flow (m)
A = cross-sectional area of the conductor perpendicular to the current flow (mm²)

Resistivity, ρ, of various materials are:

- Copper: $1.7*10^{-3}$ Ω•mm²/m
- Aluminum: $2.8*10^{-3}$ Ω•mm²/m
- Steel: $1.7*10^{-2}$ Ω•mm²/m

For a round conductor, skin current present on the transmission line is illustrated in Fig. 5.16. At very high frequencies, generally above 1 GHz, almost all of the time variant propagating electrons in the transmission line will fight each other for space to travel on, and if there is insufficient skin thickness to support the flow of electrons, losses will occur in the transmission path in addition the minimizing total amount of energy flow from source to load. The magnitude of this loss could be sufficient to cause both signal integrity problems as well as development of common-mode current due to imbalance in differential-mode signaling.

Conductors have an inductance value different from overall system-wide inductance. This inductance is called *external* inductance, which is a function of loop area enclosed by the conductor. On the other hand, *internal* inductance is *not* a function of loop area. For a round conductor, internal inductance is described by (5.7).

$$L = 0.2 \cdot l \left[\ln(\frac{4l}{d}) - 1 \right] \tag{5.7}$$

where:
L = internal inductance (µH)
l = conductor length (m)
d = conductor diameter (m).

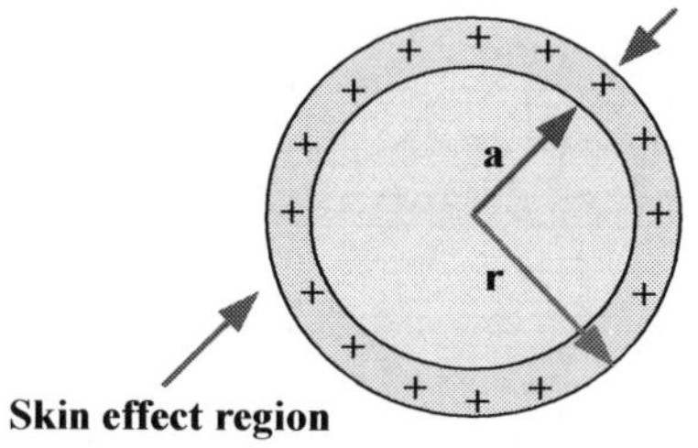

Figure 5.16 Current flow in a conductor – skin effect.

Equation 5.7 shows that inductance, L, increases linearly with length, l, while an increase in diameter, d, will reduce the total inductance logarithmically (a proportionally small degree only).

Rectangular straps and multilayer power planes have smaller inductance per unit length than that of round wire. The reason for this is that a flat strap (and extending this to a ground plane) has a larger perimeter than a round wire with the same cross-sectional area. Inductance of a ground strap is calculated by (5.8).

$$L = 0.2 \cdot s \left(\ln \frac{2s}{w} + 0.5 + 0.2 \frac{w}{s} \right) \tag{5.8}$$

where:
L = inductance of the ground strap (μH)
s = strap length (m)
w = strap width (m) [must be larger than the thickness by a factor of 10 or more].

When s/w > 4 (length-to-width ratio), Eq. (5.8) can be approximated by (5.9).

$$L = 0.2 \cdot s \cdot \ln \frac{2s}{w} \tag{5.9}$$

Equation (5.9) clearly shows that a flat strap has lower inductance than a round wire, and is more useful as a method of providing a low-impedance ground connection at higher frequencies. Extending this analysis to a solid plane that is physically large internal to a printed circuit board, we find that a plane has an impedance that is extremely small by several magnitudes compared to a wire or strap, except for the perturbations to the planes caused by annular anti-pads around vias. This is the primary reason why planes work as well as they do at higher frequencies while minimizing common-impedance coupling.

5.9.2 Avoiding a Common Impedance Path in the First Place

To reduce common-impedance coupling, care must be taken to identify all return paths during the design stage. This is best achieved when all reference connections from different systems follow a dedicated and separate path to a single-point reference connection.

Figure 5.17 illustrates a star configuration for providing power and return to various subsystems. This implementation technique require additional wiring and interconnect hardware, not to mention added cost.

To implement an improved method of common impedance grounding, functional circuits must be separated using a power distribution network for each partition in addition to the 0V reference required by logic circuitry. What this means is that we must segregate circuits by logical function. Logical function includes the following list of 0V references and does not include special circuitry that may be required or used, application dependent.

- Digital
- Analog
- Audio
- Video
- I/O Interconnects
- Control logic
- Power supply
- Network controllers

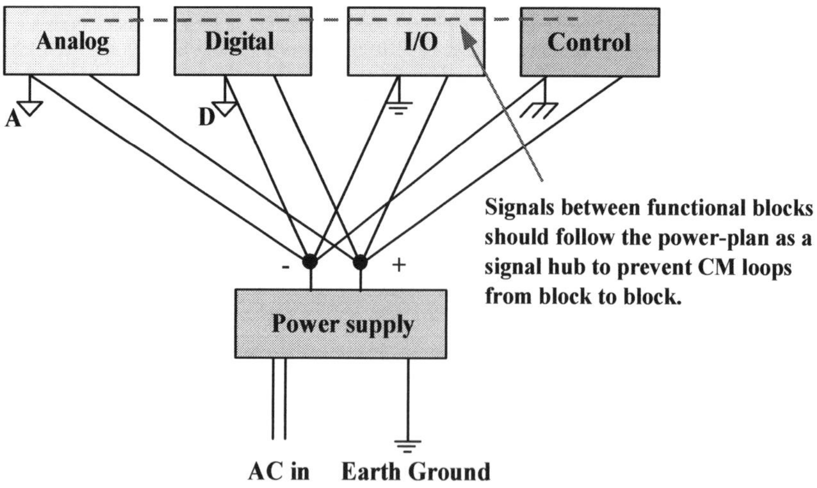

Figure 5.17 Separation of grounds to avoid common impedance coupling.

By separating noise-generating circuitry to prevent common-impedance coupling, increased noise immunity will be established between sections as long as there are no interconnects between. Each area must be connected to the primary 0V reference (Fig. 5.17), which in turn is usually connected to a safety or earth ground. This clearly shows that in reality, there is only one type of ground methodology for the entire system: single-point.

Preventing common impedance-coupling is best implemented with single-point grounding, which realistically may not be an option. As discussed in the next section, single-point grounding is best when the signal is 100 kHz or less containing lower-frequency Fourier spectra, while multi-point is preferred for higher frequency signals due to smaller loop area dimensions that minimize development of common-mode current.

What happens when a product must be multi-point grounded and common-impedance coupling is to be avoided? For the system shown in Fig. 5.18, a unit must operate in a low frequency environment that requires single-point grounding. An I/O interconnect has high-frequency noise on the cable shield as a result of exposure to externally induced high-energy radiated fields. This cable shield must be single-point grounded if frequency of operation is less than 100 kHz. For higher frequency signals above 1 MHz, *both* ends of the cable shield must be connected to the 0V-reference or chassis ground, application dependent. This frequency range is given only for guidance and will differ based on application of use, signal edge rates, frequencies required for optimal communication, along with other factors such as how either the source or load connections are bonded to an isolated ground system or earth ground. The items and other factors to consider and are presented in extreme detail within [3]. Cable shielding and whether to connect the shield on one end of both is discussed extensively in Chapter 6, as this is a system-wide design requirement.

To implement virtual grounding of a cable return wire or shield to make this transmission line appear to be a multi-point grounded cable when in fact it must be single-point, a bypass capacitor can be provided between shield and chassis ground, which is generally difficult to implement. This capacitor must be optimally calculated for a particular frequency range of interest and installed with low impedance connections (short interconnect) at the end of the cable shield which is "not at DC potential," but connected to the chassis ground reference (System 1, Fig. 5.18). This design technique is not easily manufacturable and highly discouraged.

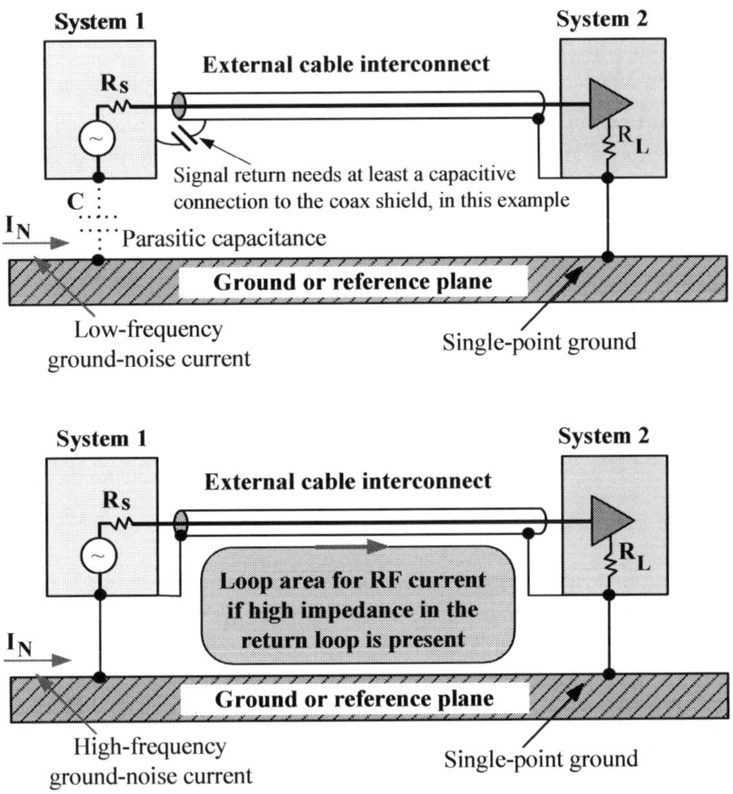

Figure 5.18 Single-point grounding for lower frequencies and multi-point for higher frequencies.

Other methods of avoiding common-impedance coupling, in addition to using single-point grounding, are available. These are use of an isolation transformer, common-mode choke, optical isolator or balanced circuitry. These options are examined later in Section 5.11, "Breaking Ground Loops."

5.9.3 Minimizing Ground Inductance

Inductance, which creates imbalance in power and return (ground) transmission line is the primary cause in allowing common-mode current to be propagated as an undesired radiated EM field. In order to reduce inductance, the transmission line must be made shorter in length or be physically wider in width. This is clearly shown in (5.10), top equation, identified as width "*w*."

Another concern is *loop area physical dimension* which is different from wire *length (l)*. When a transmission line is located physically adjacent to a reference plane, there is always finite distance spacing between the plane and transmission line, shown as variable "*h*" in (5.10) due to manufacturing requirements. The closer the physical distance between trace and plane, the lower the inductance.

Using the variables in (5.10) and (5.11), we easily calculate total inductance of the loop circuit. What we require for both optimal signal integrity and EMC is having the smallest amount of total inductance possible for each and every loop of concern.

Due to the log relationship, an increase in conductor dimensions will not significantly reduce total inductance. Therefore, a 100% increase in width will provide only a 20% decrease in inductance. The solution to minimizing ground inductance is to provide multiple parallel paths.

Length of trace

$$L_{ground} = \mu_0 \cdot \frac{l}{2\pi}\left[\ln\left(\frac{8l}{w}\right) - 1\right] \quad (\mu H)$$

Width of trace

Height above current return path

$$L_{ground} = \mu_0 \cdot \frac{1}{2\pi}\left[\ln\left(\frac{2\pi \cdot h}{w}\right)\right] \quad (\mu H)$$

(5.10)

While examining ground inductance a question commonly arises, "Can we minimize inductance to zero?" Unfortunately, no, we can never achieve zero inductance in practice. When transmission lines propagate an electromagnetic field, total inductance becomes limited by the mutual inductance (M) between the two transmission lines upon which there must be a finite distance spacing. Total inductance between conductors carrying current in the opposite direction from each other is clearly seen in (5.11):

$$L_1 = \frac{L_1 \cdot L_2 - M^2}{L_1 + L_2 - 2 \cdot M}$$

(5.11)

When conductors are identical, i.e., L1=L2, then L1 $\propto$ length and M $\propto$ spacing. Inductance also depends on the loop area between conductors. If conductors are physically close together, then:

$$L1 \approx M \Rightarrow L1 \approx L2$$

Therefore, there is a limit.

When taking into account mutual inductance and capacitance between transmission lines, Figures 5.19 and 5.20 provides an outstanding illustration on the effects of distance spacing along with their relationship to a reference plane. The potential of the plane is irrelevant. Also, internal to the charts are equations that provide these curves.

5.10 Controlling Common-Impedance Coupling in Power and Return Planes

When there are many circuits switching simultaneously with widely different voltage and current swings, logic family dependent all powered from the same distribution system, coupling of RF current may occur between components and planes. With finite impedance in planes, along with current consumed by active logic devices, a voltage drop will occur between elements. This voltage drop develops common-mode ground-noise voltage, which in turn creates undesired common-mode RF energy. Because planes are used throughout a printed circuit board assembly, ground-noise voltage present on one section of the board may propagate to other sections causing both signal quality and EMC problems between high-bandwidth and lower bandwidth elements.

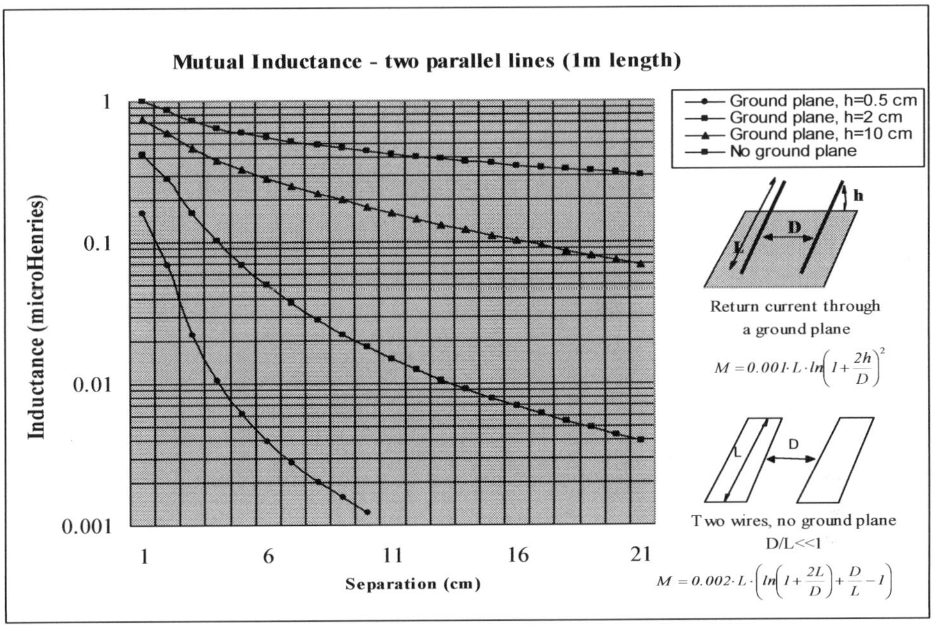

Figure 5.19 Mutual inductance between two parallel transmission lines.

Return current through a ground plane

$$M = 0.001 \cdot L \cdot \ln\left(1 + \frac{2h}{D}\right)^2$$

Two wires, no ground plane D/L<<1

$$M = 0.002 \cdot L \cdot \left(\ln\left(1 + \frac{2L}{D}\right) + \frac{D}{L} - 1\right)$$

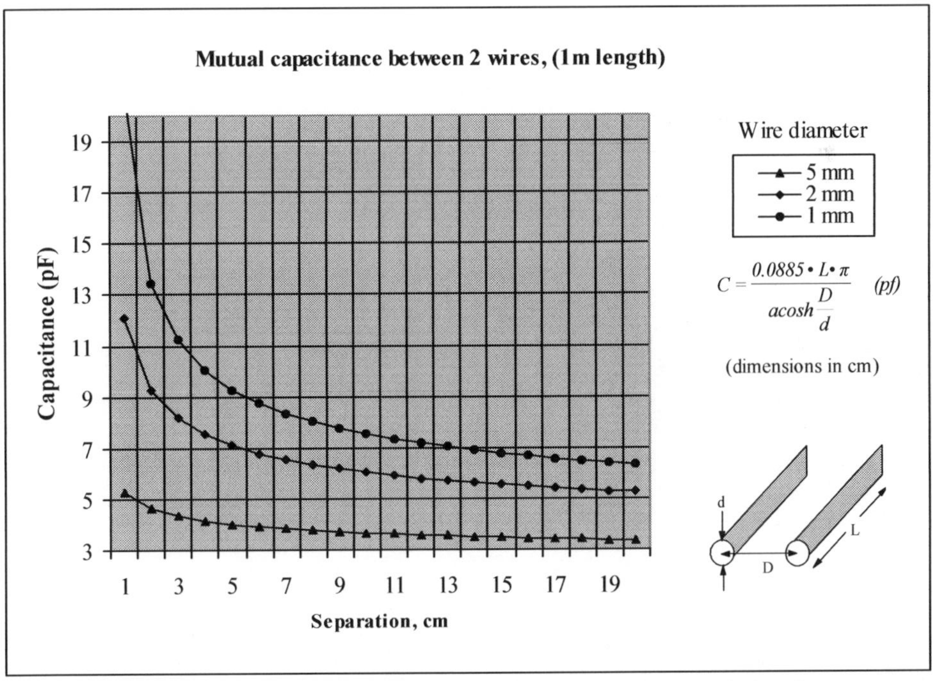

Figure 5.20 Mutual capacitance between two parallel transmission lines.

$$C = \frac{0.0885 \cdot L \cdot \pi}{\operatorname{acosh}\frac{D}{d}} \quad (pf)$$

(dimensions in cm)

Figure 5.21 illustrates the concept of common impedance coupling in power and return planes. The RF noise Device #1's reference plane is described by (5.12).

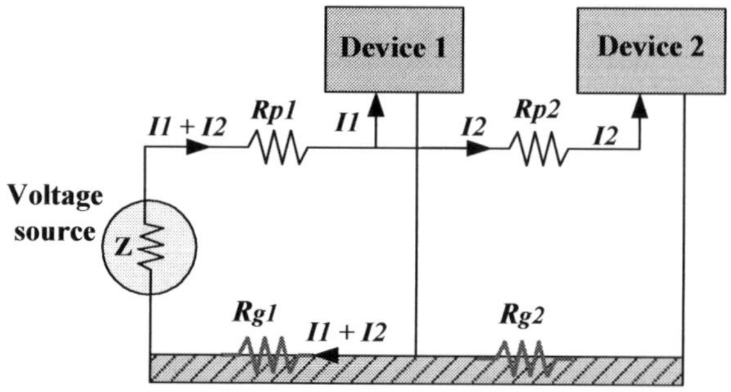

Figure 5.21 Common impedance coupling in a power and return plane configuration.

$$V_{\text{noise}} = (I_1 + I_2)\ (R_{p1} + R_{g1} + Z) \qquad (5.12)$$

If Device #2 consumes more current than Device #1 with output impedance of the source being negligible, we can easily determine total amount of RF noise voltage that is now present across Device #1 (5.13). If Device #1 is a low level analog system and susceptible to disruption, serious concerns may develop.

$$V_{\text{noise}} = I_2\ (R_{p1} + R_{g1}) \qquad (5.13)$$

When trying to minimize common-impedance coupling within any power distribution system, one should take into account the impedance presented by that power distribution network to all logic circuits which may be difficult to determine due to unknown parasitics and lack of parametric data. Depending on the design of the plane structure with regard to location in a multilayer printed circuit board, the supplied voltage and return may be implemented through use of round conductors or flat straps, if the application requires use of this method of providing a return using this configuration. The equations in (5.14) illustrates inductance for various wire configurations. Knowing the inductance of transmission lines helps designers understand why RF noise created from one device causes harmful interference. Table 5.1 details inductance of conductors operating at 1 MHz, which for technology of today is very slow [8].

$$Lo_{(round)} = \frac{\mu_o s}{2\pi} \cdot \left[\ln\left(\frac{4s}{d}\right) - 1 \right] \qquad \textit{round conductor}$$

$$Lo'_{(round)} = \frac{\mu_o s}{2\pi} \cdot \ln\left(\frac{4h}{d}\right) \qquad \textit{round conductor over a plane} \qquad (5.14)$$

$$Lo_{(flat)} = \frac{\mu_o s}{2\pi} \cdot \left[\ln\left(\frac{8s}{w}\right) - 1 \right] \qquad \textit{flat strap}$$

$$Lo'_{(flat)} = \frac{\mu_o s}{2\pi} \cdot \ln\left(\frac{2\pi h}{w}\right) \qquad \textit{flat strap over a plane}$$

where:
s = conductor length (meters)
w = width of the conductor (mm)
h = height above ground plane (cm)
d = diameter of conductor (mm)
L = inductance (Henry)
$\mu o = 4\pi * 10^{-7}$ (H/meter).

Table 5.2 Inductance of various conductors at 1 MHz.

Conductor Type	Width (mm)	Length (m)	Diameter (mm)	Height (cm)	Inductance (μH)	Reactance (Ω)
Round	–	1	1	–	1.7	11
Round above a plane	–	1	1	1	0.7	4
Flat strap	10	1	–	–	1.3	8
Flat strap above a plane	10	1	–	1	0.37	2

Inductance increases linearly with length and decreases when width becomes greater. It also becomes important to keep the length of the conductor as short (and as wide) as possible for lowest inductance. Also, the wider the physical width of the transmission line, the lower the inductance. This is the reason why planes have extremely low to minimal inductance because they are considered to be infinitely large with respect to EMC analysis.

The best way to minimize common-impedance coupling within a power distribution system is to provide separate power and return to appropriate switching devices. This works well with single- and double-sided printed circuit board assemblies. With power and return planes in a multilayer assembly, common-impedance coupling is minimized due to the much lower impedance of the power distribution system.

5.11 Breaking Ground Loops

Ground loops are a primary source of RF noise creation. RF noise is produced when the physical distance between multi-point locations are significant (>1/20 of a wavelength), and connection is made to a reference, usually at AC or chassis potential. In addition, low-level analog circuits also create RF loops. When a loop occurs, it becomes necessary to isolate or prevent RF energy transference from one circuit corrupting another. A loop consists of a signal path and part return/ground plus a parasitic path.

Figure 5.22 illustrates a ground loop within a printed circuit board mounted in a metallic chassis, where Vn represents common-mode return loss. Icm represents shunt current within the system based on V_n, created by the impedance within the RF return path (Ohm's law). Two separate grounds are provided to the same reference, one for each circuit. A difference in ground reference is present between circuits due to finite inductance in the transmission line. Unwanted noise from one circuit may thus be injected into the other as a result of this interconnect. The magnitude of ground-noise voltage, compared to the signal level in the circuit is of prime importance. If the signal-to-noise margin is affected, design techniques must be implemented to ensure optimal circuit functionality that includes decreasing total inductance in the transmission line. All components must have a solid connection to determine where 0V is located such that the voltage-level transition is appropriate for the logic family in use. For components or circuits used in a specific application that must be isolated from chassis ground, no ground loop can occur.

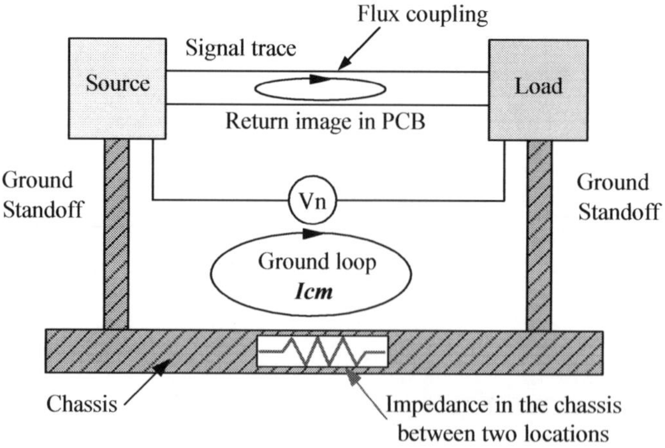

Figure 5.22 Ground loop between two circuits.

How does one avoid ground loops when a difference in 0V reference must exist? Two primary design techniques are available during the design and layout of a printed circuit board related to breaking ground loops present. Ground loops are more of a concern at lower frequencies.

- Remove one of the ground connections
- Isolate circuits using any of the following techniques:
  - Transformer isolation
  - Optic isolation
  - Common-mode choke isolation
  - Balanced circuitry

5.11.1 Transformer isolation

Figure 5.23 illustrates Fig. 5.22, electrically. When using a transformer, ground-noise voltage will be observed at only one side of the transformer. Any noise coupling that does occurs between input and output is a result of parasitic capacitance between the windings. To prevent parasitic capacitance from coupling noise from one side to the other, use of a Faraday shield built into the transformer between primary and secondary breaks up this loop, with the center tap connected to the main AC reference point or chassis ground. One disadvantage of using a transformer with a Faraday shield is physical size; an increase in printed circuit board real estate usage and additional cost. In addition, if multiple signals are to travel between isolated areas, a transformer is required for each signal line.

Advantages

- Ideal for low frequency circuits (audio, AC power), but can be used up to 1 MHz (e.g. RS-422 data bus)

Disadvantages

- Large, expensive and frequency limited (due to parasitic capacitance). The transformer may require electrostatic shielding
- Cannot be used in DC power circuits

Transformer Isolation

Common-mode noise *Vn* appears between the windings of the transformer

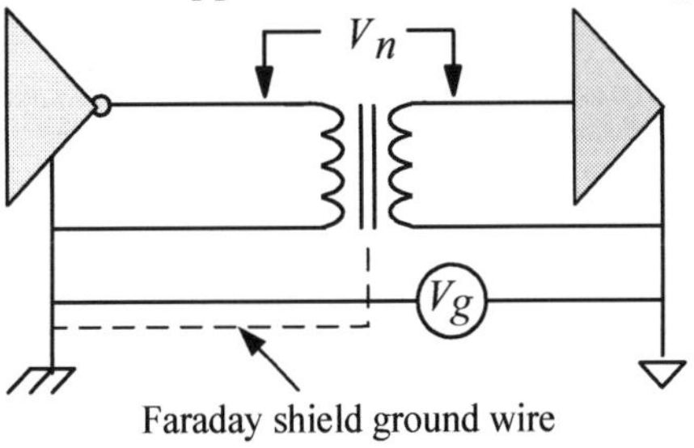

Common-mode rejection of 100-140 dB can be achieved at f = 1 kHz

Figure 5.23 Isolating a ground loop using a transformer.

5.11.2 Optical isolation

Optical isolation (Fig. 5.24) is another technique used to prevent ground loops and to minimize I_{cm}. An optical isolator breaks the transmission path completely. A continuous metallic connection thus cannot occur between circuits. A metallic connection is usually required for propagation of an electromagnetic field on a printed circuit board, but could be a wireless signal upon which optical isolation is not required. Isolators are best suited when a large voltage reference potential is present between circuits that cannot be removed by other means. Ground-noise voltage thus only appears across the input of the optical transmitter, and is best suited for digital logic designs owing to the nonlinearity of the isolator when used with analog circuitry.

Advantages

- Ideal for data/digital circuits, even at relatively high potential differences
- Small and inexpensive

Disadvantages

- Relatively large parasitic capacitance (~2 pF), which limits higher frequency performance as signal propagation and parametric requirements will be affected, such as edge rate rounding as well as increased time in signal propagation
- Used primarily in digital circuit due to non-linearity
- Cannot be used on power supply lines (AC or DC)

Optical Isolation

Common-mode noise appears between the photo-diode and photo-transistor which is removed during signal propagation

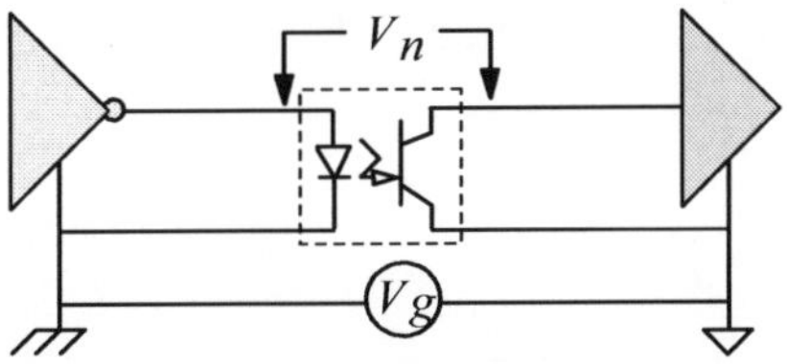

Common-mode rejection of 60-80 dB can be achieved

Figure 5.24 Isolating a ground loop using optical isolation.

5.11.3 Common-mode choke isolation

Use of common-mode chokes (Fig. 5.25) is another technique to break up RF loop currents between circuits. The advantage of using common-mode chokes is, of course, to remove undesired common-mode energy on balanced differential pair transmission lines. A common-mode choke passes the desired DC level of the differential-pair signal while attenuating higher-frequency AC components that may also be present within the transmission line. A common-mode choke has no effect on the differential-mode intelligence which is what we want to propagate, not residual common-mode energy. Multiple windings may be wrapped around the same core structure, increasing impedance but over a narrower bandwidth of frequencies.

Common-mode chokes are generally used when differential-pair signals propagate across a boundary such as into I/O interconnects. If one transmission line is imbalanced with respect to the other as differential-pair signals go off the board, a small amount of common-mode RF current may be established. Common-mode chokes removes this undesired current.

Advantages

- Does not affect differential-mode signal transmission propagation
- Generally found in I/O circuits for cables
- Versatile in frequency response
- Available in almost every type of configuration and for almost every possible application

Disadvantages

- Can be physically large and bulky
- Takes up real estate on a printed circuit board and adds cost
- May not have enough insertion loss to remove all common-mode noise present

Common-Mode Choke Isolation

Common-mode noise present on the signal and return lines is Removed, permitting only differential-mode signal propagation.

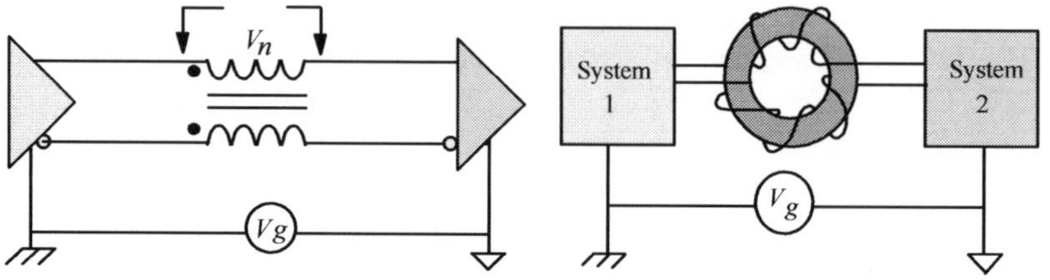

Common-mode rejection exceeding 80-100 dB can be achieved at high frequencies.

Figure 5.25 Isolating a ground loop using common-mode chokes.

5.11.4 Balanced circuitry

Balanced circuits use differential pair transmission lines to send a signal between circuits. By using differential transmission paths, drive current in both lines are supposed to be equal (Fig. 5.26). The primary advantage of using differential signaling is when transmission lines travel between two systems or assemblies that are at different 0V potential without a common 0V reference. If there is no possible means for a 0V reference to be established across the finite impedance between circuits, development of common-mode EMI is prevented. All networking and telecommunication circuits that are routed external to systems or assemblies use twisted pair wire with no ground or 0V reference. An example of balance circuitry is Ethernet and USB.

Twisting wires remove common-mode current through the process of flux cancellation, since signals are supposed to be equal and opposite. An example of differential-mode cabling is Ethernet, where wire lengths can be very long, physically with no common-mode energy present unlike USB with shorter cable lengths.

Differential-mode signaling requires components to have a high level of common-mode rejection. This means they should not have any undesired common-mode currents established within the transmission line. The input circuitry of the device should be able to remove common-mode noise, thus ensuring that only differential-mode signaling is proper. Many component manufacturers provide a Common-Mode-Rejection-Ratio (CMRR) number in their data sheets. CMRR is defined as the ratio of:

$$\frac{\textit{common - mode voltage, } V_{cm}\textit{, applied to both inputs required to generate output voltage, } V_o}{\textit{differential voltage, } V_{dm}\textit{, applied between the inputs to generate output } V_o\textit{.}}$$

When using balanced circuit isolation, the center tap of the isolation transformer if one is provided, is generally connected to chassis ground through a capacitor to ensure there is no DC voltage short circuit. If one side of a differential line is out of balance, phase-wise or due to a loss within the transmission line routing, common-mode current present will be shunted to ground, ensuring that the secondary has only a true differential-mode signal.

Advantages

- Rejects only common-mode (ground loop) noise with little effect on functional differential signals (signal-mode sensitive filter)
- Can be used in several circuits simultaneously
- Versatile in frequency response

Disadvantages

- May be large in size and frequency limited (due to parasitic capacitance) which can cause functionality problem at higher frequencies, requiring a second wire.

Balanced Circuit Isolation

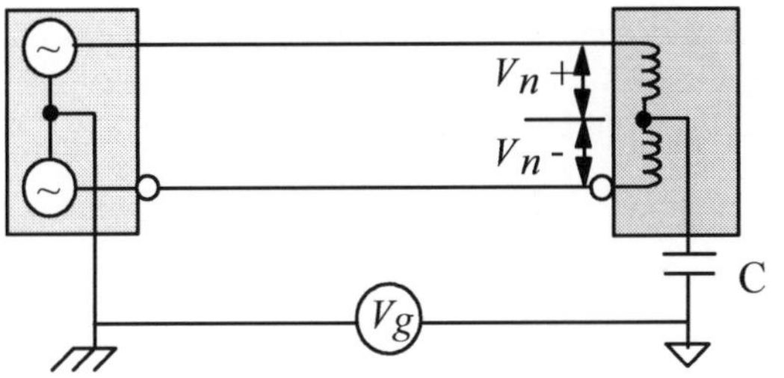

Figure 5.26 Isolating a ground loop using balanced circuit isolation.

CMRR identifies how much common-mode noise will be rejected from entering the input of a receiver. The higher the balance between the differential pairs, the greater the amount of undesired common-mode RF current is rejected (Fig. 5.27). At higher frequencies, achieving a large CMRR value may be difficult due to the complexities of controlling all the parasitic and distributed impedances that may be present.

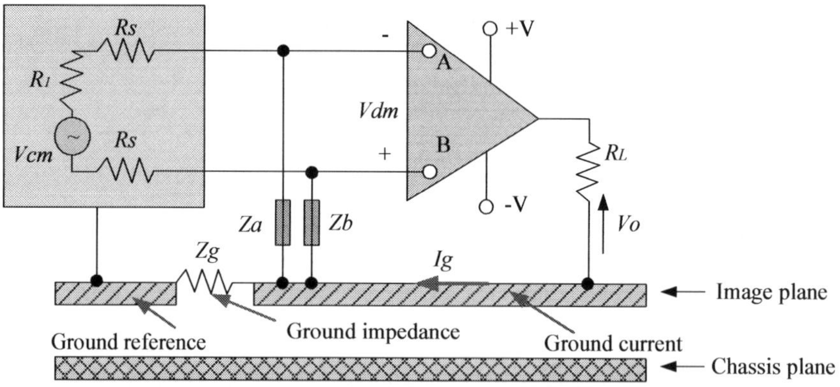

Figure 5.27 Circuit representing common-mode-rejection-ratio.

Common-Mode-Rejection-Ratio (CMMR) is mathematically defined as (5.15).

$$CMRR = 20\log\left|\frac{V_{cm}}{V_{dm}}\right| \quad dB \quad (V_0 = constant) \tag{5.15}$$

Using the circuit of Fig. 5.25, we calculate CMMR as (5.16).

$$CMMR = 20\log\left|\frac{R_1(Z_b - Z_a)}{(Z_a + R_1)(Z_b + R_1)}\right| \quad dB \tag{5.16}$$

In Fig. 5.27, observe the location of both image and chassis plane. The differential mode transmission line system is referenced from the 0V-image plane, not chassis plane. Any I_{cm} developed between source and load must flow in the 0V (image) reference. Termination resistors, Z_a and Z_b, must be chosen with a tight tolerance to assure impedance matching between the two transmission lines is perfect. If an impedance imbalance is present, I_{cm} is increased. Generally, termination resistors are incorporated into the package of digital devices, input side only.

When using (5.16), the tolerance rating of the resistors is the critical parameter. Series resistors (R_s) are provided to match the transmission line impedance to ensure functionality of the circuit, commonly called series termination.

When dealing with differential-mode circuits to minimize I_{cm}:

1. The impedance control of the signal lies only in the image plane.
2. Signal flux is bound to the internal image plane, not chassis plane. The chassis plane is too far away, physically, to be of any significant value for flux cancellation.
3. The chassis plane only shunts common-mode loss occurring within the image plane.

5.12 Resonances When Using Multi-Point Grounding

A common architectural concept for many designs is to have a printed circuit board secured to a metal mounting plate or chassis. A resonance may be inadvertently developed between the power and/or return planes and chassis housing depending on layer stackup and routing topology. This is due to parasitic capacitance between an element at potential and a second element at 0V or ground. Figure 5.28 shows both capacitance and inductance present in a printed circuit board that is screw secured to a mounting panel or metal enclosure/chassis. Depending on physical distance spacing between mounting posts, which is at a fixed dimension that corresponds to one element of a dipole antenna, along with both inductance and capacitance of the assembly resonant at a particular frequency, undesired RF loop currents may be generated. These loop currents are call eddy currents and propagate on the inside of the metal assembly. When eddy currents are present, these currents may couple either by radiated or conducted means to other circuits nearby, the chassis housing, internal cables or harnesses, peripheral devices, I/O or even into free space.

Problems that arise in printed circuit boards using multi-point grounding may have resonances that occur between all ground stitch locations and the AC reference or chassis plane. While the AC reference or chassis plane may be at 0V potential, this AC-based reference may be completely different from the DC 0V reference required for digital or analog circuitry. This difference between reference levels is more apparent when higher frequency, faster edge rate signals are present.

Depending on physical distance spacing between ground stitch locations, a resonance condition will develop, depending on spectral excitation. This resonance is present because there is parasitic capacitance and interconnect between the power and return planes, in addition to capacitance and inductance induced by the mounting ground stitch standoff mounting post shown in Fig. 5.28.

Figure 5.28 illustrates a printed circuit board's image plane secured to a metal mounting plate with both capacitance and inductance shown. Capacitance is between the power and return planes in addition to parasitic capacitance relative to the external metal structure, either physically above or below the board mounted in a metal enclosure. The planes themselves have finite impedance between ground stitch locations. We can easily determine the self-resonant frequency of the power and return planes through simulation or measurement.

Use of a network analyzer will provide a quick way of determining the *actual* self-resonant frequency between ground stitch locations. Multiple measurements will be required since the self-resonant frequency of the printed circuit board is dependent on the inductance of the planes.

When a low-impedance 0V reference plane is provided in a multilayer printed circuit board, or a chassis ground stitch connection provided between 0V reference and metal chassis it becomes important, like single-point grounding, that any bond connection be kept as short as possible to minimize development of common-mode current across an inductive interface. In most applications when mounting a printed circuit board in a metal enclosure, connecting the reference plane to chassis ground is always through use of the angular ring of the through-hole screw connection on the bottom side of the printed circuit board assembly. The angular ring is connected by multiple thermal connections (spokes) to the internal reference plane, or 360 degree connection in certain applications.

This is the best means of having a low-impedance bond connection between the 0V reference plane of the printed circuit board and chassis ground. When this occurs, we essentially have a single-point grounded system at chassis potential, which then is usually connected to the third wire AC mains ground terminal, if one is provided.

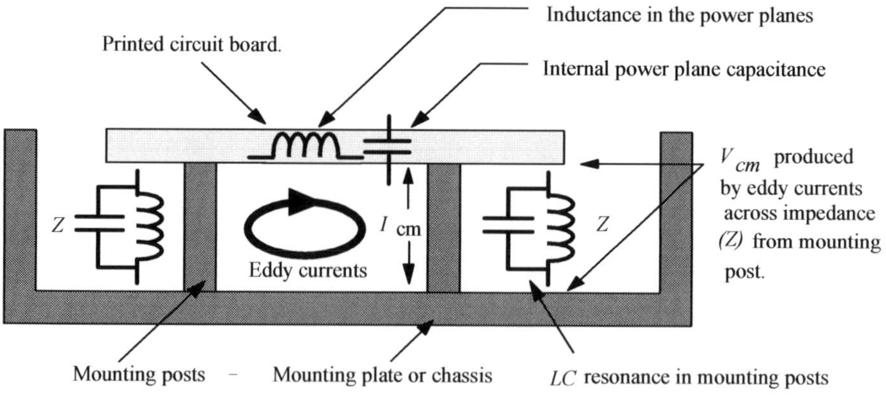

APPLICATION MODEL OF MULTIPOINT GROUNDING

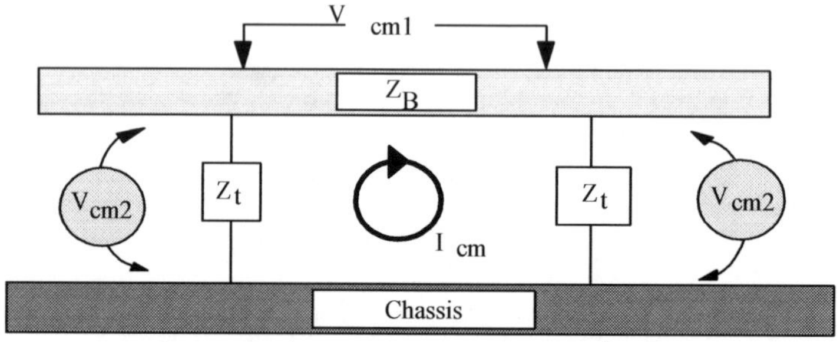

V_{cm2} is reduced by the mounting posts (ground stitch locations).
Resonance is thus controlled, along with enhanced RF suppression.

ELECTROMAGNETIC MODEL OF MULTI-POINT GROUNDING

Figure 5.28 Resonance in a multi-point ground to chassis.

Since the printed circuit board's power and return planes are resonant at various frequencies, the same analysis is applied to the metallic enclosure used to secure the board. This metallic enclosure may be a chassis for a motherboard, a mounting plate used in a card cage assembly with a backplane, a shield partition between two boards to prevent radiated noise coupling between the two, or other applications not identified in this short list. Again, inductance will exist between in mounting standoffs. Now that we have identified inductance, what about parasitic capacitance?

Because there is a finite distance between the printed circuit board and the metallic structure there is a transfer impedance. For example, the printed circuit board can be considered as the positive plate of a capacitor at voltage potential, and the metal enclosure the negative plate, with air being the dielectric media between the two plates.

In addition to overall inductance of the metal enclosure which is extremely small, and parasitic capacitance between the printed circuit board and mounting plate using standoffs to secure the board to the chassis (generally pem studs press-fit into the metal), inductive behavior under certain conditions of use is minimal. If any inductance is present, common-mode currents can be established and propagated as eddy current which will find a means to radiate EMI through a localized antenna structure.

The explanation of why standoffs are inductive does not have to do with the physical construction of the standoff, metallic wise, but with the screw used "if and only if energized with RF current." Screws contain inductance that may be several orders of magnitude greater than inductance inherent within the printed circuit board, or total parasitic inductance of the overall assembly, not related to the screw secured mounting.

The reason why screws are inductive is best illustrated in Fig. 5.29. It is impossible to model screw inductance using SPICE or a field solver to acquire 100% accurate results because of a large number of parameters that cause this inductance to exist, and which cannot be identified or quantified for computational analysis. Some of these parameters include material composition, the number of threads that makes actual physical contact with the standoff in intermittent locations, thread spacing, pitch of the threads, plating material provided for the screw and standoff, compression strength, and length of the screw from top to bottom that actual touches the standoff through its threads.

Screws contain a helical thread. Only the top portion of the screw thread, and then only a small portion of the thread, namely 10-20%, actually makes physical contact with the standoff. We cannot guarantee that all threads will make 100% solid bonding contact with the standoff's sidewall in a 360 degree manner. Tolerances in manufacturing of both standoff and screws, depending on number of dendrites present in the metal, means only incidental contact will occur.

The standoff must be physically larger in diameter than the screw diameter to allow the screw to be inserted. As a result we will always have incidental contact between *some* of the threads, *not* the entire length of the screw. This is observed in Fig. 5.29.

A helical thread functions the same function as a helical antenna, but only if energized with RF current. A voltage potential is thus developed between the bottom and top of the screw. This voltage potential difference exacerbates creation of RF current.

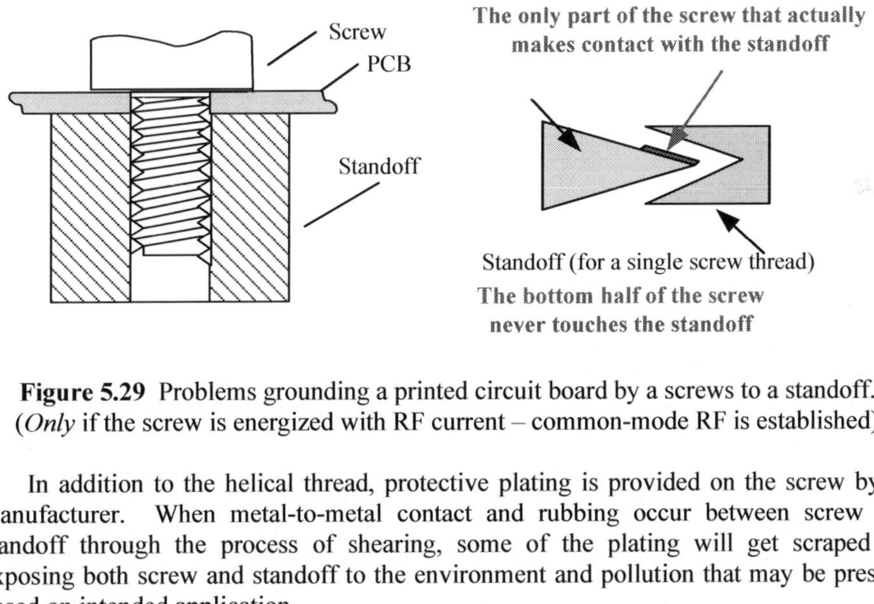

Figure 5.29 Problems grounding a printed circuit board by a screws to a standoff. (*Only* if the screw is energized with RF current – common-mode RF is established)

In addition to the helical thread, protective plating is provided on the screw by its manufacturer. When metal-to-metal contact and rubbing occur between screw and standoff through the process of shearing, some of the plating will get scraped off exposing both screw and standoff to the environment and pollution that may be present, based on intended application.

Galvanic corrosion may occur making the screw nonconductive (an insulator) in extreme conditions. If the intended application is to have a low-impedance common-mode ground path, one cannot exist if there is corrosion. As such, a screw must only be used *only for compression purposes* between board and enclosure. Screws must not be relied upon to transfer RF currents to chassis ground. This type of connection is illustrated in Fig. 5.30 showing both correct and incorrect means of screw securing a printed circuit board. This incorrect method occurs because the designer does not understand loop current and transfer impedance, and will ground the 0V plane(s) to chassis by a highly inductive trace through the screw on the top side of the board with a 0-ohm resistor or ferrite bead. Mounting pads for surface mount components are generally provided so one can experiment with grounding the 0V reference plane with these devices. This type of grounding reference planes to chassis ground guarantees development of EMI! In fact, if this implementation is used and the printed circuit board fails radiated emissions testing, remove the screw. It could be the screw that is radiating, nothing else!

To properly shunt RF current to chassis ground in an efficient manner which is low inductive, connect the 0V plane(s) to the bottom angular ring of the through-hole via and let the angular ring sit on top of the standoff, with the screw providing only compression.

Large mounting pads provided on the bottom of the printed circuit board that overlaps the standoff walls helps make the desired low-impedance ground connection, not the actual screw. The standoff is usually installed in the chassis with a good bond connection by the sheet metal fabricator. Thus, if a low-impedance path to ground is required for the printed circuit board, this is done through the walls of the standoff, not screw threads.

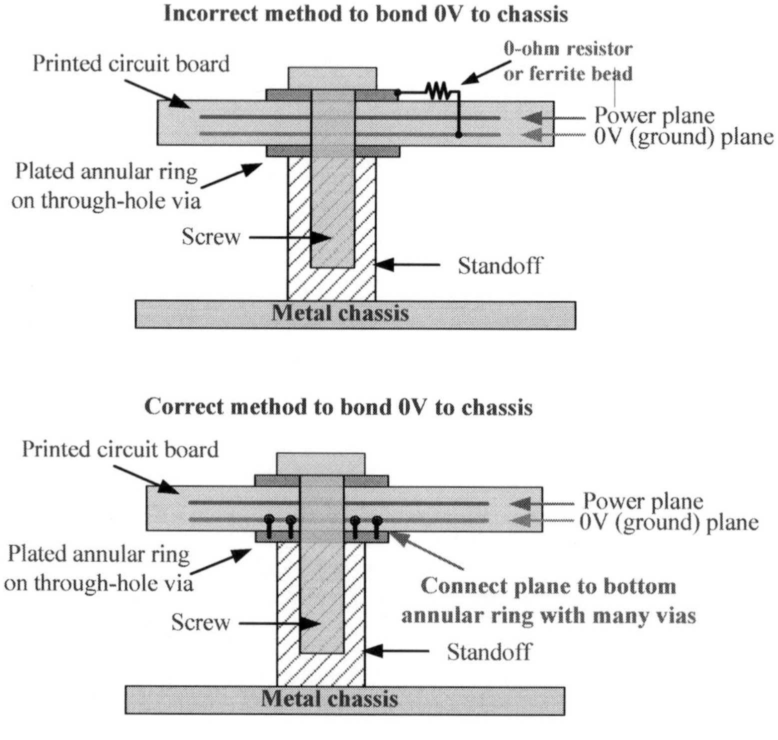

Figure 5.30 Mounting a printed circuit board to an enclosure using screw and standoff.

5.13 Signal and Ground Loops (Not Eddy Current Loops)

Signal and ground loops is a major contributor to the propagation of undesired RF energy. RF current will return to its source through any path or medium possible: components, wire harnesses, ground planes, adjacent traces and free space. This current is created between source and load due to inductance in transmission lines when there is a voltage potential difference between elements. Path inductance causes magnetic coupling of RF currents to occur, thus increasing RF losses.

One important design considerations for EMI suppression on a printed circuit board is minimizing or preventing undesired signal return currents propagating in an unexpected loop, and there many in every printed circuit board layout. Analysis must be made for each and every ground stitch connection (mechanical securement between the printed circuit board and chassis ground) relative to RF currents generated from noisy electrical components. Higher speed logic devices must always be located as close as possible to a ground stitch to minimize formation of loops in the form of eddy currents to chassis ground, and to break up the dipole antenna structure present within the 0V plane. This design requirement is now examined.

An example of multiple loops that could occur, for example a computer with PCI adapter cards and single-point grounding, or multiple connections to a single point reference (a.k.a. multi-point) is shown in Fig. 5.31. As observed, multiple signal-return loop areas exist. Each loop creates a distinct electromagnetic field spectra. RF currents from each loop will create an electromagnetic radiated field at a unique frequency depending on the physical size of the loop antenna structure. Containment measures such as a metal enclosure must now be used to keep these propagating currents from coupling to other circuits or radiating to the external environment as EMI. Internally generated RF loop currents are thus to be avoided.

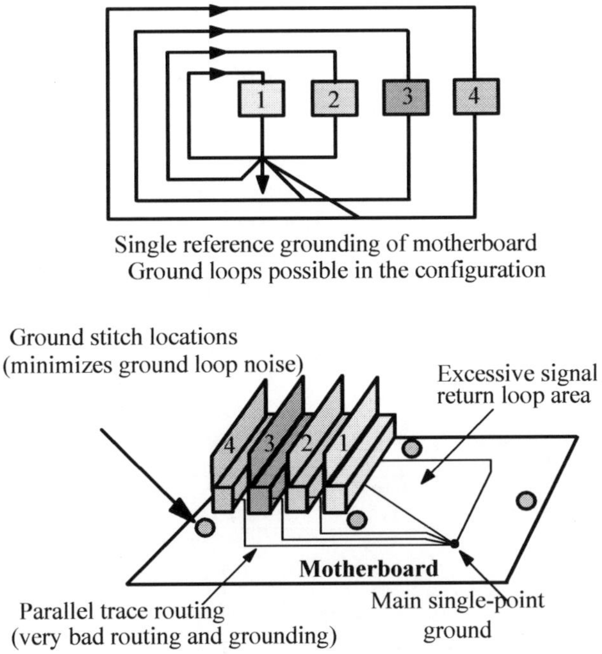

Figure 5.31 Ground loops within a printed circuit board assembly.

To expand on the concept of loop area shown in Fig. 5.31, we have Fig. 5.32 which shows the loop area between two components. Depending on circuit configuration, this may be acceptable if operation is low frequency. If there are higher frequency components and the return path is a wire or transmission line with any inductance, common-current will be developed. If the return path is a plane physically located next to signal routing layer in a multilayer stackup assembly, this is highly desired as there is now minimal inductance in the transmission line return path..

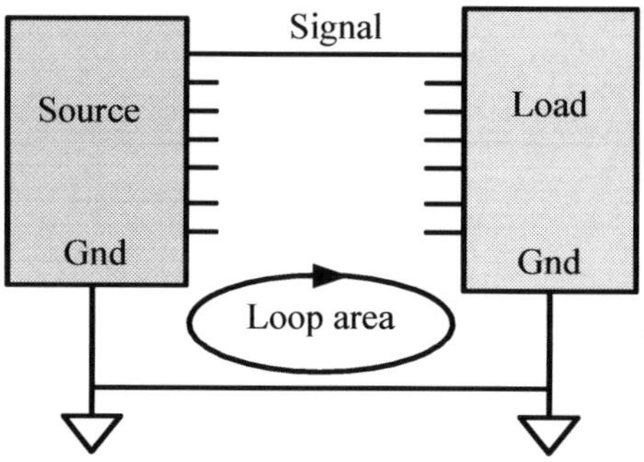

Figure 5.32 Loop area between components.

REFERENCES

1. Montrose, M. I. 1999. *Printed Circuit Board Design Techniques for EMC Compliance-A Handbook for Designers*. Hoboken, NJ: John Wiley & Sons/IEEE Press.
2. Montrose, M. I. 2000. *EMC and the Printed Circuit Board-Design, Theory and Layout Made Simple*. Hoboken, NJ: John Wiley & Sons/IEEE Press.
3. Joffe, E. & Lock, K. S. 2010. *Grounds for Grounding-A Circuit-to-System Handbook*. Hoboken, NJ: John Wiley & Sons/IEEE Press.
4. W. Michael, United States Patent #4,145,674.
5. Paul, C. R. 2006. *Introduction to Electromagnetic Compatibility*. 2nd ed. Hoboken, NJ: John Wiley & Sons.
6. Ott, H. 2009. *Electromagnetic Compatibility Engineering*. Hoboken, NJ: John Wiley & Sons.
7. Bogatin, E. 2010. *Signal and Power Integrity Simplified*, 2nd ed. Upper Saddle River, NJ: Prentice Hall.
8. Coombs, C. F. 1996. *Printed Circuits Handbook*. New York: McGraw-Hill.
9. Hartal, O. 1994. *Electromagnetic Compatibility by Design*. W. Conshohocken, PA: R&B Enterprises.

6 Shielding, Gasketing and Filtering Made Simple

6.1 The Need to Shield

Electromagnetic shielding reduces RF field amplitudes propagating in free space by blocking fields with a barrier made of conductive and/or magnetic materials. Shielding is typically applied to enclosures and cable shields to isolate electrical circuits, systems and transmission lines from either propagating undesired RF energy into the environment and conversely, keeping unwanted externally induced EMI from causing operational perturbations.

Shielding reduces coupling of electromagnetic and electrostatic fields between two elements. A conductive enclosure that is used to prevent field transfer is commonly called a Faraday cage. The amount of field reduction depends upon material used, thickness, size of the shielded volume, the frequency of the RF energy, and impedance of the electromagnetic field impinging on the shield with respect to the shield impedance. Also of interest are size, shape and orientation of apertures in a shield to an incident electromagnetic field.

Shielding and filtering are complementary. Filtering is discussed later in this Chapter. The shield can be an all-metal enclosure if protection down to low frequencies is not needed. For higher frequency protection (>30 MHz), almost any thin metallic material will be adequate. Shield performance can become negligible if its integrity is breached with excessive slots, openings or cable penetrations under certain conditions such as size and number and their relationship to where both desired and undesired RF energy exist.

When designing a system, a choice must be made between using a plastic or metal enclosure. For many products this decision is known ahead of time, allowing engineers to implement efficient methods of EMC suppression. For low cost consumer products, these are generally plastic due to cost and lighter weight. When designing a system, a decision must be made to shield or not to shield, although the severity of the requirement specifications may force the decision. The following gives guidance.

- If predicted fields will exceed limits through imbalance and common-mode conversion, shielding is essential.
- If the printed circuit board layout requires dispersed interfaces, shielding will probably be essential.
- If the printed circuit board layout allows for concentrated interfaces, a ground plate may be adequate that acts as an RF return (common-mode image) path and flux cancellation.
- Consider shielding only critical circuitry with a localized cover, such as the receiver section of a telecommunication device located adjacent to the transmitting section to prevent crosstalk.

To summarize, at a high level:

- Shields are intended to control or minimize the propagation of electromagnetic fields between two regions internally among circuits or circuit board segments, or externally to the environment.
- Shielding is the extent of attenuation provided by a metallic surface or structure between two regions in space.

6.2 Basic Shielding Equations

Shielding effectiveness (*SE*) is defined as the ratio of impinging energy to residual energy, or the part that propagates through the shield. The basic equation describing *SE* is (6.1).

$$SE(dB) = SR(dB) + SA(dB) + SB(dB) \tag{6.1}$$

where:
SE(dB) = total contribution from reflection at each surface
SR(dB) = reflection loss
SA(dB) = absorption loss (function of frequency and material)
SB(dB) = loss due to multiple reflections (usually negligible and very small).

Equation (6.1) appears to be simple, which it is. Equation (6.2) details the real variables, assuming we need to actually calculate all factors during simulation analysis, and not just understand basic shielding theory.

$$SE(dB) = 20\ log\ (e^{\gamma\text{-}l}) + 20\ log\ \left(1 - \Gamma \cdot e^{2\gamma - l}\right) + 20\ log \left(\frac{1}{\tau}\right) \tag{6.2}$$

where:
γ = propagation coefficient (= *SR*)
Γ = reflection coefficient (= *SA*)
τ = transmission coefficient (= SB)
l = thickness of the shield barrier.

Shielding effectiveness is a dimensionless unit of measurement expressed in *dB*. The unit value of both electric and magnetic fields are removed and converted to a logarithmically expressed ratio. When dealing with logarithmic mathematics in *dBs*, we are concerned with "*power*" and its derivatives of voltage and current. Electric and magnetic fields are voltage and current elements. We need both time variant voltage and current for electromagnetic field propagation. Therefore when dealing with logarithms, we multiply measured units by 10 to acquire units of power, whereas when analyzing electric and magnetic fields, the factor of 20 is used.

Equation (6.3) describe electric and magnetic fields with regard to shielding effectiveness, both voltage and current. Voltage times Current equals Power (*P=VI*).

$$\text{SE(dB)} = 20 \log \frac{\text{E (without shield)}}{\text{E (with shield)}} = \frac{E_0}{E_1} \quad \text{(for E fields)}$$

$$\text{SE(dB)} = 20 \log \frac{\text{H (without shield)}}{\text{H (with shield)}} = \frac{H_0}{H_1} \quad \text{(for H fields)} \tag{6.3}$$

To help illustrate how shielding affects signal propagation, refer to Fig. 6.1. The "without shield" could be a printed circuit board installed in a *plastic enclosure*, whereas "with shield" is a *metal barrier*. As observed, significant field reduction occurs if the right type of metallic material that reduces signal propagation is used to create a shield partition, or Faraday barrier.

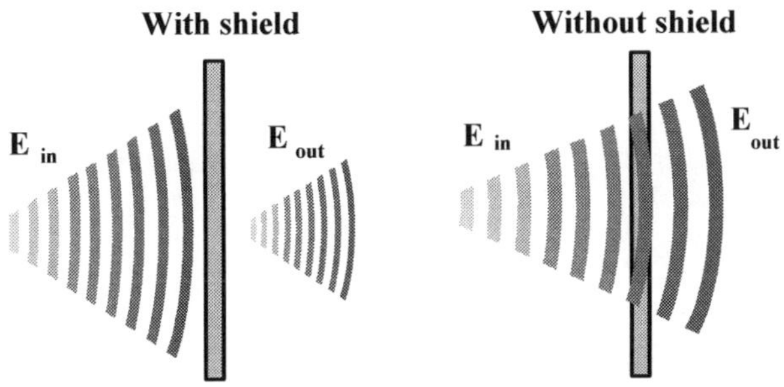

Figure 6.1 Shielding effectiveness example.

If shields were perfect barriers total power (*P*) on the secondary, or outer portion of the barrier, would approach zero. In most applications, barriers have discontinuities such as slots and openings for ventilation, cables and user interfaces. In practical applications, all shields are merely an attenuator of RF field intensity with the majority of field reduction based on absorption and reflections characteristics of the material.

6.3 Theory of Shielding Effectiveness – Made Simple

To explain how shielding works in simplified form, we convert a shield partition to a transmission line model, identical to that used for signal propagation on a printed circuit board. It is easy to visualize a printed circuit board trace and its relationship to sending an electromagnetic field between location. We also know that the wider the trace width and/or the smaller the dielectric separation between trace and reference plane, the lower the impedance. Conversely, when a trace has a smaller width or the distance spacing increases from a reference plane, impedance increases.

How does transmission line theory relate to shielding theory? In Fig. 6.2, a shield barrier is shown with an incident or reflected wave on one side, the left. Any residual electromagnetic field that propagates internal to the barrier (providing the thickness is very small) may allow an attenuated transmitted signal to exit on the opposite side, but at a much lower amplitude of power. This attenuated transmitted wave is generally low enough in amplitude to not be a concern related to emission and immunity protection.

There are two primary aspects of shielding that attenuates signals, absorption and reflection. In Fig. 6.2, an electromagnetic field hits the first boundary condition and is reflected back. This reflected wave loses very little energy, thus both incident and reflected waves still exist inside the assembly. If the reflected wave phase adds with the incident wave or other waves present, total RF intensity internal to the system could be increased.

On the bottom half of Fig. 6.2 there is a transmission line representation similar to a printed circuit board trace. The transmission line impedance on the left side is at a fixed impedance (free space is 377 Ohms). When the transmission line becomes smaller in width its impedance increases (Z_m >> 377 Ohms). With an increase in transmission line impedance there is a reduction of the electromagnetic field, per Ohm's law. On the right side of the shield partition we return to a normal transmission line structure (377 Ohms). The loss that occurs due to Z_m is known as absorption loss.

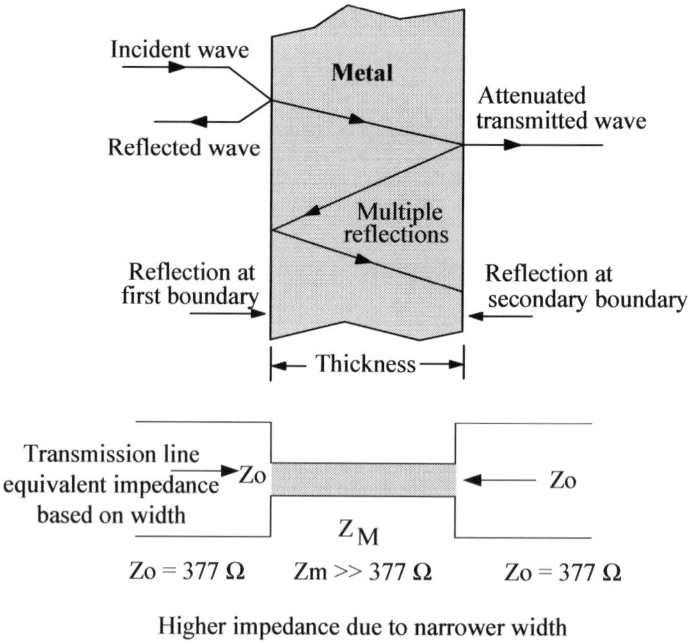

Figure 6.2 Shielding theory illustrated as a transmission line.

If we visualize a shield as a transmission line in simplified form, the concept of reflection and absorption loss becomes obvious.

6.3.1 Technical Explanation-Shielding Theory

Shielding theory is based on two fundamental mechanisms, reflection and absorption loss. Figure 6.1 showed these mechanisms by placing a metallic barrier in the path of an electromagnetic wave propagating with orthogonal ***E*** and ***H*** vectors. The ***E***- and ***H***-field components are related by Maxwell's equations, and when combined together in the far-field they create a plane wave. This mathematical relationship is described by (6.4).

$$Z = E\ (V/m)\ /\ H\ (A/m) \tag{6.4}$$

In free space, wave impedance is calculated per (6.5).

$$Z_o = E/H = \sqrt{\mu_o / \varepsilon_o} = \sqrt{\frac{4\pi 10^{-7}\ H/m}{\frac{1}{36\pi}(10^{-9})\ F/m}} = 120\pi \text{ or } 377 \text{ ohms (exactly } 376.99 \text{ ohms)} \tag{6.5}$$

A metallic barrier however has much lower impedance (6.6):

$$Zs = \sqrt{\omega\mu / \sigma} \quad ohms \tag{6.6}$$

where:

ω = $2\pi f\ (Hz)$
μ = $\mu_r \mu_0$ (permeability)
σ = conductivity [1/Ωm or "mhos/m"].

Equation (6.6) implies an impedance mismatch is present for the electromagnetic field, which causes reflections at the boundary. The remaining field that is not reflected at the barrier is either absorbed internally or continues to pass through, depending on thickness of the metal barrier. The electric-field is predominantly reflected when it encounters material with low impedance. The magnetic-field component becomes reflected when it enters a material containing a higher impedance. Between the two edges of the barrier surfaces, there can be multiple reflections, which may be overlooked if the absorption loss in the barrier is at least 10 dB or greater.

6.3.2 Shielding Effects

We now analyze how a Faraday's shield (in reality a Gaussian structure) affects field propagation. Without having to resort to extensive field theory (i.e., *EMC Made Simple*©), we take the example of a thin, conductive spherical device and place it in an electric field (Fig. 6.3). According to Gauss, the field inside the sphere will be zero everywhere on the shield. A metallic sphere of high conductivity will eliminate any internal electrical field because induced charges on one side tend to generate a second electric field that cancels the original one on the other side.

The reason why this occurs is not because the sphere has absorbed the field, but because the ***E***-field present caused electrical charges of opposite polarity along the boundary of the sphere to be created (Gauss's Law) on the other side. Within a hollow conductor, when there is an electric field present, charge is redistributed very rapidly over the surface so that the field lines end on surface charges. These electrical charges generate an electric field that tends to cancel the original field located inside the sphere. Because electrons can move easily along any conductive surface, the thickness of the boundary condition is not in itself a significant factor.

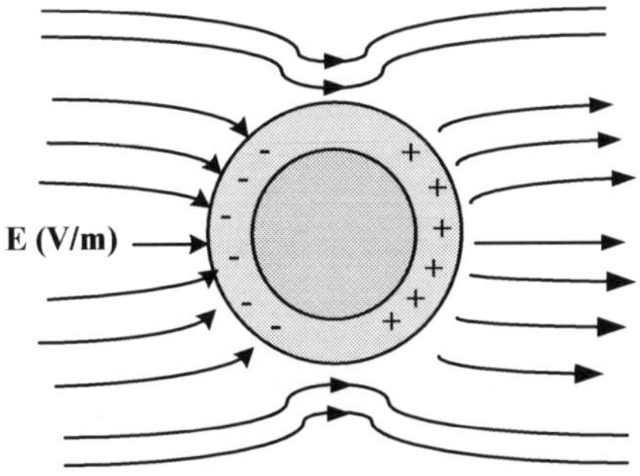

Figure 6.3 Electric field effects on a boundary condition.

There is no field within the conducting walls and hence, no field inside the hollow space. Charges inside the hollow conductor can be separated to give an electric field. No external electric field can influence the electrical conditions inside. The inside of the conductor is thus shielded from outside influences.

Magnetic-field attenuation can however be achieved by means of a boundary condition made of magnetic material combining high permeability (μ>>1) with sufficient thickness to attract the material's magnetic field by providing a low-reluctance path (Fig. 6.4). Alternatively, a thin shield made of a conductive material with low permeability can provide some level of effective shielding for magnetic fields assuming there is adequate skin depth presented by the thickness, frequency and conductivity of the material (Fig. 6.5). This is due to the fact that an alternating magnetic field will induce so-called eddy currents in the barrier, assuming the shield has adequate conductivity.

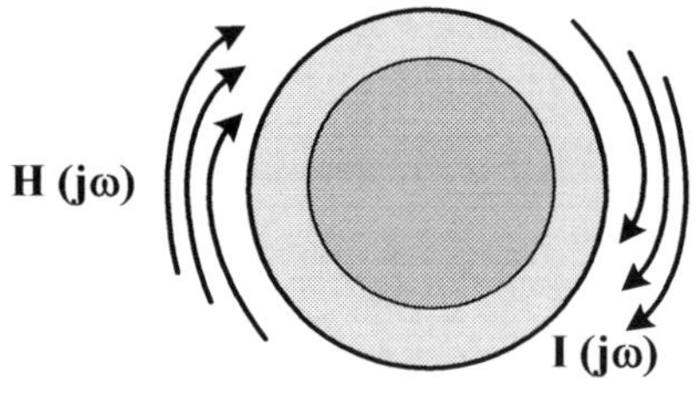

Figure 6.4 Magnetic field effects on a boundary condition.

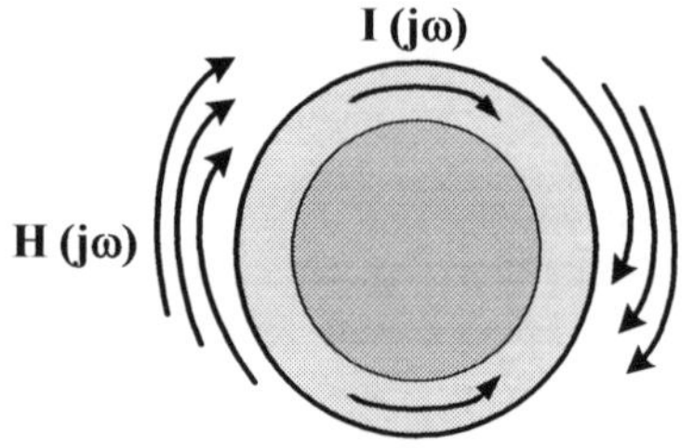

Figure 6.5 Magnetic field producing eddy currents within a boundary condition.

Eddy currents will themselves create an alternating magnetic field of equal and opposite orientation inside the sphere. This effect will increase as frequency increases, resulting in high shielding effectiveness at higher frequencies.

Lower-frequency magnetic fields are more difficult to shield or prevent field propagation through. Whereas absorption typically calls for the installation of thick shields using magnetic materials, shields based on the induced-current principle may be reasonably effective at power-line frequencies. As a result, aluminum screens may be used to protect against 50/ 60-Hz magnetic fields generated by transformers.

Any opening in a shield will limit its effectiveness. Shielding theory assumes that RF currents will flow as long as there are no obstacles in their path such as an aperture or ventilation slot. It is essential that any and all openings be arranged in such a manner as to minimize their effect on current propagations (Fig. 6.6). Apertures have high frequency resonances so induced RF current flowing on the barrier can cause the aperture to act as a transmitting antenna.

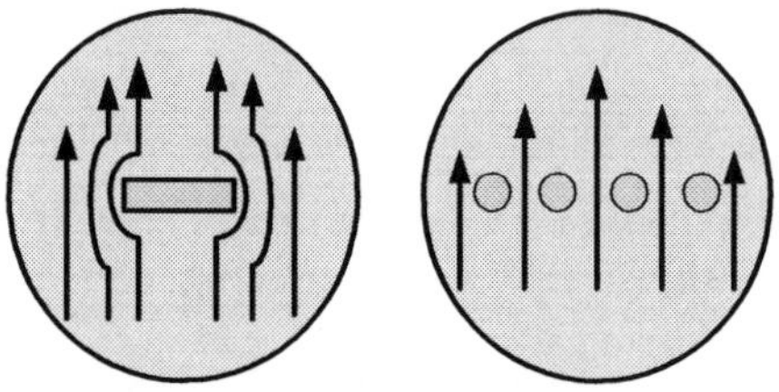

Figure 6.6 Electromagnetic fields propagating around openings in a shield partition.

6.3.3 Near-Field Conditions

Plane-wave conditions assume that the electromagnetic field is fully developed in free space ($E/H = Z_0 = 377\Omega$), which is true if the distance from the radiation source is a significant distance away, wavelength wise. This significant distance is called the far-field region. The amplitude of the ***E***-field and ***H***-field, respectively, decreases in the far-field as $1/r$. Details why are provided in Chapter 1.

In the near-field region, by contrast, the ratio between ***E*** and ***H*** is complex and varies with distance from the source. A higher-impedance source antenna (i.e., dipole antenna) will produce a near field dominated by E ($E/H >> Z_0$), while a lower-impedance antenna (i.e., a current loop) will yield a near field dominated by H ($E/H << Z_0$). At distance $\lambda/2\pi$, the wave impedance will approach Z_0 and the field will start to decrease linearly with distance when $r = \lambda/2$.

In the near field, shielding effectiveness must be designed separately for both electric and the magnetic field component since we may not have knowledge of which field is dominant in the reactive field region. This is why we cannot always use data given by shielding manufacturers since they are published for far field conditions and not for the near field, which is the primary area of interest for almost all product applications. Shield barriers are usually very close, physically, to printed circuit boards and are generally in the near field for most frequencies of interest, not the far field, unless the system operates in the upper GHz range where wavelengths are extremely small.

6.4 Losses Achieved with Shielding Material

There are two primary areas associated with reducing field propagation across a barrier, or preventing an electromagnetic field from either entering or leaving a shielded enclosure. As mentioned earlier, enclosures must have openings for interconnects and human interfaces which minimizes shielding effectiveness, unless special design techniques are implemented to ensure RF leaks do not occur.

Absorption (SA) and reflection (SR) loss are the two primary areas of concern. There is a third loss which rarely affects total shielding effectiveness, and that is if the barrier has any thickness that is significant to the wavelength of the signal impinging on the barrier. This third loss is called "re-reflected" loss (SB).

We now examine these three loss mechanism in shield barriers as well as skin depth and skin effect.

6.4.1 Reflection Loss

Reflection loss is the relationship or ratio of the impinging wave impedance to barrier impedance. The impedance of a barrier is a function of conductivity, permeability and frequency (the combination of these parameters describe skin depth). Material with high conductivity includes copper and aluminum which has a higher ***E***-field reflection loss than lower conductivity metal such as steel.

Reflection loss increases with an increase in frequency, at least for electric fields depending on the conductivity of the material at the electromagnetic field energy level. Losses also increase for any magnetic field that may be present. In the near field, or if the distance between source and antenna is less than $\lambda/2\pi$, this distance will affect reflection loss. The ***E***-field impedance is high in the near-field which means reflection loss is high. Conversely, ***H***-field impedance is low in the near field and thus reflection loss is low. When the barrier is in the far-field, the incident field is the item of concern and its impedance is constant which means the distance from source to antenna is irrelevant.

Reflection loss occurs within every transmission line, which results in part of the energy being reflected back to the source. This occurs under the following conditions:

1. At a discontinuity or impedance mismatch in any transmission line (shield barriers are modeled as transmission lines), the reflection loss is the ratio of the incident power to the reflected power, expressed in dBs.
2. In an optical fiber interconnect, the reflection loss that takes place at any discontinuity is called the refractive index, especially at an air-glass interface such as the fiber endface. At the endface, a fraction of the optical signal is reflected back toward the source due to impedance mismatches. This reflection phenomenon is also called "Fresnel reflection loss", or simply "Fresnel loss."

The equations associated with reflection loss are provided in (6.7). Field impedance is the primary element of concern. Figure 6.7 illustrates what a reflected wave looks like after it encounters a shield barrier.

$$
\begin{aligned}
&Z_o = \sqrt{\mu_o / \varepsilon_o} && \mu = \mu_o = 4\pi \times 10^{-7} \ H/m \\
& && \varepsilon_o = 8.85 \times 10^{-12} \ F/m \\
&Z_s = \sqrt{2\pi f \mu_r \mu_o / \sigma} && \sigma = 5.8 \times 10^{7} \ S/m \\
&R(dB) = 20 \log(Z_o / 4Z_s) \\
&\textit{In free space, } Zo = 377\Omega\textit{, therefore} \\
&R(dB) = 20 \log \frac{94.25}{|Z_s|}
\end{aligned}
\qquad (6.7)
$$

where:
Z_0 = impedance of the wave prior to entering the shield
Z_s = impedance of the shield
f = frequency.

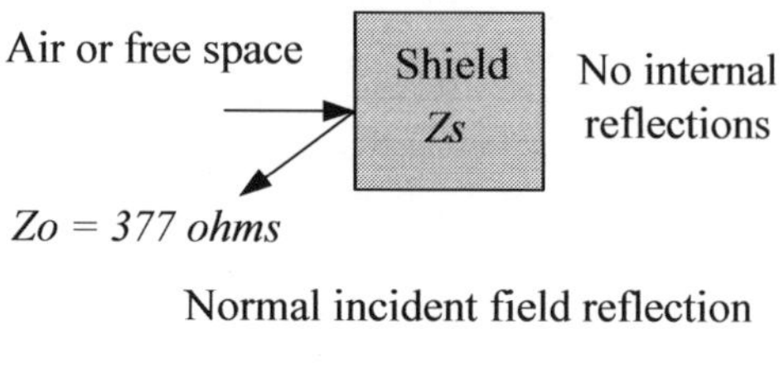

Figure 6.7 Reflection from a shield barrier.
(The amplitude of the signal decreases when reflection occurs)

To summarize, reflection loss is:

- High for electric fields
- High for high-frequency magnetic fields
- Low for low-frequency magnetic fields
- Decreases with frequency

Reflection Loss in a Shield Barrier

Within a shield barrier, reflection loss minimizes field propagation from one side to the other. Based on material composition and thickness with respect to skin depths, the attenuation of a signal will vary with frequency, distance from source and type of wave; ***E***-, ***H***-field or plane.

Within Fig. 6.8, three curves (electric, magnetic and plane wave) are illustrated. These curves show the frequency that makes separation between source and barrier equal to $\lambda/2\pi$, where λ is the wavelength of the signal. The point source produces both electric and magnetic fields. Reflection loss for a practical source is where the electric and magnetic field lines merge. In Fig. 6.8, when the distance from the source is 30 meters, the plane wave begins to form at 1.6 MHz. At a 1 meter distance, the plane wave begins at 300 MHz ($\lambda=c/f$).

Reflection loss of the electric field decreases with frequency until separation distance is $\lambda/2\pi$. Beyond this point the reflection loss is the same as for a plane wave. On the other hand, the reflection loss of a magnetic field increases with frequency before the loss begins to decrease at the same rate as the plane wave.

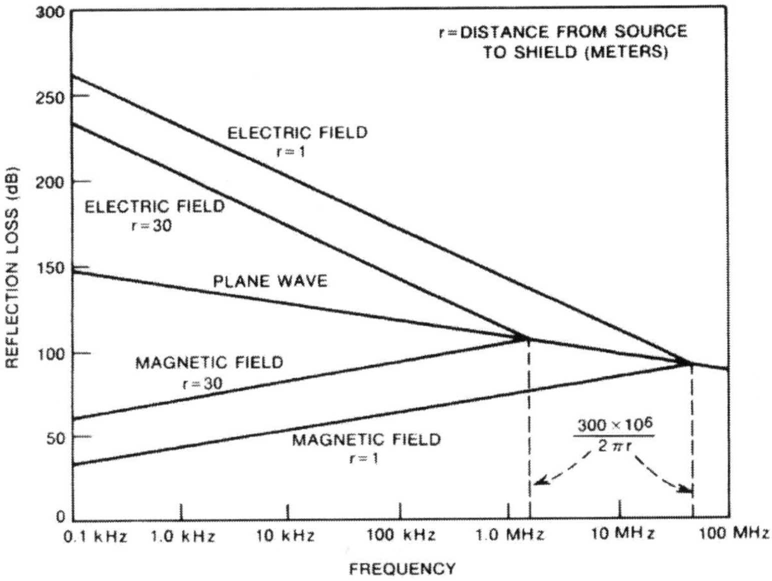

Figure 6.8 Reflection loss in copper shield relative to frequency and field composition. (Source: Electromagnetic Compatibility Engineering–H. Ott)

6.4.2 Absorption Loss

Absorption loss is defined as loss within a shield barrier caused by the dissipation or conversion of electromagnetic energy into another form of energy as a result of its interaction with a material's medium. This dissipation is usually is in the form of heat.

Absorption loss increases with an increase in frequency of the electromagnetic wave, barrier thickness, barrier permeability and conductivity. Absorption loss is independent of field characteristics and dependent on the type of field present and thickness of the shield only.

Absorption loss primarily depends on thickness of the barrier along with skin depth, discussed in the next Section. Skin depth is dependent on material properties (relative conductivity and permeability at a frequency of interest). For example, steel offers higher absorption than copper with the same thickness. At higher frequencies, absorption becomes the dominant mode increasing exponentially with the square root of frequency.

When an electromagnetic field passes through a medium its amplitude decreases exponentially due to ohmic losses and heating of the material per (6.8):

$$E_1 = E_0 e^{-t/\delta} \quad and \quad H_1 = H_0 e^{-t/\delta} \tag{6.8}$$

where:

$E1\ (H1)$ = wave intensity at a distance t within the media

δ (skin depth) = The depth where the field/surface current is attenuated to e^{-1} (37%, or approx. 9 dB) from its value on the surface of the barrier.

An illustration of absorption loss is provided in Fig. 6.9. Here we see an incident field approaching a shield barrier. At a specific distance inside the barrier ("Distance from edge"), a majority of the field strength present is absorbed and the amplitude of the signal decreases quickly (the slope in the right illustration). It takes very little material thickness to achieve high levels of absorption loss. One skin depth, or a very small amount of metal thickness, decreases the signal amplitude to approximately 37% of its original amplitude, or 9 dB (8.78 dB exactly). This number, calculated by (6.9) is due to skin effect loss in the material.

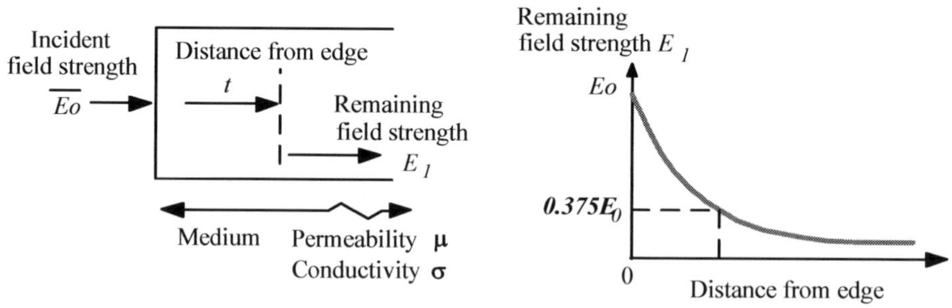

Figure 6.9 Absorption loss within a shield barrier.

$$A = 20\left(\frac{t}{\delta}\right)\log(e) \quad dB,$$

$$A = 8.69\left(\frac{t}{\delta}\right) \quad dB \tag{6.9}$$

If we double the thickness of the shield, we double the loss. The amplitude of the signal will then decrease with frequency significantly. In the range of 30-100 MHz, absorption loss will predominate for this reason.

Absorption loss is not a constant value for all types of metal barriers with different metallurgical compositions. Figure 6.10 show absorption loss of copper and steel with different thicknesses. Steel has a significantly greater amount of ferrous oxide, a magnetic material that absorbs magnetic field propagation more than copper. Also at lower frequencies, skin depth level is extensive, relatively speaking, but at higher frequencies it takes less metal thickness to achieve the same amount of field reduction.

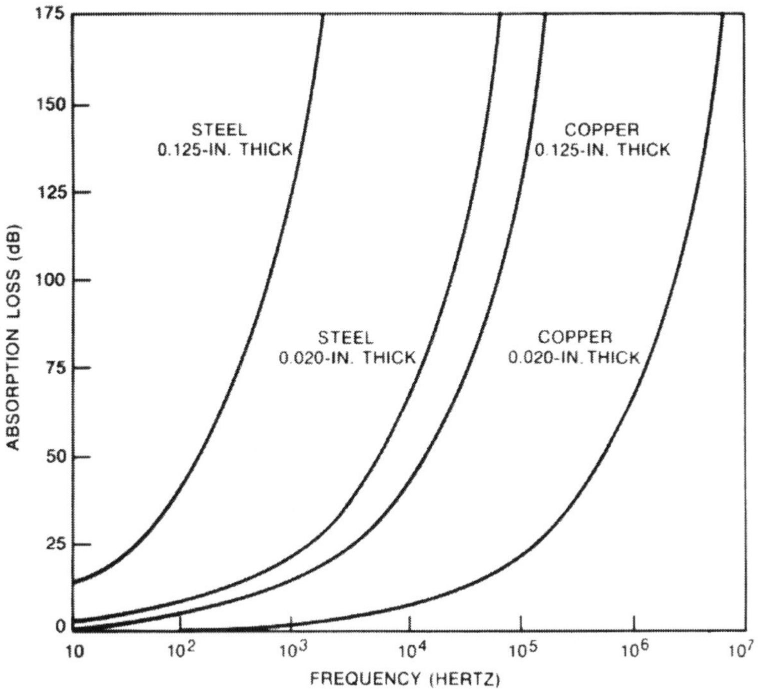

Figure 6.10 Absorption loss for various types of metal and their thickness. (Source: Electromagnetic Compatibility Engineering–H. Ott)

6.4.3 Skin Effect and Skin Depth

The topic of skin depth previously presented is the primary factor in determining total absorption loss within a metal barrier. What exactly is skin effect and the physical thickness of a metal barrier required for optimal performance?

Skin effect is the tendency of an alternating electric current (AC), or RF field, to become distributed within a metallic transmission line or structure such that the current density is largest near the surface. This causes the current density of the magnetic field to decreases as the distance within the barrier becomes greater from the surface boundary.

Electric current flows mainly on the "skin" of the conductor, or between the outer surface boundary and an internal level which is called the skin depth layer. The result of having skin effect in a shield causes the effective impedance of the conductor to increase at higher frequencies, and will occur when the skin depth distance becomes smaller in thickness. Skin effect also reduces the effective cross-section area of the conductor and is due to opposing eddy currents induced by a changing magnetic field impinging on the barrier.

The internal magnetic field propagating within a conductor is in addition to the magnetic field present between both signal and return path. These internal magnetic fields create induced current loops within the transmission line. These current loops produce their own magnetic fields. These new magnetic fields oppose the initial magnetic field from the source, which effectively pushes current carrying electrons toward the surface of the conductor extremities.

At very low frequencies close to DC, both current density and resistance of the material are evenly distributed across the cross-section geometry in typical conductors. As frequency increases, current distribution starts to change due to both changing electric and magnetic fields. This current density concentrates at the boundary of the transmission line. This effect often happens at a medium frequency in the order of 1-5 skin depths. The medium frequencies are also known as the transition frequency range. At medium and higher frequencies, where conductor thickness is greater than five skin depths, current density is assumed to be concentrated at the conductor surface.

At 60 Hz in copper, skin depth is about 8.5 mm. At higher frequencies skin depth becomes much smaller. Because the interior of a large conductor carries very little current, tubular conductors such as a metal pipe can be hollow on the inside which saves weight and cost as it propagates an electromagnetic field. This is how waveguides operate at high frequencies, with no metal internal to the transmission line structure.

The distance required for a wave to be attenuated to $1/e$ or 37% of its original value within the shield is defined as *one skin depth* calculated by (6.10).

$$\delta = \sqrt{\frac{2}{\omega\mu\sigma}} \ (meters) = \frac{2.6}{\sqrt{f\mu\sigma}} \ (inches) = \frac{6.6}{\sqrt{f\mu\sigma}} \ (cm) \qquad (6.10)$$

where:

$\omega = 2\pi f$ (frequency in Hz)

$\mu = \mu_o = 4\pi \times 10^{-7}$ H/m

$\sigma = 5.8 \times 10^{7}$ S/m.

An example of absorption loss per skin depth using (6.10):

$$A(dB) = 20 log[E_o/E_{(x)}]$$
$$at\ x = \delta,\ E(\delta) = E_o/e$$
$$A(db) = 20\ log(e)$$
$$A(db) = 8.7\ dB\ /\ skin\ depth$$

If wall thickness = 0.3 mm
Skin depth = 6.6 μm
A(dB) = 8.7 dB x thickness/skin depth
A(dB) = 395 dB

Table 6.1 illustrate skin depth for various metals relative to the frequency of the electromagnetic field that encounters this shield barrier. It is easily observed that as frequency gets higher, it takes almost no metal thickness to reduce an electromagnetic field 37% from its maximum amplitude, or 9 dB.

Table 6.1 Table of skin depth for various metals.

Frequency	Copper inches (mm)	Aluminum inches (mm)	Steel inches (mm)	Mu-metal inches (mm)
60 Hz	0.335 (8.45)	0.429 (10.89)	0.034 (0.86)	0.014 (0.36)
100 Hz	0.260 (6.60)	0.333 (8.4)	0.026 (0.66)	0.011 (0.28)
1,000 Hz	0.092 (2.34)	0.105 (2.66)	0.008 (0.20)	0.003 (0.08)
10,000 Hz	0.026 (0.66)	0.033 (0.083)	0.003 (0.07)	--
100,000 Hz	0.008 (0.20)	0.011 (0.27)	0.0008 (0.020)	--
1 MHz	0.0026 (0.066)	0.003 (0.07)	0.0003 (0.007)	--
10 MHz	0.0008 (0.020)	0.001(0.03)	0.0001 (0.0025)	--
100 MHz	0.00026 (0.0066)	0.0003 (0.007)	0.00008 (0.002)	--
1000 MHz	00.0008 (0.0020)	0.0001 (0.003)	0.00004 (0.001)	--

6.4.4 Reflections Internal in Thin Shields

The re-reflection loss factor ***SB*** is insignificant for the majority of applications when absorption loss is greater than 10 dB. This re-reflection coefficient is however important for very low frequencies. Re-reflection loss is strongly dependent upon the magnitude of absorption loss in any material. Just as a reflection occurs at the first air-to-metal boundary, a similar reflection occurs at the second metal-to-air boundary. With an absorption loss value greater than 9 dB or one skin depth, the re-reflection term can be ignored in the shielding effectiveness equation (6.1).

Multiple reflections refer to what happens at various surfaces or interfaces within the shield barrier. An example of a shield barrier with a *small surface area* is porous or foam material. An example of a shield with a *large interface area* is composite material containing a conductive filler or solid metal.

If the shield is physically thin relative to the frequency present, the reflected wave from the secondary boundary is re-reflected off the first boundary and then returns to the secondary boundary to be reflected again. This effect can be neglected in the case of a thick shield due to high absorption loss, known as the factor "***SB***" in the Shielding Equation (SE) for total effectiveness of the structure (6.1). Multiple reflections are thus illustrated in Fig. 6.11.

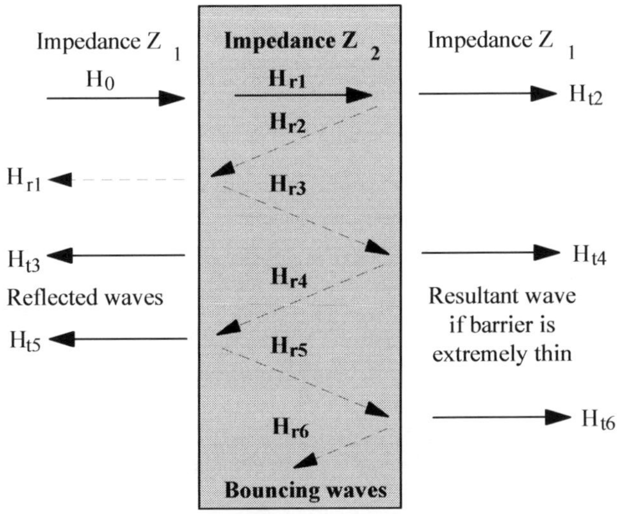

Figure 6.11 Multiple reflections internal to a shield barrier.

The re-reflection loss correction factor (***SB***) for thin shields affects magnetic fields only, and is clearly seen in Fig. 6.12 which illustrates the ratio of shield thickness to skin depth and how much attenuation occurs for various thickness of a material.

To review, the shielding effectiveness equation (6.1) is given again below:

$$SE = SA + SR + SB$$

where:

SE = total shielding effectiveness
SA = absorption loss
SR = reflection loss
SB = re-reflection loss.

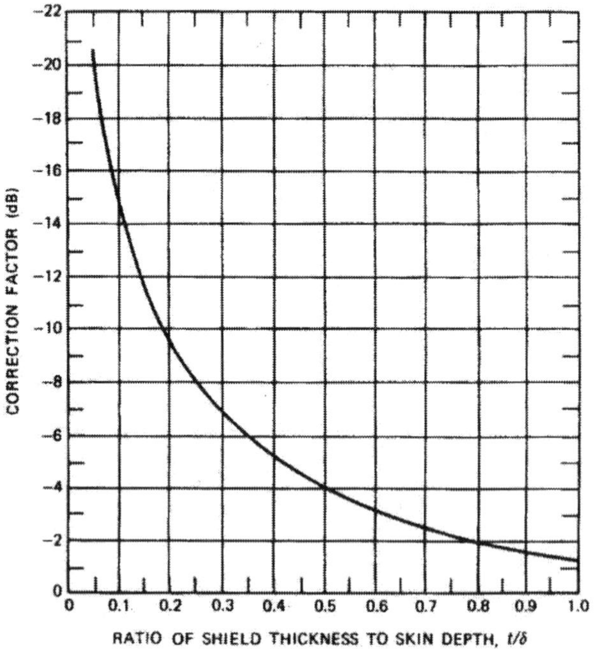

Figure 6.12 Re-reflection loss correction factor (*B*) for thin shields/magnetic fields. (Source: Electromagnetic Compatibility Engineering–H. Ott)

6.4.5 Composite Absorption and Reflection Loss

We now analyze effects of both absorption and reflection in a shield barrier at the same time, ignoring the re-reflection loss coefficient which means the material has a thickness greater than several skin depths at all frequencies of interest.

In the 30-100 MHz range, absorption loss is greater than reflection. Any solid shield thick enough to be practical provides more than adequate shielding for the emission levels of commercial digital equipment.

- At ***low frequency***, reflection loss is the primary shielding mechanism for the electric field.
- At ***high frequency***, absorption loss is the primary shielding mechanism.

In Figure 6.13, lower frequency signals are shown from 10 Hz to 10 kHz, illustrating how reflection and absorption losses occur at the same time, with the vertical axis as total shielding effectiveness in dBs. As frequency increases, RF energy reflected off the barrier decreases linearly. What is easily noticed is that the absorption losses start to increase quickly above 100 kHz. At approximately 5 MHz, almost all attenuation and shielding effectiveness is due to absorption (reflection has little affect starting at this frequency).

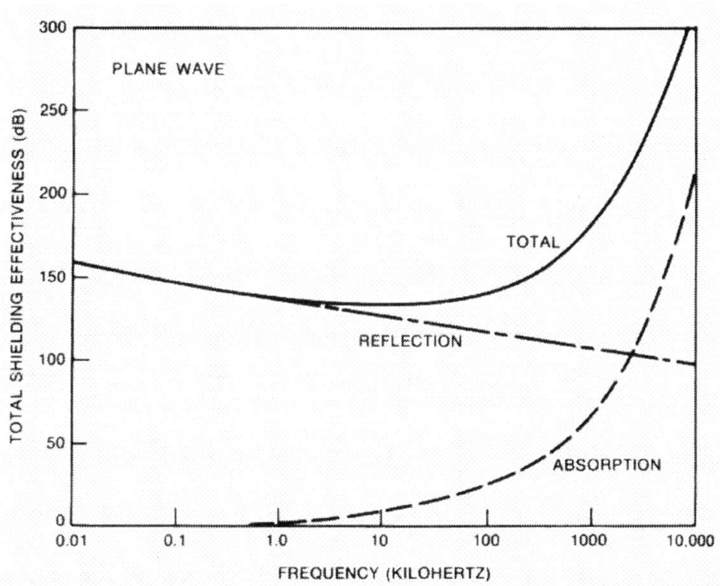

Figure 6.13 Shielding effectiveness of a 0.02-inch thick copper shield in the far field. (Source: Electromagnetic Compatibility Engineering–H. Ott)

Figure 6.14 shows a 0.020 inch (0.51mm) thick solid aluminum plate with two types of fields and a plane wave. Remember, plane waves are created from field impedance equations (6.3) and (6.4), or $Z = E / H$. Since the plot (x-axis) goes from 100 Hz to 10 MHz, this means that the antenna is located in the far field or greater than 30 meters away because the plane wave impedance is still in the reactive or near-field region. If we were in the far field, the plane wave would have a fixed impedance of 377 ohms, which is above the top axis of 300 ohms and thus not shown.

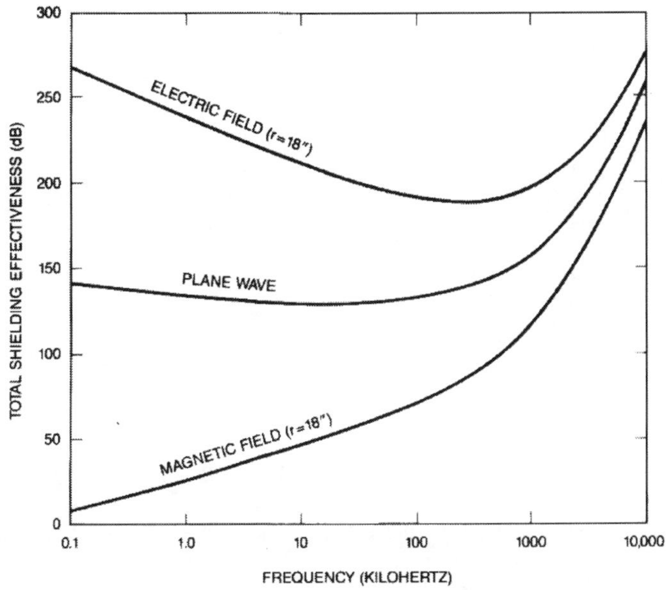

Figure 6.14 Electric field, plane wave and magnetic field shielding effectiveness. (Source: Electromagnetic Compatibility Engineering–H. Ott)

Regardless of the plot's relationship to free space, the item to note is how the impedance of both ***E*** and ***H*** fields behaves simultaneously in the near-field, at lower frequencies.

6.5 Apertures in Shield Barriers

There are applications where openings in contiguous shield barriers cannot be avoided. Apertures allow a radiated field to propagate from one side to the other causing either an emission or immunity situation to develop. The most common openings are those used for air ventilation and interconnects to the outside world. Fans are frequently used to move hot air inside the assembly to the environment. These openings are frequently in the shape of one large or many small holes.

Figure 6.15 illustrates various types of holes or openings within a shield barrier and how RF current flow propagates on the skin of the metal. When currents are induced into the shield, these currents and associated fields generate scattered fields, which counteract or reduce the effects of the incident field by the process of reflection. The reflected field is of a polarity such that it tends to cancel the incident field in order to satisfy the boundary condition related to the total electric field tangent to the barrier. In a perfect conductor, the total electric field is zero if there are no disruptions in current flow.

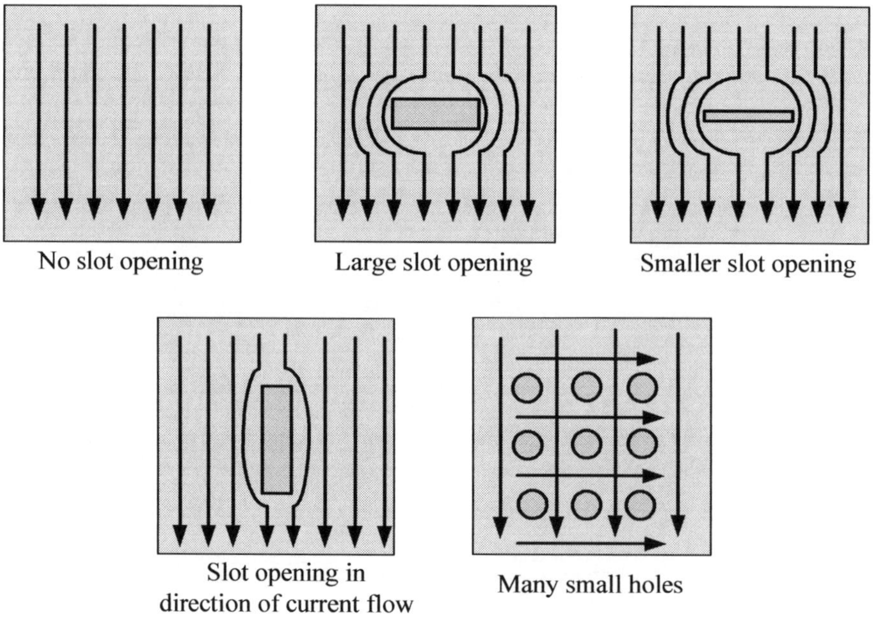

Figure 6.15 Slots in a shield barrier and effect of induced current flow.

If there are slots in the barrier, current flow will be interrupted and shielding effectiveness reduced. If we orient the slot parallel to the direction of the induced current, the opening will have much less effect on shielding effectiveness. Since it is not possible, or very difficult to determine the relative polarization of induced current flow, a larger number of smaller holes are thus preferred. Many small holes provide as much ventilation as one long slot by perturbing the induced current less, thereby reducing degradation of shielding effectiveness caused by the aperture.

Within Fig. 6.15, we see that with no openings, current flow is not affected and maximum shielding effectiveness is present. With the horizontal slot configuration, current disruption is identical regardless of slot size in that one dimension (height). The same is true for the vertical slot with current disruption but at a much lower level. Since it is nearly impossible to know the direction of current flow in the metal ahead of time, we have to assume that any polarity or direction of a slot will disrupt current flow significantly.

In Fig 6.15, the holes present less disruptive current propagation to result in improved shielding effectiveness while permitting air flow, although the total physical area of the slot and holes are identical to a single large hold.

The physical dimensions of the slot (width, length or diameter) may be at a resonant frequency of an RF signal. It is the physical linear dimension of the slot based on wavelength that permits RF energy to pass through, although current disruption may be present. If the wavelength is greater than ½ of the frequency present, or any harmonic of the wavelength slot, this opening becomes an efficient antenna with maximum radiation capabilities.

The linear length of the slot or opening is more important than thickness of the metal in determining radiated emission potential of the gap due to absorption loss of the material. The total area of a circle has no effect on the wavelength of the signal that gets propagated thorough it. It is the "*diameter*" of a circle that is the critical dimension, not "*radius.*" For rectangular slots, it is either the horizontal or vertical dimensions that determine antenna efficiency. Very small holes are self-resonant in the upper GHz range, which is why they are preferred for regulatory compliance as well as to minimize risk of fire by acting as a fire shield barrier per international product safety requirements.

Apertures affect magnetic field leakage more than electric fields since magnetic fields propagates in loops. Each line of current in the figure is magnetic field. As a result, emphasis must be given to minimize magnetic field leakage. For almost all applications of use, and if signals are greater than 30 MHz, absorption provides adequate attenuation of the magnetic field. To ensure maximum efficiency of magnetic field attenuation, any barrier greater than several skin depths should be adequate.

Penetration of fields through apertures that are small compared to a wavelength of a propagated field ($\lambda/2\pi$) will not occur. For slots with a dimension equal to or less than half a wavelength, shielding effectiveness will be defined by (6.11):

$$SE = 20\log\left(\frac{\lambda}{2\ell}\right) \tag{6.11}$$

where:
λ = wavelength of the signal
ℓ = maximum linear dimension of the opening.

The maximum amount of leakage from an aperture is dependent on: 1) the maximum straight line linear dimension compared to wavelength, "not" area of the opening; 2) the electromagnetic field and wave impedance present; and 3) the frequency of concern along with its relationship to the self-resonant frequency (linear dimension) of any aperture or opening.

6.5.1 Single Apertures

To illustrate how field leakage or penetration occurs through a single opening in a shield barrier based on the physical dimension of a signal's wavelength, we now examine Fig. 6.16. In this illustration we have both electric and magnetic fields shown. Shields contain field intensities produced from circuits which are generally in the near field, thus plane wave propagation is not shown [1].

Electric fields propagate using the dipole antenna model. Magnetic fields require a loop antenna for efficient field propagation. Once an electromagnetic field penetrates the shield, we observe a plane wave in the far-field. In Fig. 6.16, a wire or metallic item is physically located external to the shield. This metallic item could be another printed circuit board, cable assembly or another shield barrier. Between the shield barrier's driven elements, or in this case from the incident electric field through the aperture to the external element which in this example is a 0V reference or ground, direct or indirect field may propagation occur.

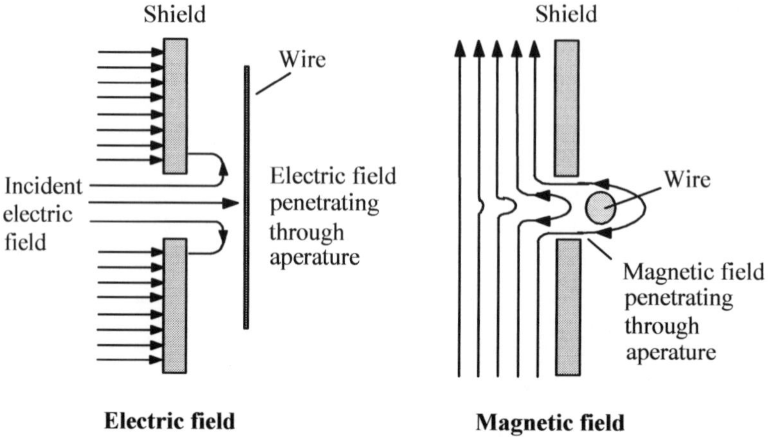

Figure 6.16 Penetration of fields through apertures large compared to the wavelength.

For every electric field line there is a reflected wave back to the source, not shown in Fig. 6.16. Only the incident electric field is illustrated (arrows going to the right, not the left which do exist). Reflection loss will keep the energy from causing harmful disruption to the outside world. For the magnetic field, absorption loss is the key element since the field propagates on the skin of the enclosure as eddy currents. Depending on thickness of the material, the signal is attenuated except at the point of an opening where the field has the opportunity to exit, find a victim to encircle and then return back to its original path.

To acquire a better understanding on how field propagation travels through an opening and its effects in the far field, Figures 6.17 and 6.18 illustrate an important principle in field propagation. Figure 6.17 shows a simplified example on how reflected waves respond when they hit a shield barrier. Since we never know what the incident field impingement angle will be in a real product, there may be thousands of reflections simultaneously. The reflected wavefront is sent back to the inside of the system. Depending on the linear dimension of the opening, a portion of the electric field will propagate through this hole or aperture. Magnetic field lines are not shown in this drawing to simplify the illustration.

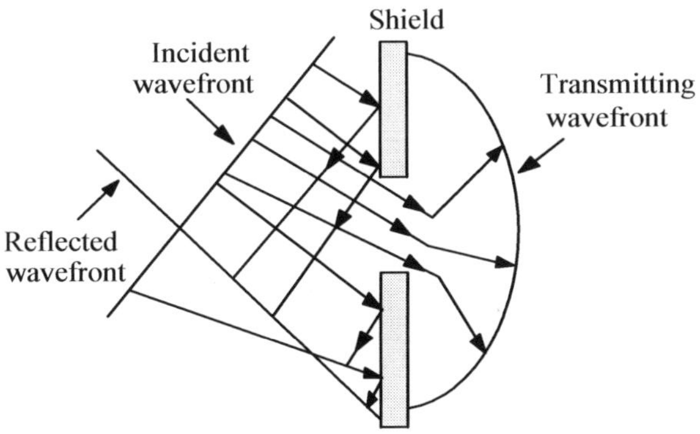

Figure 6.17 Penetration of fields through holes in thin barriers.

When we combine both magnetic and electric fields at known distance from the source generator, we now have an *electromagnetic* field. Once in free space this electromagnetic field is observed as a plane wave but only in the far field. In reality, electric and magnetic field do not propagate separately based on either the dipole or loop antenna model. In free space we have a plane wave that is described by Huygen's principle, created at the barrier opening.

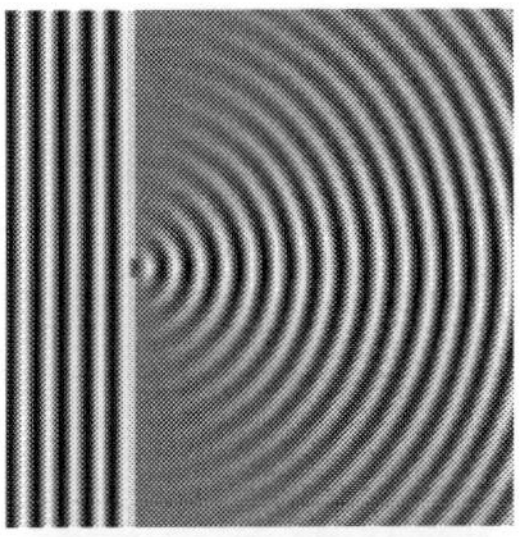

Diffraction of a plane wave when the slit width equals the wavelength

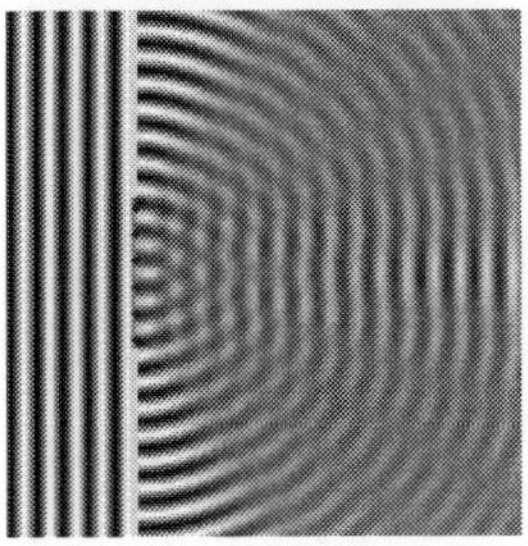

Diffraction of a plane wave at a slit whose width is several times the wavelengths

Figure 6.18 Huygen's principle - Diffraction of plane wave at opening at a barrier. (Courtesy – Wikipedia. Huygen's Principle, Illustrations in the public domain)

Christiaan Huygen's principle was based on the analysis of light waves in 1678 but is applicable to the field of electromagnetics as discovered by Augustin-Jean Fresnel in 1816, who explained both rectilinear propagation of a wave and diffraction effects. This principle is a method of analysis applied to problems of wave propagation both in the far-field limit and in near-field diffraction.

1. Each point or particle on the primary wavefront acts as a source for secondary wavefronts, which sends out a disturbance (waves) in all direction in a similar manner as the original source.
2. The new position of the wavefront at any instant of time is given by the forward envelope of second wavelets at that instant in time.

To summarize, in Figure 6.18, we see magnetic field lines on the left side of the shield barrier as was illustrated in Fig. 6.16. At the opening or aperture, field lines are converted to rectilinear field propagation. The only way a rectilinear field can be established is with both magnetic and electric fields, but only if the opening is several times the wavelength of the signal present.

For reasons discovered by Huygen, it is best not to allow any opening larger than about $\lambda/30$. For effective screening of electromagnetic fields with frequencies up to 1 GHz, apertures should be no bigger than 1 cm. A good rule of thumb to follow is to avoid openings larger than $\lambda/50$ to $\lambda/20$ at the highest frequency of operation, including the 10th harmonic for a signal of concern. Multiple apertures in close proximity, however, will reduce shielding effectiveness hence the density of the apertures is an important consideration. See Section 6.5.2 below.

To determine shielding effectiveness versus frequency for openings in a barrier, Table 6.2 provides linear dimensions of an opening whether it is a slot (width or length), or circle (diameter, not radius or perimeter). It is easily noticed that small openings provide significant shielding effectiveness for lower frequency signals but decrease in effectiveness when the frequency becomes greater. This is due to the wavelength of the signal based on the linear dimension and current disruption that occurs in the shield material. If there were no openings in the barrier, attenuation of 300 dB is easily achieved. Figure 6.19 illustrates Table 6.2 in graphic form to show the linear decrease in shielding effectiveness based on the physical length of the slot and frequency of concern. Equation (6.11) provides the numerical analysis for the data in Table 6.2. When using (6.12) select the largest dimension whether it is in the vertical or horizontal polarity

Table 6.2 Shielding effectiveness vs. frequency and maximum length for single aperture.

	Aperture Length				
***f* (MHz)**	**0.5" (1.2cm)**	**1" (2.54cm)**	**3" (7.62cm)**	**6" (15.24cm)**	**12" (30.48cm)**
30	52 dB	46 dB	36 dB	30 dB	24 dB
50	47 dB	41 dB	32 dB	26 dB	20 dB
100	41 dB	35 dB	26 dB	20 dB	14 dB
300	32 dB	26 dB	16 dB	10 dB	4 dB
500	27 dB	21 dB	12 dB	6 dB	<1 dB
1000	21 dB	15 dB	6 dB	<1 dB	<1 dB

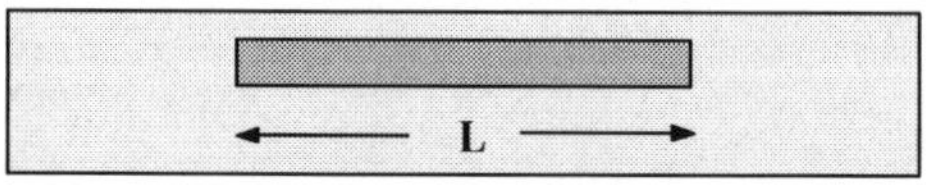

$$S = 20 \log \frac{150}{f(\text{MHz}) * L\,(\text{m})} \qquad (6.12)$$

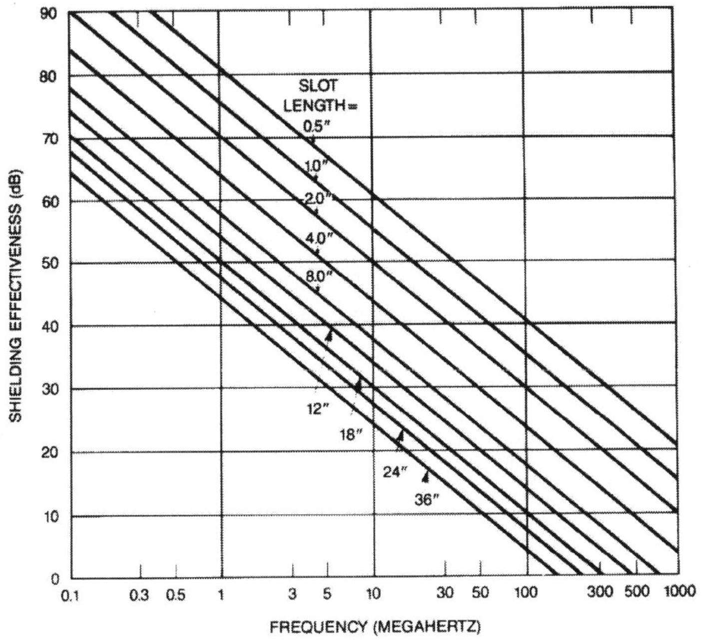

Figure 6.19 Shielding effectiveness vs. frequency and maximum slot length for one aperture.
(Source: Electromagnetic Compatibility Engineering–H. Ott)

6.5.2 Multiple Apertures

When there are more than one aperture, shielding effectiveness decreases based on the number of openings, the frequency involved and physical spacing between slots [1].

When using opening for the purpose of air ventilation, shielding effectiveness is based on the thickness of the shield barrier and distance spacing between holes defined by (6.13) and illustrated in Fig. 6.20:

$$SE(dB) = 20 \log (s/2d) \quad [for\ d > t] \qquad (6.13)$$

where:
SE = total shielding effectiveness
s = spacing distance between holes
d = diameter of the hole
t = thickness of the barrier.

For an array of closely spaced opening (Fig. 6.21), reduction in shielding effectiveness is proportional to the square root of the number of openings (n) calculated by (6.14). Based on the number of openings, shielding effectiveness (SE) decreases as the square root of the number of apertures multiplied by -20log, or the number of apertures multiplied by -10log. The negative sign indicates SE decreases. These equations provide insight into how much degradation occurs along with an example on how to calculate required SE level before holes are provided.

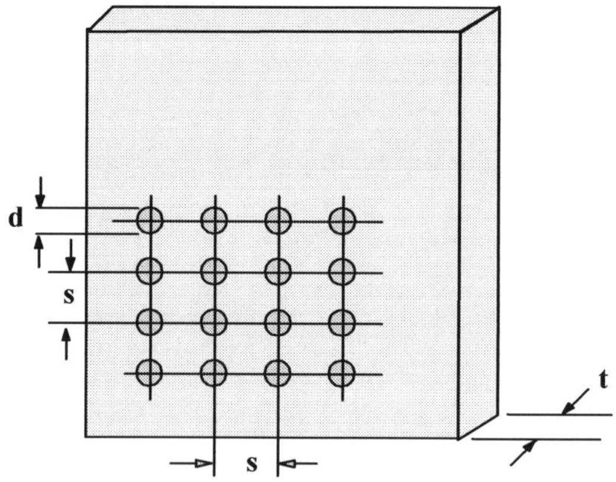

Figure 6.20 Penetration of fields through holes in a thin barrier based on linear dimensions.

$$SE = -20 \log \sqrt{n} \quad \text{or} \quad SE = -10 \log n \tag{6.14}$$

N	SE (dB)
2	-3
4	-6
6	-8
8	-9
10	-10
15	-12
20	-13
25	-14
30	-15
40	-16
50	-17
100	-20

Example:
A product requires 20 dB of shielding effectiveness at 100 MHz. There are 30 slots all of the same length. How long can one slot be?

- Thirty slots will reduce shielding effectiveness by -15 dB
- Therefore each slot must provide 35 dB of shielding effectiveness
- To provide 35 dB of shielding effectiveness at 100 MHz, each slot can be one inch (2.54cm) long (Table 6.2).

With regard to total net shielding effectiveness, if there is a linear array of equal size holes, this is equivalent to the shielding effectiveness of only one of the holes plus the total shielding effectiveness from having multiple apertures, described by (6.15).

$$SE = 20 \log \left(\frac{150}{f\, l \sqrt{n}} \right) \tag{6.15}$$

where:
SE = net shielding effectiveness
fs = frequency of the signal (MHz)
l = length of the slot (meters)
n = total number of holes.

In many applications, air ventilation may be implemented on more than one side of an enclosure for both inlet and outlet in order for fans to work efficiently. If air ventilation holes are located throughout the system, overall shielding effectiveness is not decreased any more than that of a single width opening. Electric and magnetic field propagation internal to the system hits all internal sides of the enclosure simultaneously. The magnitude of RF energy that happens to be present on the outside portion of the barrier is that which leaks out through the holes.

Equation (6.14) is for a "linear array" of openings in "one dimension," not a two-dimensional array as shown in the Fig. 6.21 (vertical and horizontal). This means that only the "top row" of holes affect (decreases) shielding effectiveness. The second, third and fourth rows do not decrease shielding effectiveness since these additional holes do not provide additional current diversion after the first row of holes which have already diverted the current. The distance spacing between openings is so small, wavelength wise, that the diverted current remains diverted until the last row of holes is passed.

When applying (6.14), remember this equation is valid only for a single row of holes in the matrix. This row can be in either in the horizontal, vertical or diagonal direction. This is the number used as "*n*" in (6.14). If there is an unsymmetrical matrix of holes, the array with the largest number is the one to use. For example, if there is a 2x10 matrix of holes, use the number 10 for "*n*."

6.5.3 Slot Antenna Polarization

We know a slot antenna will radiate RF energy based on the discussion above. Huygen's Principle tells us a propagating electromagnetic wave will occur based on the polarity of the magnetic field present on the inside of this barrier. This was clearly illustrated in Fig. 6.18. The magnetic field lines on the left travel in the vertical axis, yet the electromagnetic wave is horizontal. What caused this a shift in polarization to occur?

A slot antenna produces a radiation pattern with the same magnitude as its complementary antenna except with a 90° shift in polarization [6, p. 307], Fig. 6.21.

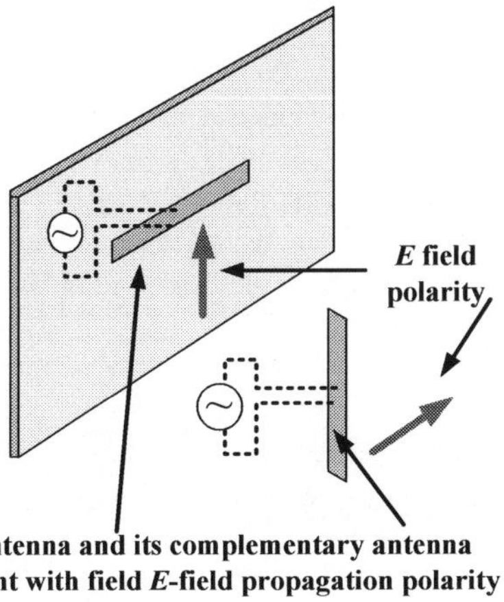

Figure 6.21 Slot antenna cross polarization effect on a radiated signal.

With this knowledge, when testing products for EMC compliance and observing a signal that requires identification as the source of the transmitted signal, if the antenna is polarized in the far field vertical, then examine or look for cables, slots or other apertures that are in the horizontal polarity. The same is true for reversed polarity. Understanding cross-polarization patterns can save significant time in isolating a radiating source when troubleshooting a system that needs to comply with EMC test requirements.

6.5.4 Waveguide Below Cutoff

If an opening or aperture is required in a shield barrier, attenuation of the signal can be achieved if the opening is designed to be a waveguide below cutoff. What this means is that the opening becomes a transmission line that prevents signals from propagating from one side to another below a specific frequency. This specific frequency is called the cutoff frequency and becomes an attenuator for all unwanted signals below this point. Attenuation of the waveguide is a function of physical length, "*t*" [8]. At the cutoff frequency and those frequencies above this level, attenuation of RF field propagation is typically minimal.

Conductors should never be passed through waveguides below their cutoff frequency as this compromises their effectiveness. Waveguides below cutoff can be applied to plastic shafts (e.g. control knobs) so that they do not compromise shielding effectiveness when they exit or penetrate an enclosure or shield barrier.

A window with multiple waveguides (e.g., a "honeycomb" configuration) can be an effective tool for preventing propagation of an electromagnetic field from one side of a barrier to another such as air vents used in an anechoic chamber. In such installations, no external RF field will be able to penetrate the waveguide for any distance if the diameter of the opening is small compared to the wavelength of interest. Shielding effectiveness at a distance d [m] behind a hole of diameter D [m] can, for frequencies with a wavelength $\lambda \gg D$, be approximately calculated as:

$$
\begin{array}{c}
20 \text{ dB at } d/D \approx 1 \\
40 \text{ dB at } d/D \approx 2 \\
60 \text{ dB at } d/D \approx 5
\end{array}
$$

If a hole in a shielded room is required to allow a tube to pass through (such as water or an air inlet or exhaust outlet), a mechanical waveguide may be provided as a filter for electromagnetic waves with frequencies well below its cutoff frequency. This may take the form of a circular, honeycomb or rectangular tube (Fig. 6.22). The signal will be attenuated and decrease with frequency, eventually becoming zero at the cutoff point. The equations to determine this cutoff frequency are provided in (6.16).

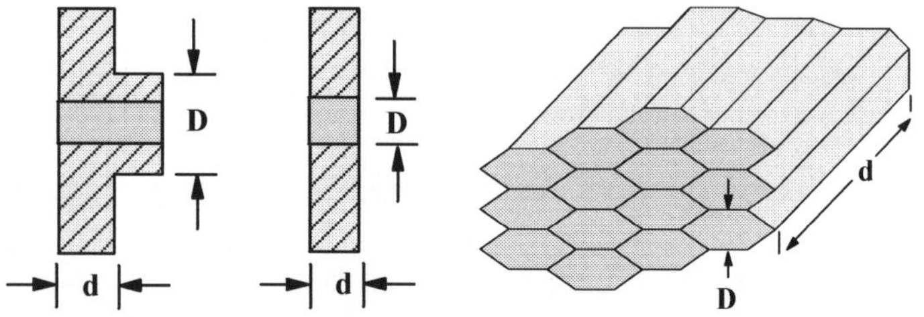

Figure 6.22 Typical waveguide configurations and dimensions to calculate cutoff frequency.

For a round waveguide, cutoff frequency (inches)

$$f_c = \frac{6.9 \times 10^9}{d} \quad Hz \ (inches) \qquad f_c = \frac{17.5 \times 10^9}{d} \quad Hz\,(cm) \ \ (f \ in\,GHz)$$

For a rectangular waveguide, cutoff frequency (inches)

$$f_c = \frac{5.9 \times 10^9}{l} \quad Hz \quad [l = largest\ cross\ section\ dimension\,(inches)]$$

Magnetic field shielding - round waveguide

$$S = 32\frac{t}{d} \quad dB$$

Magnetic field sheilding - rectangular waveguide

$$S = 27.2\frac{t}{d} \quad dB \tag{6.16}$$

A waveguide with length three times greater than diameter has over 100 dB of shielding effectiveness.

A sample list of waveguide below cutoff applications includes the following:

1. Honeycomb panels and cooling vents – most common application.
2. Passing control shafts through metallic panels.
3. Passing fiber optic cable into a sealed chamber or enclosure.

Illustrations of typical waveguides used as air vents are shown in Fig. 6.23.

Photo courtesy Tech-Etch

Figure 6.23 Examples of honeycomb waveguides for air ventilation.

6.5.5 Waveguides or Transmission Lines Between Enclosures and Systems

Waveguides are frequently used to carry non-electrical services and metal-free fiber optic cables through the wall of a shielded enclosure. What is critical is that *no metallic conductors or cables are passed through the waveguide except fiber optic!* Any penetration by a transmission line that is not filtered at the "wall" (boundary) destroys waveguide properties and renders it useless for shielding integrity.

When a cable such as external AC/DC power, remote control operation, video monitoring equipment or an Ethernet cable to a router or simulator outside of the chamber is passed through a tube or pipe, any RF energy present outside the chamber will enter and cause reduction of any shielding effectiveness in addition to injecting ambient noise into the chamber. The opposite is true for systems in the chamber that may corrupt equipment located outside the chamber, especially if radiated immunity testing is occurring.

It is for this reason, to prevent field propagation from either entering or leaving the chamber, a bulkhead panel with appropriate interface connectors are used generally coaxial based. Many chambers have a small support room adjacent to the large chamber with a waveguide that allows cables to pass between rooms (e.g. coaxes, antenna control, video interface if used, and similar support instrumentation). This control room is itself a Faraday structure when its door is closed, no external RF noise either enters or leaves the chamber.

What happens when a cable, control knob or other item must travel between rooms? It all depends on where the two facilities are located in relationship to each other. If physically together, such as test chamber and control room, it is easy to have cables pass through a bulkhead. If not physically together but located a distance apart, a waveguide or Faraday shield structure is required. Waveguides are a hollow metal structure where RF energy propagates between systems does so through the process of skin effect, with RF energy on the inside barrier of the hollow waveguide, traveling on the first skin of the metal.

The usual form of construction is a simple metal tube welded into the wall in the same manner as a cable connector. It does not matter whether the tube projects into or out of the enclosure as long as the thickness and diameter of the tube is calculated for cutoff at a particular frequency per (6.15). A shielded cable can penetrate the chamber through a waveguide below cutoff as long as the shield of the cable is bonded by a high-quality, low-impedance connection to the waveguide on *both* sides of the transmission line to establish a Faraday shield around the signals in the shielded transmission line.

6.6 Apertures in Shield Barriers

Shielding involves three primary areas of concern; reflection, absorption and re-reflection. We already examined slots, or apertures, required for air ventilation or interconnects that includes cables. Also, both single and multiple apertures were discussed as well as waveguides that permit cable penetration within a particular frequency range of interest.

We now examine what happens when a shield barrier is violated. Example of problems areas or concern are shown in Fig. 6.24. Knowing which one to address first when an EMI event develops, either emissions or immunity, is critical to achieving EMC.

6.6.1 Proper and Improper Shield Penetrations

When considering shield or barrier penetrations, one seeks to keep electromagnetic energy either inside or out. The smallest amount of impedance in the bond connection for interconnects (connectors), or a poor seam weld, can be difficult to troubleshoot or impractical to fix in addition to permitting significant RF energy to propagate.

As discussed in Chapter 5, in addition to Reference [5], it is critical to ensure the flow of RF current travels to a desired 0V reference. There is no such thing as a true "ground" in any electrical product. In reality, what we call ground is actually a 0V reference tied to a single potential point. The word *ground* when used with regard to shielding applications may refer to an [implied] earth ground. In this configuration, the system may be connected to the third wire ground (green/yellow) within an AC mains cable. If the cable does not have a third wire ground such as a two-wire AC mains plug, or battery operated, then the system operates off of a primary single-point 0V reference, not a virtual earth ground connection.

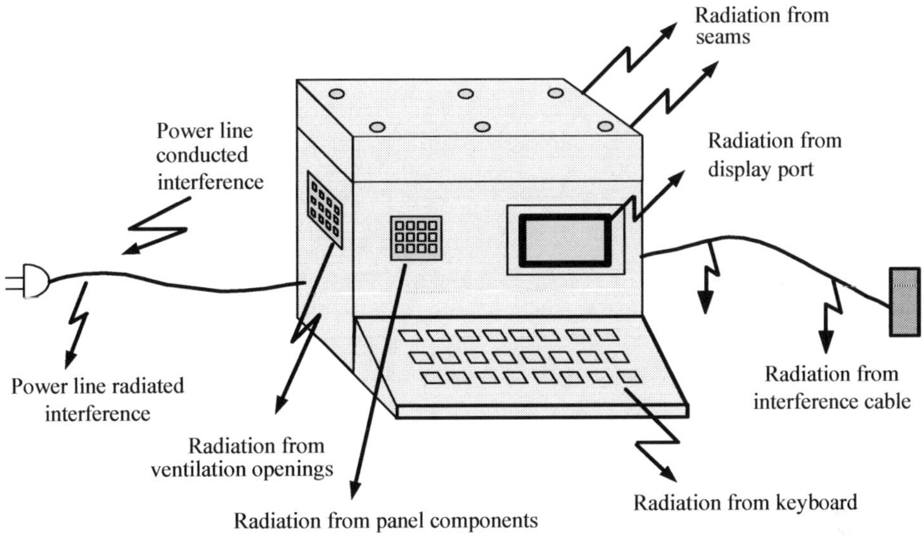

Figure 6.24 Possible shielding integrity violations from an enclosure or assembly.

Electromagnetic fields that propagate on the skin of metal enclosures with RF energy are called eddy currents. As discussed in the Section 6.5.1 (Single Apertures), a combination of magnetic field currents (eddy currents), and internal electric field propagation, creates a plane wave at the barrier junction based on Huygen's Principle. This plane wave will propagate as a radiated field. To prevent field propagation, the shield barrier must be as solid as possible with only required air vents and other openings designed not at a frequency that is resonant with any harmonic or wavelength of all signals present within the assembly.

Another area of concern lies with connectors, either metal or plastic. Figure 6.25 shows both proper and improper means of connector use within a shield barrier. When a high quality, or low impedance bond connection is made from the shield of a cable, or the housing of an I/O filter (e.g. feed-through capacitor), interface devices (buttons, indicators, display modules, shafts), all must have consideration made during the design and implementation stage to ensure proper usage and maximum benefit.

Shielded connectors used with interfaces such as Ethernet (RJ-45), HDMI and USB have a metal shell housing with spring clips designed to make contact with the thin edge of the barrier opening, generally the thickness portion of the metal.

When incorporating a connector with spring fingers into a metal bulkhead, these fingers might not make a quality low impedance bond connection to the metal housing due to contamination from oil on fingers, paint, anodization, varnishes or other non-conductive material that may be present on the surface of the metal (increasing conductivity). In addition, these fingers may not have sufficient compression force to ensure low-impedance bonding due to poor mechanical design or assembly. Without low impedance bonding, common-mode currents will be established across what is believed to be a quality ground connection, which it is not due to high transfer impedance.

The best means of securing a connector to a shield barrier is with a solid 360 degree continuous connection, which means using a serrated ring or another means of creating a secure bond connection by penetrating through any non-conductive plating or paint overspray. Another way of making a solid low-impedance bonding of a connector housing that does not have spring fingers, or clips, is use of conductive gaskets, discussed later in this Chapter, Section 6.9.

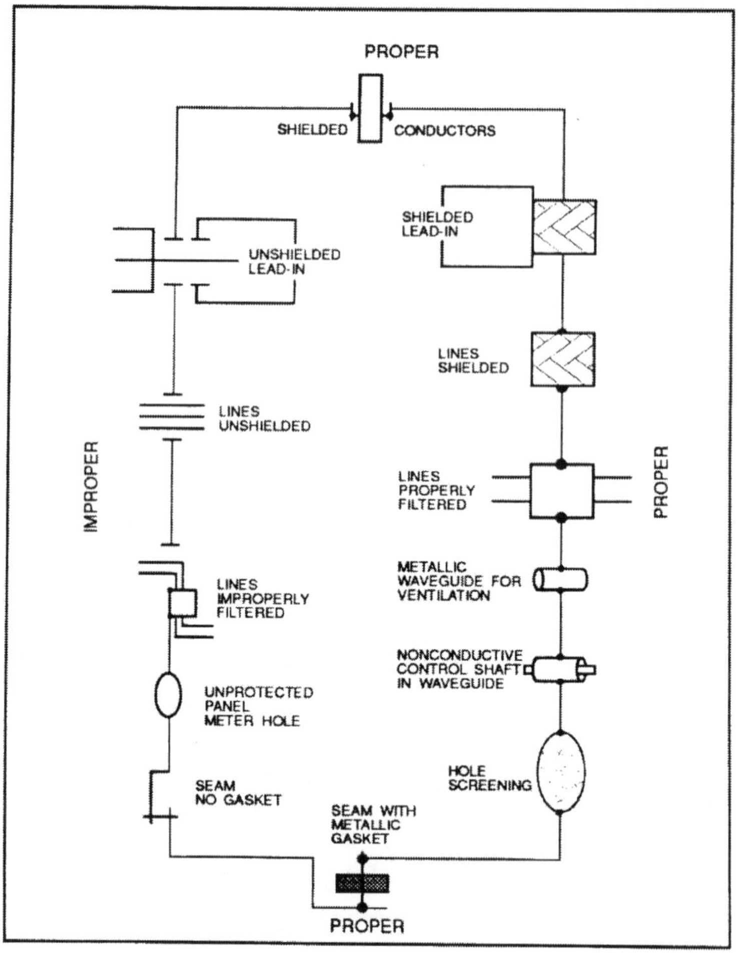

Figure 6.25 Proper and improper application of connectors penetrating a shield barrier. Courtesy: Elya Joffe

What is not generally recognized by electrical and mechanical engineers is that certain grades of metal, at least those which are low cost and contain a dull appearance, and not a bright finish, are usually compromised in terms of conductivity. Some metal may be actually "non-conductive!" Take your fingernail, a knife or sharp object and scratch the surface of the metal. If a bright line appears that is a different color than the rest of the metal, then we know the metal is compromised by poor plating and could be non-conductive, which is disastrous if one applies a conductive gasket to a varnished, non-conductive metal surface.

To validate this discovery, take an ohmmeter with buzzer or speaker mode selected. Do not worry about any reading on the meter. We only want to "hear a sound" on the speaker when the tips of the leads are touched together. What we are doing is making a test fixture to ensure DC continuity is present when the leads are touched together. After making sure the buzzer is working, take both leads and using gravity, drop touch both to the metal. Do "*not apply pressure under any condition*" to the probe tips or have them in a *vertical* orientation. Your hand should be several inches away from the tips. The probe tips should lie nearly horizontal or flat on the metal. Touch the metal with the edge of the probe, *not the end of the tip*. If the meter does not buzz or make a sound when both leads are physically touching the metal, this tells us that the metal is non-conductive. Move the leads around in a circle to ensure that physical contact is made on both tips without applied force.

Now take the probes and make them vertical and with pressure, push the tips down and penetrate the non-conductive plating. If the meter now buzzes we have DC continuity. Repeat the gravity drop test by moving the probes along the metal in circles (both leads simultaneously) and then vertical with pressure. This is an easy test to prove that the plating on the metal is either conductive or non-conductive (*EMC Troubleshooting Made Simple*)! Aluminum oxide is a hard or tough plating process when applied properly. It takes a lot of pressure to break through the plating to achieve conductivity, unlike copper oxide which is very soft and almost malleable. It is thus easier to break through a copper oxide coating to achieve full conductivity especially if low impedance bonding is required between top and bottom covers, or spring fingers on interconnects requiring low impedance bonding.

The reason why this plating is non-conductive is that *some vendors of sheet metal*, instead of providing a 0.002 inches [0.005mm] plating, may plate the metal to 50% of the required thickness. Chemical bath plating takes time by immersing metal into a chemical bath and allowing an electrolytic charge to deposit on the material. By leaving the metal in the bath for a shorter period of time for full plating, they save time and money. They essentially plate only to 0.001 inches [0.0025mm] thickness (50%), enough to prevent visible corrosion from occurring. This is a *questionable process from vendors*, who then dip the half-plated metal into a plastic varnish to achieve a total finished thickness of 0.002 inch, which takes only a few seconds to do. Varnish is less expensive that chemicals and to the untrained eye, the surface still looks conductive. The dull finish appearance on the metal may be the non-conductive varnish!

Trying to make an [assumed] low-impedance compression bonding to varnished coated metal using a flat washer or spring clip will ensure RF field propagation and development of common-mode currents. Much time will be spent trying to solve a mechanical or electrical design problem related to grounding or shielding when in fact, varnish is the culprit!

6.7 Cable Shield Grounding and Termination

A shielded cable generally has electromagnetic shielding in the form of a wire mesh or braid, aluminum foil wrapped in longitudinal or spiral form, ferrite compound, a combination of materials or other metallic elements that surrounds an inner core conductor (the transmission path or signal). The inner core conductor can be a single transmission line, multiple wires or twisted pair. The shield impedes the escape of any RF propagating field present within the cable dielectric (the material between the transmission line and shield) to the environment. This RF field is both electric and magnetic traveling in TEM mode (transverse electromagnetic). In addition, cable shielding also prevents undesired radiated fields within the environment of use from entering the core conductor and causing harmful disruption to the electromagnetic wave that defines a signal traveling down the assembly. Some cables have two or more shields for specific applications, sometimes with the same shield structure, or different such as a second layer of braid or foil. This application is for use in environmentally hostile environments, or requires a rigid protective cover. An extensive discussion on cable shielding for EMC is found in [10].

To provide an optimal cable shield from radiating common-mode RF energy, or preventing externally induced fields from causing harmful interference to internal signal lines, we must first understand different types of shield structures and whether we need to ground this shield at one side or both.

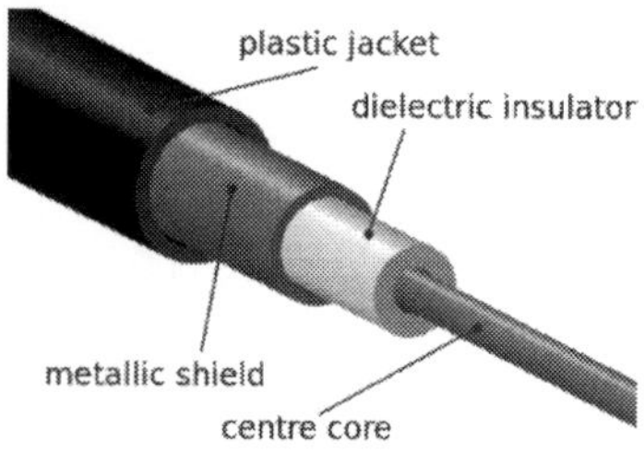

Figure 6.26 Overview on how a shield appears on a cable assembly. (Photo courtesy – Internet, owner unknown)

6.7.1 Types and Applications of Cable Shields

There are different type of shield structures for cables, each intended for a specific application and environment of use. Environment includes internal or external applications, oil or water exposure, excessive high and low temperatures, extremely high radiated electromagnetic fields, thermal consideration based on power handling capabilities (voltage and current), the need for rigid versus flexible, and many more environmental conditions of use.

Shield coverage is generally selected based upon required application and cost. The degree of need depends on frequencies of concern and noise susceptibility of the signal(s) within the cable. Termination is also critical because if not done properly, more problems will develop at the connector interface than any other part of the system.

For the majority of uses, shield effectiveness is best if the shield is intact without damage, no broken strands, or flaking foil or braid that has been stressed to the point of separation. In addition, method of securement to the connector shell housing and cleanliness is also critical for optimal performance.

The following are the most common shield structures found in commercial products, although there are many other special configurations of shield coverage possible that include triple and quad. Photographs of these common shields are shown in Fig. 6.27 while Fig. 6.28 shows the magnitude of benefit due to the number of shield layers provided.

- **Served shields** are a spiral-wound wrap with groups of many small-gauge wire strands surrounding the insulation of the conductor(s). These wire strands are loosely wrapped. They are easy to unwind and terminate but prone to be relatively high inductive as they are coiled around the cable. The high inductance causes them to not be suitable for high frequency use. This type of shield is the most flexible of all shielded cable assemblies and often used in audio applications. Served shields are usually soldered or crimped to a lug or termination post; best applicable to low-frequency signals.
- **Braided coax shields** are usually terminated by crimping or clamping and are occasionally soldered or terminated with a heat-shrink shield to protect a pigtail, if provided, commonly found in aircraft wiring harnesses. Braided non-coax shields can be soldered if the conductors are able to exit the shield barrier

through an opened space in the braid, or terminated again with a heat-shrink protected pigtail.

- **Metalized "foil" shields** consist of a metallized flexible wrap (mylar, polyimide, aluminum, etc.) spiraled around the conductor(s) and is very thin on the order of 0.0003 inch (0.0076 mm). Foil coverage can approach 100% although its resistance is far greater than other shield designs, and due to skin effect because of the thin application of the metallization, causes them not to provide significant performance at higher frequencies. Foil shields are sometimes used in combination with braided shields for enhanced shielding performance. Since metalized foil shields are usually aluminum in construction they cannot be crimped or easily terminated, therefore a "drain wire" is always provided that makes contact with the foil. This drain wire can then be soldered to a termination point. This pigtail is highly inductive and can create serious EMI problems, both emissions and immunity at higher frequencies if not terminated properly within a metalized connector housing.
- **Braided shields** are woven over and under each other to form a tight but flexible cylinder. This may result in the need to unbraid or loosen the weave in order to terminate the shield to a low-impedance bonding connection. Braided shields are easy to terminate onto coaxial connectors or to specifically designed connector back-shells to preserve the shield all the way to the connector body. Braided shields are anywhere from 82-95% optical coverage.

  There are several forms of braided shields. One is a group of small-gauge wires known as "carriers" or "bands" laid side-by-side similar to a ribbon cable with multiple wire (known as "strands). Braids can also be a "strip braid" using solid ribbons of conducting material that provides a more uniform inner surface to a coaxial conductor which has an advantage at very high frequencies. When combined with other shield designs, braided shields form a very effective EMI barrier.
- **Solid shielding** consists of a metal tube that is either rigid or semi-rigid. The most common material used is copper or aluminum. This shield surrounds the dielectric and center conductor of the transmission line. Semi-rigid coaxial cables using solid shielding can be formed manually or with a tube bending tool. Coverage is 100% and resistance is low, therefore this is the best shield possible. Common applications include short coaxial jumpers inside instruments or ground-based antenna feeds subject to extensive or harsh environmental conditions. Solid shields are usually soldered or clamped for termination purposes since soldering aluminum is not practical, low cost or easily performed.

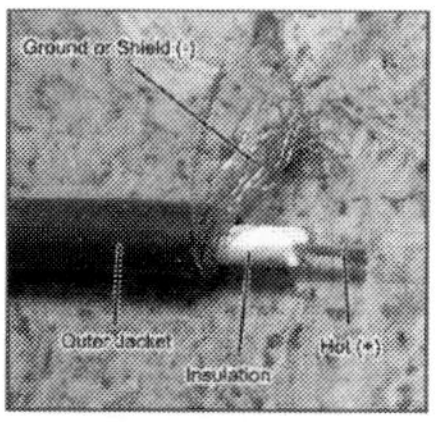

Served shield

Braided coax shield

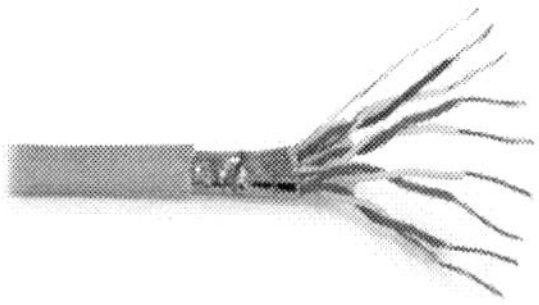

Foil shield

Foil shield with drain wire

Braided shield over foil with twisted pair

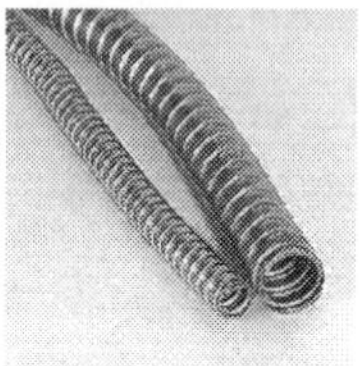

Solid shield (applied over cable)

Figure 6.27 Various types of shields used for cable assembly.
(Photo courtesy – Internet, owner unknown)

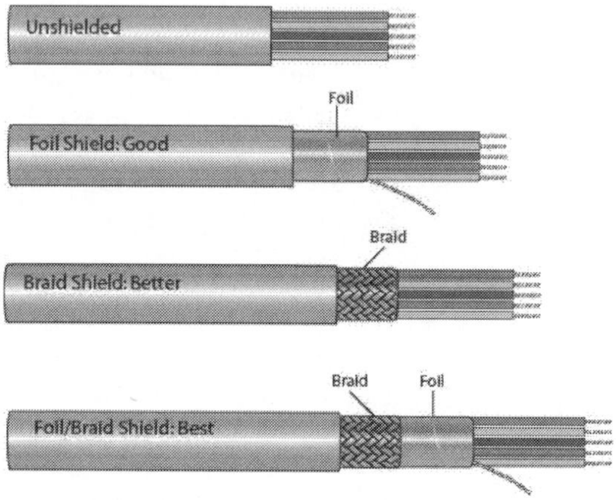

Figure 6.28 Summary of shielded cable configurations and protection levels.
(Courtesy: Alpha Wire International)

6.7.2 Grounding the Cable Shields–One End or Both

A frequent question asked by engineer is "Do we ground the shield of a cable at one end, both ends or let it float? We summarize each methodology with both advantages and disadvantages as "*Understanding Cable Shield Termination Methodologies Made Simple*".

Before an answer to the question can be given, a review of antenna configurations must occur. There are two basic types of antennas; dipole and loop. Chapter 1 provides details on how these perform based on both electric and magnetic field structures.

With regard to the electric field, this depends on the dipole antennas model that contain two parts, a driven element and return. Between the driven element and return is capacitance created by a dielectric which could be free space or physical material (e.g., polyester) between the center wire and shield. This is shown in Fig. 6.29, left side drawing. Dipoles are efficient antennas at higher frequencies, generally above 1 MHz due to the wavelength of the signal being propagated, and physical length of the elements being resonant at a particular frequency. This interconnect could be a coaxial cable connected at only one end, or the driven element left unconnected at the load.

With regard to the magnetic field, a loop antenna will exist between two circuits or interfaces where the driven element of the cable is the source path in the right side illustration of Fig. 6.29, and the shield is signal return, assuming a coax is used for signal propagation.

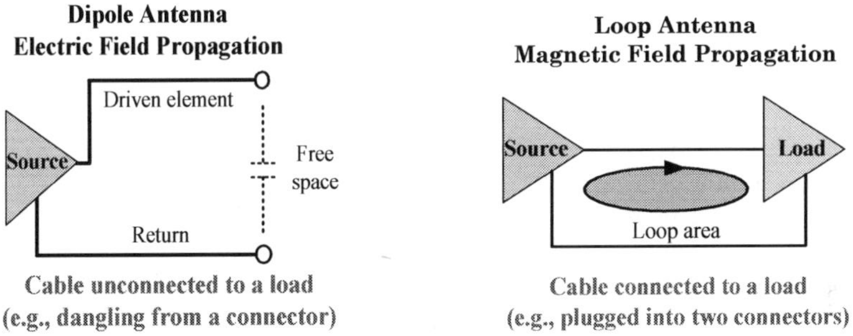

Figure 6.29 Basic antenna structures associated with cables and interconnects.

<u>Understand the context in which this discussion is being presented in a visual manner for ease of understanding EMC Made Simple</u>

Imagine a cable or two-wire transmission line (includes coaxes) plugged into a connector driven by logic circuitry (e.g. two signal paths; driven element and return). Do not terminate this cable but leave it unconnected at the far end. Electromagnetically, we created a highly efficient dipole antenna that will propagate RF energy based on the frequency of the signal and physical length of the antenna (self-resonant frequency; $f=c/\lambda$). If there is RF current driven into the cable we have an emissions problem. Conversely, any RF energy present in the environment such as commercial broadcast signals will find this antenna and may cause undesired RF to couple to the transmission line system causing harmful disruption to logic circuits. At lower frequencies, generally below 1 MHz, dipole antennas are inefficient radiators due to their physical size with regard to the wavelengths of signals present.

Loop antennas on the other hand are highly efficient at lower frequencies and for this discussion, assume 1 MHz as the high end of the magnetic field frequency range. The exact cutoff frequency is irrelevant for this visualization and discussion on how loop antennas works.

As seen in Fig. 6.29 right side, a signal is driven into the transmission line (cable) from a source element to load. An electromagnetic field (e.g., clock signal) propagates from source to load and then returns through a 0V reference creating a loop structure (magnetic field antenna). This is called differential-mode signaling. The full loop connection is always made through interconnects instead of free space capacitance.

Taking our two wire transmission line model, represented as either a dipole or loop antenna, a shield is now added to the wire assembly. For purpose of *discussion* and *visualization*, assume the cable is coaxial with a center conductor and shield. Do we ground the shield at one end or both? The answer is "*it depends*."

It depends on factors such as the frequency sent through the transmission line, environment of use (noisy electromagnetically or cleanroom quality), mechanical requirements (flexible or rigid), and whether the return signal is isolated from the shield as a separate wire, or the shield "*is*" the DC return path. To extend this visualization further, instead of having only a single center conductor we can have alternatively an unlimited number of transmission lines as the center conductor that is protected by an overall [Faraday] shield.

Regarding physical connection of the signal return path (e.g., a separate wire in the cable assembly) there are several methods of transmission line implementation, whether to connect signal return to the cable shield and then cable shield to a separate chassis ground (essentially grounding or shorting out DC return to chassis ground), or let the shield connect to chassis ground and be isolated from DC signal return (e.g. dual shield configuration. This isolated shield is such that the DC return is a separate conductive wire and not part of the overall outer shield covering).

1. **Signal return wire isolated from cable shield.** The signal return is a separate transmission line internal to the cable assembly isolated from the overall shield (not connected), or a true two-wire assembly covered with a separate overall shield. We use this type of application when the signal return cannot be connected to chassis or shield ground (e.g. video and differential pair signals, audio, RS-232, etc.). If the signal return path is accidentally connected to the shield of the cable assembly (chassis ground), shorting of the DC signal return to chassis ground may cause a functional problem since the 0V reference of the signal may be significantly different from the 0V potential of the chassis. This potential difference creates a large inductive loop as well as common-mode RF current across this impedance interface, previously discussed in Chapter 5. Across this impedance, with fixed voltage potential and common-mode current, this source of impedance become a radiating antenna structure.

2. **Signal return wire connected to cable shield.** The DC signal return path is now connected to the cable shield (RF return) making this a true, two-wire transmission line identical to that of a coaxial configuration. With this type of application, the signal return propagates on the inner portion of the shield and undesired RF energy from the external environment is on the outer half. This is due to skin effect and the frequency of the propagating field, but in most cases the thickness of a braid is many skin depths thick thus we have two separate transmission line systems that just happens to be sharing the same conductor.

   Both AC (external) and DC (internal) currents are present and do not know about the existence of each other, nor do they care. The shield of the cable is then connected to chassis or earth or chassis ground which means the DC potential of the digital signal is at chassis potential and hopefully, there is little impedance difference between circuits and the return plane in relation to chassis ground to guarantee high quality signal integrity. This type of connection is best used for lower frequency applications.

We now examine how cable assemblies with a shield behaves when we ground either one end or both. The key element to keep in mind are the following:

1. Frequency of concern.
2. Physical length of the transmission line acting as an efficient antenna structure.
3. If the cable assembly behaves as a dipole or loop antenna.

Grounding the shield at one end only

In Fig. 6.30, the cable shield is connected at only one side at the source, chassis ground. The DC signal return is not illustrated herein, only the signal line and coaxial shield. If there is no chassis ground or metal enclosure provided, then the following discussion is modified as if the signal return path is the now same as shield ground, which must be connected at both ends in order for optimal signal functionality.

Dipole antennas are efficient radiators at higher frequencies. A 1 meter long cable is resonant at 300 MHz. A 10 meter cable is resonant at 30 MHz. The cable shield at the source in Fig. 6.30 is an inefficient dipole antenna if its length is physically long ($\lambda=c/f$), therefore the answer to this question is "it depends" on the length of the cable with respect to frequency. The signal line is the driven element while the shield is the return. Thus, dipole antennas are inefficient radiators if signals are lower in frequency which means the cable has to be very long, but is very efficient in propagating an electric field if electrically short in length.

With regard to the large loop antenna that propagates magnetic fields, again physical size plays a major factor in determining if the cable shield must be grounded at one end or both. With the shield grounded at both ends, this structure forms an excellent loop antenna. If the loop is small, then magnetic fields will not efficiently radiate lower frequency signals . Thus the answer to the question is again, "it depends" on physical length.

- Dipole antenna model: Radiates an electric field but only if the physical length is short. If the cable length is long, this dipole is now an inefficient radiating structure.
- Magnetic loop antenna model: Loop currents cannot flow and is an inefficient radiator at all frequencies.

Grounding the shield at both ends

Grounding the shield on both sides removes one half of the dipole antenna structure (ground side) which means it now an inefficient radiator at all frequencies for all *electric fields* as easily observed in Fig. 6.31. However, magnetic field propagation now occurs due to loop current flow (Ampere's law). With a physically large loop area due to a long length cable, the radiated signal is generally low such as audio, well below frequencies used for telecommunication purposes and generally not a concern for EMI compliance. A physically small cable length with shield grounded at both ends is thus inefficient as a radiating structure due to the impedance of the signal with regard to free space.

Essentially, grounding the shield at "***both***" ends creates a Faraday shield like structure between source and load (Fig. 6.31) which prevents *electric field propagation* (shorts out the dipole model), but will permit *magnetic fields* to flow as a result of the loop.

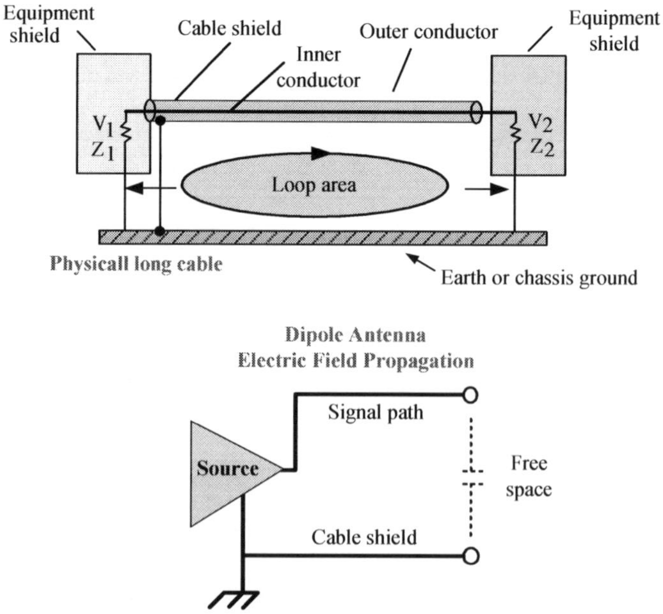

Figure 6.30 Connection of cable shield at one end–*best for lower frequency signals.* The dipole antenna model is an inefficient radiator at these lower frequencies

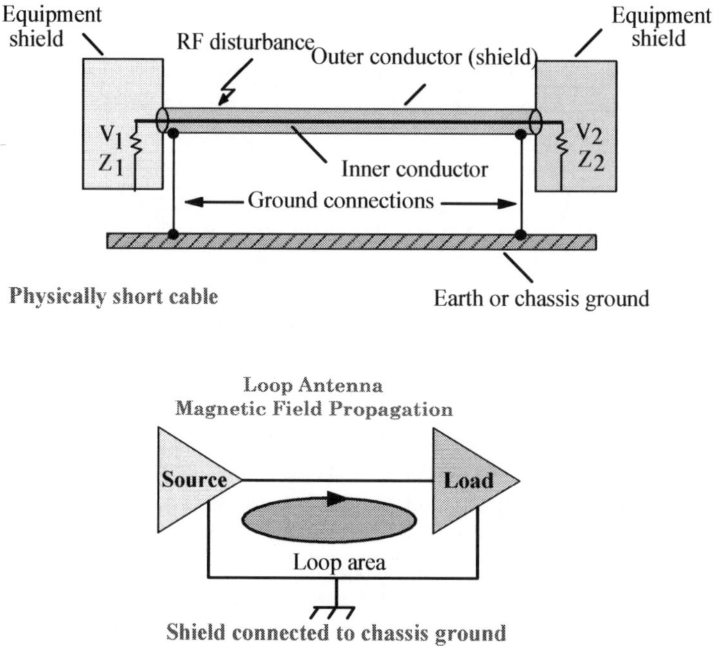

Figure 6.31 Connection of cable shield at both ends–*best for higher frequency signals.* The loop antenna model is an inefficient radiator at higher frequencies

To summarize:

Advantages of grounding a cable shield to the chassis or enclosure

- With both ends grounded, ***E***-field excitation at lower frequencies may occur but does not exist at higher frequencies due to the Faraday shield effect and grounding out of the return element of the dipole antenna structure.
- Magnetic fields could radiate if there is un-cancelled RF loop current on the shield, but due to the generally large size of the loop, this antenna model is usually an inefficient radiator at higher magnetic field frequencies.
- Having both ends grounded avoids resonances at higher frequencies for both ***E***- and ***H***-field excitations.

Disadvantages of grounding a cable shield to the chassis or enclosure

- Ground loops can occur if poor bonding is made between the shield of the cable and earth or chassis ground, especially if signal return is connected to shield ground and not as a separate transmission line path.
- If the cable length is physically short and poor bonding of the shield occurs, the interconnect cable will radiate as an efficient dipole antenna.
- If the cable shield is connected between two systems at different AC potentials (not equipotential connected) hazardous currents may be developed between the system which could be fatal but only if one physically touches the shield of the cable and is also in contact with a different "ground" reference. This is why unshielded twisted pair cables are used for long distance applications; minimize hazard of electric shock.

6.7.3 Cable Shield Termination Overview–System Level

When dealing with system level interconnects, Fig. 6.32 shows several common applications. In the top illustration, a Faraday shield configuration is present since the signal transmission line is completely contained within a metallic barrier or cable shield. Parasitic capacitance between printed circuit board and shield to chassis ground causes RF coupling to occur between assemblies, but rarely does so since the electromagnetic field propagated on the inside ½ portion of the shield (inner skin effect) and the outer ½ portion (outer surface skin effect) that carries environmental RF energy do not interact with each other.

Electromagnetic fields propagate in loops through RF parasitic coupling. Parasitic coupling to chassis ground from cables generally happens at higher frequencies, usually above 1 MHz. Above 1 MHz, depending on the physical length of the cable relative to the wavelength of a transmitted signal, common-mode current present may couple from the shield to chassis ground using the dipole antenna model.

In the bottom illustrations of Fig. 6.32, a pigtail is utilized to connect the cable shield to chassis ground. Pigtails are created by soldering a wire to the shield (usually aluminum foil or conductive thin film of metalized mylar) and securing it to the connector housing (if metallic), or a physical ground stud at the chassis enclosure either internal or external to the box. Pigtails always have (perhaps excessively) high impedance due to self-inductance of the wire ($L[dI/dt]$), which can become very large causing common-mode current to be established across the pigtail. This RF energy is observed as radiated EMI and may increase emissions by up to 40 dB!

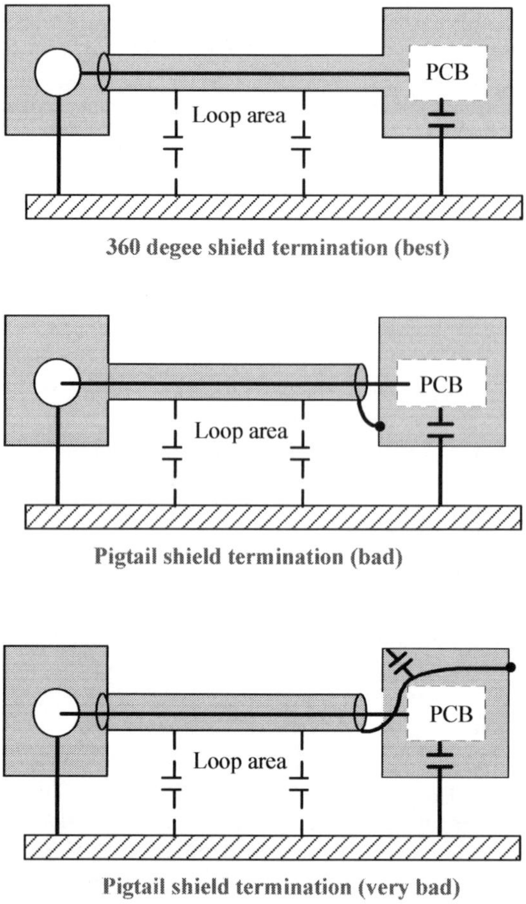

Figure 6.32 Connection of cable shield - system level.

If the pigtail is physically bonded to the chassis on the outer portion of the enclosure, several items of concern must be noted.

- Depending on the length of the cable relative to the signal frequency and wavelength ($\lambda = c/f$), we now have loop current flow if the pigtail is connected at both ends. At lower frequencies, magnetic fields will exist.
- If one end of the cable (receiver side) is not connected to chassis ground or floating, the cable shield will emulate a dipole antenna and become an efficient antenna at higher frequencies for electromagnetic field propagation.
- Common-mode current will be established across the high inductance of any pigtail and will find a way to propagate or receive unwanted EMI.
- If there is any external RF current at high levels within the environment in which the system is operated within, this RF energy can couple into the cable shield and may cause functional disruption of internal signals through the process of both inductive and capacitive coupling.

6.7.4 Implementing a Cable Shield into an Assembly

If specific directions are not provided by the cable manufacturer then:

1. Twist both signal and return conductors tightly. This enhances common-mode flux cancellation by minimizing the loop area between the two transmission lines (lowers total mutual loop inductance, *L[di/dt]*).
2. Coaxial cables are less efficient at low-to-mid frequencies if the shield is used to carry both signal return and interference currents (both common-mode and environmental). Change to a different type of cable assembly (multi-conductor with shield) using a dedicated signal return wire within the jacket.
3. The drain wire that is sometimes provided in a foil-screened cable is not designed to be used as a ground connection, but instead may be provided by the manufacturer to easily rip open the cable jacket for easy access to the internal wires (called a rip-cord) or used as a spacer to fill up the area of the wire to ensure the shield surrounds a tightly woven cable assembly.
4. If a foil shield is part of the cable, it must be bonded 360° to the connector housing without the drain wire used as a pigtail. Use the conductive side of the foil (the outer portion of a foil shield is non-conductive). If a drain wire must be used, ensure this is not exposed to the environment or bonded to the shell of a metal connector, unless there is a second shield barrier provided over the solder connection fully encapsulating all wires, include the drain.
5. Braid shields are electrically superior to overall spiral-wrap foil since the spiral-wrap configuration will separate if the cable is bent or stressed beyond a specific bend radius.
6. A tightly woven braid shield has greater optical cover and thus enhanced performance with a lower transfer impedance.
7. Braid and foil shields provided together allows for greater shield coverage and lower transfer impedance due to optical coverage of the braid (woven material) however, skin effects may limit improvement at frequencies lower than a few MHz.
8. Longitudinally wrapped foil is better than spiral foil, at the expense of cost and flexibility.
9. Individually twisted pair wire within a shielded cable assembly may need additional individual shields to prevent capacitive crosstalk between wire pairs (signal integrity concerns; also the impedance of the wire pair is significantly higher than if non-shielded).
10. Multiple isolated shields are better than non-isolated, in general.
11. Thicker, heavier shields perform generally better at the expense of a larger cable diameter; minimal flexibility, and additional weight along with cost.
12. The best possible double-screen cable is one routed in a dedicated metal conduit (pipe), grounded 360° at both ends with a low impedance bond connection.
13. Low capacitance and/or high-velocity factor cables are better at carrying data signals for longer distances.
14. For higher-rate data signals, the characteristic impedance of the signal lines and their connector interface (pins inside the connector and to printed circuit board traces) must be within a tight impedance tolerance. This can be expensive to implement due to the high cost of impedance controlled connectors.
15. A shielded cable subjected to tight bend radii or repeated flexing can destroy shield integrity by opening up gaps in the shield coverage.

6.7.5 Terminating a Cable Shield Properly

In Fig. 6.33, an example on providing a solid, low impedance bond for a cable shield is shown. The key item is that the connection occurs within a Faraday shield structure (metal connector shell assembly-upper left illustration) or to a metal ground plate within a power distribution box. Both connections are made using a ferrule secured to the main ground reference.

With a typical cable connector such as D-sub, shielded RJ-45, USB and the like, the shield must be made to the connector housing without use of a drain wire or pigtail unless the distance of the pigtail wire is "extremely" short. Again, the bond connection of the shield must be made with a low impedance connection which means soldering.

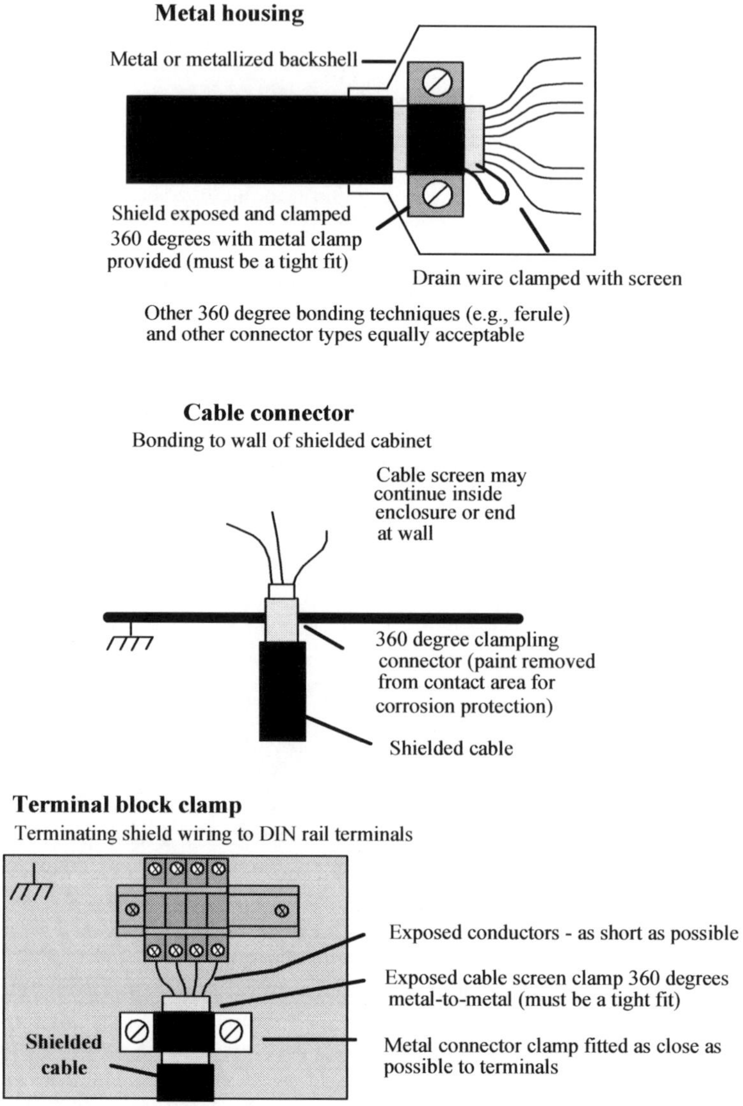

Figure 6.33 Terminating cable shields.
Courtesy: Elya Joffe

6.7.6 Aspects to Consider When Specifying a Shielded Cable

The following are to be consider on which type of cable shield to use, application dependent.

- **Mechanical compatibility:** The cable shield's outer diameter must match the connector's housing's physical dimensions, often within tight tolerances if a ferrule is used.
- **Electrochemical compatibility:** The material used for the shield must be compatible to prevent corrosion between dissimilar material such as aluminum coming into physical contact steel, which are electrogalvanic dissimilar.
- **Ease of assembly:** High workmanship standards are required for professional quality of the cable assembly. Can the shield type chosen be easily grounded or opened up to allow access to inner conductors without damage?
- **Conductivity:** Full low-impedance connection must be made across joints, which directly affects shielding effectiveness.
- **Physical manipulation of the cable:** Excessive bending and movement may damage the shield.
- **Environment of use:** The shield must be able to retain quality performance under harsh use, repeated handling, flexing, sheer and stress damage due to installation in metal conduit, moisture, ingress, outgassing and the like.

6.8 Shielded Compartments

There are various methodologies when implementing a shield barrier on a printed circuit board. These methodologies include use at the enclosure, functional and component level. At the enclosure level we create a highly efficient Faraday shield (metal box) sometimes containing air vents (slots and apertures) or openings for cable and display penetration. At the functional or component level, we keep radiated fields from causing harmful interference to other sections of the printed circuit board or active circuits located in close proximity. This on-board protection is required for self-compatibility, not just to comply with regulatory requirements. Conversely, radiated fields external to the shield are attenuated from entering the system to minimize operational disruption .

An example of the need for self-compatibility is ensuring the transmitter of a wireless device does not interfere with the receiver section or other processing elements located within an intentional transmitting devices such as WiFi or GPS.

.

The need for providing a shielded compartment is required for achieving:

- Isolation of "noisy" from "sensitive" circuits and components by using a barrier partition to separate areas into functional sections on a printed circuit board (may also be accomplished using a via fence or split plane,s if no metal cover can be provided as a shield partition).
- Maintaining bonding and grounding integrity for:
  - Bonding of internal structural members to chassis ground.
  - Grounding of circuits and components to an appropriate 0V reference.
- Internal cable layout, routing and shielding adjacent to the metal chassis to prevent internal radiated noise coupling (provides for flux cancellation with the metal enclosure acting as the RF return path).

The need for ensuring isolation between components or functional subsections is because we must protect sensitive or RF noisy circuits from disrupting other functional areas, electromagnetic wise. High intensity radiated fields lead to EMI interactions between circuits. Isolation between high- and low-level analog components is required along with separation between high- and low-frequency bandwidth devices. To achieve high levels of isolation the system should be divided into compartments as shown in Fig. 6.34.

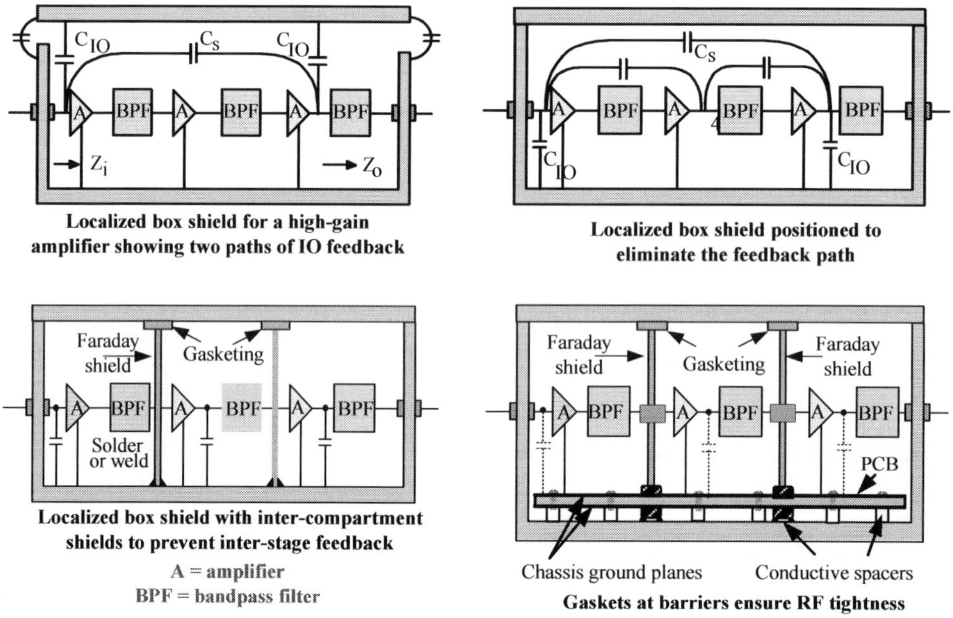

Figure 6.34 Barrier partitions with shielding protection within an enclosure.

When a printed circuit board's interface penetrates the barrier between compartments, isolation may be compromised. One way to minimize this effect is to use both top and bottom chassis planes on the printed circuit board bonded to the enclosure housing using a low impedance bond method (conductive gasket or screw/standoff). At penetration points, a gasket may be inserted along with filters, if necessary, in order to ensure no undesired coupling occurs between sections.

When using assemblies or partitions within metal enclosures:

- All assemblies should be bonded to the enclosure using high quality methods such as standoffs connected to the 0V reference, or another ground reference with the standoff making connection to the annular pad of the mounting via on the printed circuit board bottom layer.
- Use a direct, continuous metal to metal contact with the best means of securement possible (low impedance bonding).
- When not feasible, only one common grounding point shall be implemented (single-point ground) such as chassis ground, ESD ground, analog ground, digital ground or 0V reference.

When using assemblies without a metallic enclosure (e.g., a plastic housing)

- Assemblies without a metallic enclosure shall be grounded to an appropriate location according to the "grounding tree" methodology (Chapter 5), usually through the motherboard where one ground location is used for implementing the "system's central reference point."

- Grounding of circuits and systems must be in accordance with the grounding scheme/tree chosen. Attempt to identify potential ground loops before implementation between printed circuit board and enclosure.

6.8.1 Board Level Component Shields

With high technology products, many are now becoming essentially a system-on-system design operating in the GHz range. With multiple systems or modules located on a printed circuit board, localized shielding of each element may be required. These elements include for example; transmitting and receive sections, processors, memory arrays, wireless modules (GPS, WiFi, Bluetooth, etc.), high definition video control logic, high performance audio processing, and other elements that may be easily subjected to RF field disruption either from within the assembly or the external environment.

In addition to keeping RF fields from coupling to components and transmission lines, it sometimes become necessary to reduce or minimize undesired radiated common-mode current from digital switching elements in order to comply with regulatory limits, especially those used on printed circuit boards installed in a plastic enclosure which provides minimal shielding protection. Localized shields on printed circuit boards are usually inexpensive and may provide sufficient shielding effectiveness, better than the use of metalized enclosures or the use of aluminum shields.

Examples of component shields used on a printed circuit board are shown in Fig. 6.35. These shields come in both single and two piece assemblies. Securement is by soldering or using a side support structure upon which a top enclosure snap fits onto a vertical support, if a two piece assembly. A two piece assembly allows easy access to components for repair, replacement, upgrading of a module, or ease of manufacturing.

Almost all wireless systems contain both transmit and receive sections. Receivers must be very sensitive to low levels of RF energy for signal reception and can be easily disrupted by undesired external fields. These undesired external fields come from either the transmitter section, the CPU and clock generation circuitry, video display drivers, network interfaces or a host of other digital switching elements located in the near vicinity. Only a specific signal at a particular frequency from a carefully designed and tuned antenna is permitted to be picked up by the receiving section. The same is for the transmitting section and its' antenna. The transmitter could cause harmful interference to other sections of the system or electrical devices nearby and needs to be shielded, except for the physical antenna element.

What is discussed above is self-compatibility. Component shields guarantee functional operation is ensured between functional sections, or to meet regulatory compliance requirements both emissions and immunity.

6.9 Gaskets-Application and Implementation

Conductive gaskets make a low impedance bond between two conductive surfaces to ensure undesired RF energy does not leave or enter an enclosure or shield barrier. An ideal shield barrier contains a continuous seal between two mating surfaces with no apertures or openings, and is sometimes called a Faraday cage or Faraday shield. Samples of shield designs are found in Fig. 6.35.

Courtesy-Tech-Etch

Figure 6.35 Printed circuit board component shield samples.

Gaskets are used for either temporary or semi-permanent sealing applications between joints of metallic structures.

Temporary applications

- Securing access doors to enclosures, cabinets or equipment
- Mounting cover plates or panels for equipment maintenance, alignment or other purposes

Semi-permanent applications

- Mounting either screen or conducted glass windows to housings containing electrical or electronic test equipment
- Mounting honeycomb or other ventilation covers to enclosure cabinets or equipment
- Securing parallel members of equipment housing to a frame structure using machine screws

Metallic enclosures utilizes two or more pieces. Designers put a printed circuit board, power supply and other elements in an enclosure first before adding or installing a top cover. Even if a hinge is used to secure the cover or door, there are still two pieces that need to be connected together. Securement can be made by a temporary locking mechanism, rivets or screws. If screws or rivets are used, the distance spacing between each should be spaced based on a simple calculation and not symmetrical, which mechanical designers prefer for ease of manufacturing and symmetry. The physical distance between screws or rivets must be calculated to be at a distance less than $1/20\lambda$ (or $\lambda/20$) of the 10th harmonic of the highest generating frequency of the system where λ = wavelength of the RF field ($\lambda=c/f$).

Any straight line linear dimension between chassis securement screw locations permits undesired common-mode RF energy to propagate both to/from the system. Magnetic field current propagating on the surface of the enclosure (skin effect) will find this slot, and with the electric field bouncing around both inside and outside, an electromagnetic field is created and launched into free space (per Huygen's Principle).

The maximum physical distance spacing between screws, rivets or locking mechanism is calculated by (6.16). If the distance spacing between securement points is greater in distance relative to the slot antenna's efficient resonant frequency, then a gasket will be required to seal the gap. It is easier to use an inexpensive gasket for optimal shielding protection than to many install screws, for example less than ½ inch (1.3 cm) apart, which is a typical dimension for an efficient slot antenna operating in the GHz range.

Based on antenna theory, shielding effectiveness is easily calculated and is most efficient when its maximum linear dimension is equal to ½ wavelength, defined as 0 dB relative to shielding effectiveness. As the aperture becomes smaller, radiation efficiency decreases at a rate of 20 dB per decade, or increases at the same rate if the slot becomes physically larger. For an aperture with a maximum linear dimension equal to or less than ½ wavelength, shielding effectiveness in dB is equal to (6.17)

$$S = k\ log\left(\frac{\lambda}{2l}\right) \tag{6.17}$$

where:
S = total shielding effectiveness (dB)
k = 20 for a slit or 40 for a round hole
λ = wavelength (meters)
l = longest linear dimensions of the aperture

Equation 6.16 can be rewritten as (6.18).

$$S = 20\ log\left(\frac{150}{f_{MHz}\ l_{meters}}\right) \tag{6.18}$$

The determine maximum linear distance between securement locations before shielding effectiveness starts to decrease is provided by (6.19).

$$l_{meters} = \left(\frac{150}{10^{\frac{S}{20}}\ f_{MHz}}\right) \tag{6.19}$$

6.9.1 Material composition and performance

When selecting a conductive gasket for a particular application, one must choose material based on performance characteristic and environment of use. The type and thickness of the housing material can have a significant impact on performance. For example, protective coatings (such as chromate on aluminum) will create a thin non-conductive layer causing EMI by preventing an effective low-impedance bond connection to be established. When this situation occurs (e.g., non-conductive layer), a gasket that can pierce the insulated coating such as oriented wire or knitted wire mesh will be required.

The mechanical strength of the enclosure requiring a gasket can also affect EMI performance. For example, a flexible metalized plated plastic box may result in having many inadvertent slot antennas throughout the frequency spectrum if the gasket does not compress to an appropriate level and recover to its original dimensions at the same rate when the box is opened (non-compressed state). This is called compression set if the gasket remains deformed when compression is removed.

Due to skin effect of a propagating field on a conductive surface, the electromagnetic field must transfer into the gasket and then retransfer to a second surface. This means that full conductivity must be present throughout the entire skin of the gasket. A composite material having a conductive filler or element with a small unit size chemical structure (microns) is more effective than material having conductive fillers with a large unit size in comparison, relative to a micron.

For effective use of the entire cross section of a filler compound unit for shielding, the unit size of the filler should be comparable to or less than the skin depth of the field for optimal performance. Filler material that is very small is preferred for higher performance but is not commonly available at low cost, especially when the filler size physically decreases to near atomic levels. Metal coated polymer fibers or particles are thus used as fillers for this reason, namely cost. Metal coated polymer fibers or particles also suffer from the fact that the polymer interior of each fiber or particle does not contribute to overall shielding effectiveness if each particle is not physically touching each other in a tightly controlled manner.

An elastomer is a flexible and resilient material but is unable to provide quality shielding performance unless coated with a conducting substrate (e.g., a metal coating or metallization). The elastomer comes with a conductive filler during the extrusion process (typically metal particles). This type of coating suffers from poor wear resistance. Conductive filler when used in this manner of manufacturing results in a decrease in resilience, especially if a high filler volume fraction usually required for sufficient shielding effectiveness is used. As resilience decreases, it becomes more severe as the filler concentration increases. The use of a filler material is however effective even at a low volume fraction and is highly desirable for optimal performance for many applications.

6.9.2 Common Gasket Material

There are many gaskets varieties, all based on geometry and material selection. There are four principle categories of shielding gaskets, among many other types detailed below: knitted wire mesh, beryllium copper or other metal spring figure composition, conductive particle filled elastomers, and conductive fabric or cloth over foam. Each has advantages and disadvantages depending on application. Regardless of the type of material selected, important factors to considered are: RF impedance ($R+jX$, where R=resistance and jX=inductive reactance); effectiveness; material compatibility; compression forces, compression compressibility, compression range, compression set; and environmental sealing.

Other factors to consider when choosing a gasket includes:

- Desired frequency of protection
- Operating environment
- Load and forces applied to the material
- Corrosive considerations
- Attenuation performance
- Storage environment
- Cycle life
- Exposure to nuclear, biological or chemicals
- Electrical requirements
- Space and weight considerations
- Material thickness and alloy used
- Recyclability
- Flammability requirement for product safety concerns
- Fastening/mounting methods
- Cost

The following are common material used for EMI protection. Each type is dependent on application, environment of use, mechanical design constraints and other items in greater detailed within Tables 6.3 and 6.4. Figure 6.36 shows a sample of typical gaskets types, however many designs applications are customized and not all varieties are shown.

- ***Aluminum foil***: Excellent for electric fields but provides minimal shielding effectiveness for magnetic fields.
- ***Knitted wire mesh***: Tin-plated, copper-clad or steel-knitted wire mesh, that include monel and aluminum. Knitted wire mesh is a cost-effective approach for low cycling applications requiring different forms and shapes. This material is designed to provide shielding for enclosure joints, door contacts and cables. This mesh has the ability to penetrate plating on metal to enhance low impedance bonding and offers high shielding effectiveness over a broad frequency range, available in many sizes, shapes and knit construction.
- ***Oriented wire mesh***: Oriented array of wires of monel or aluminum are impregnated into a solid or sponge silicone, designed to be used in military, industrial and commercial application requiring shielding and grounding in conjunction with environmental sealing, or repeated opening and closing of access doors and panels using high compression forces. The stiff wires in the mesh penetrates plating on metal which enhances low impedance bonding.
- ***Conductive elastomer:*** This silicone based elastomer provides high shielding effectiveness and improved corrosion resistance for both environmental sealing and shielding. Conductive elastomers provide shielding effectiveness up to 120 dB at 10 GHz with a wide choice of profiles to fit a large range of application. Conductive fillers include, but are not limited to the following material: Carbon; passivated aluminum; silver plated aluminum/copper/glass/nickel; nickel-coated carbon; silver and silicon/florosilicone rubber.
- ***Metal strips***: [Tin-plated] beryllium copper designed to be placed between two flat surfaces (e.g., metal case and its top cover, or for doors used with anechoic chambers). Beryllium copper is a highly conductive, corrosion-resistant spring material. Tin plating may be used to lower contact resistance to other metal surfaces and is corrosion compatible with aluminum in the presence of moisture and salt spray. Note: Beryllium in its native chemical composition when extracted from the earth is extremely toxic, but once treated or processed becomes safe to handle.
- ***Conductive coating/paint***: Silver, nickel or copper applied to a surface to enhance conductivity and provide a pre-defined level of shielding effectiveness.
- ***Conductive cloth over foam***: Conductive cloth over foam is low cost material that is extremely flexible and thin, excellent for I/O connectors that must be bonded to a panel and which can take significant compression. Conductive cloth offers very small size stitching of metalized nylon. This cloth is extremely effective in shielding as it offers a smooth, soft surface. Holes can be punched into the material if the foam is rigid. This type of material comes in every conceivable size and shape including large size pieces that can be utilized on the walls of a room if a shielding material is required (e.g., a shielded tent for EMC testing). There is another type of conductive cloth which is similar to a sheet of fabric without the use of foam material for applications requiring low compression force.
- ***Form in Place***: This type of conductive material, similar to a conductive elastomer, is dispensed onto any conductive surface of an enclosure that requires environment sealing, has complex or rounded surfaces, or miniature devices requiring a precision gasket to protected again EMI fields and the environment.

Table 6.3 Properties-common types of RF gaskets and fingers.

Type	Comments
Gasket–cloth over foam	Conductive cloth over open cell foam. Very compressible. Wide variety of form factors (rectangular, P-shape, hinged, with/without adhesive, etc.). Related varieties include plated rubber foam. Wide contact area slows onset of galvanic problems.
Gasket–loaded rubber	Includes very high-performance types; military heritage-combines hermetic seal with wide contact area. Limited material compressibility (unless hollow extrusion; wide variety of extruded or cut shapes possible). Can be expensive, sometimes intolerant of rubbing contact.
Gasket–mesh	Wire and wire over foam types. Not currently popular in commercial applications.
Spring fingers–strip (Beryllium copper)	Relatively noble spring metal (e.g., beryllium copper, sometimes with nickel or tin plating) can lead to galvanic issues. High conductivity. Generally good tolerance to wiping contact on insertion along finger orientation. Relatively small area of contact-line or multiple points. Wide variety of shapes. Highly compressible. Can easily be damaged with shearing thus can only be used in a compression state.
Spring coil	Continuous flat coil set in a groove. Spring finger variation.

Table 6.4 Characteristics of common gasket materials.

Material	Chief Advantages	Chief Limitations
Compressed knitted wire	Most resilient of all metal gaskets. Available in variety of thickness and resiliencies, used in combination with neoprene and silicone.	Not available in sheets. Must be 0.04" (0.10cm) or thicker. Subject to compression set.
Brass or beryllium copper	Best break-through for corrosion protection films.	Not truly resilient or generally reusable. Can be easily damaged.
Oriented wires in rubber silicone	Works as both a fluid and RF seal. Effective against corrosion if ends of wire are sharp.	Might require wider or thicker size gasket for same effectiveness, which reduces with mechanical use.
Aluminum screen impregnated with neoprene	Combines fluid and RF seal into one material. Thinnest gasket possible.	Very low resiliency (high flange pressure required).
Soft metals	Cheapest for small size use.	Cold flows. Low resiliency.

Metal over rubber	Takes advantage of the resiliency of rubber.	Foil cracks or shifts position. Generally low insertion loss with poor RF properties.
Conductive rubber (carbon filled)	Combines fluid and RF seal into one material.	Provides moderate insertion loss.
Conductive rubber (silver filled)	Combines fluid and RF seal into one material. Excellent resilience with low compression set. Reusable and available in almost any shape.	Not as effective as metal for use with magnetic fields. May require salt spray environmental protection.

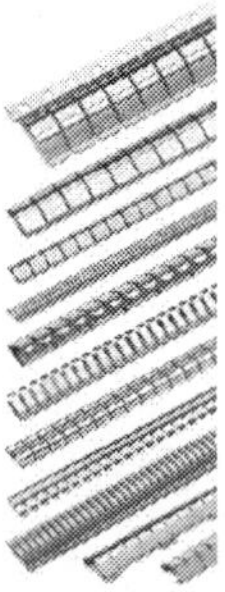

Fingerstock Metal Strips

Knitted Wire Mesh

Conductive Elastomer

Conductive Foam/Fiber

Metalized Fabric

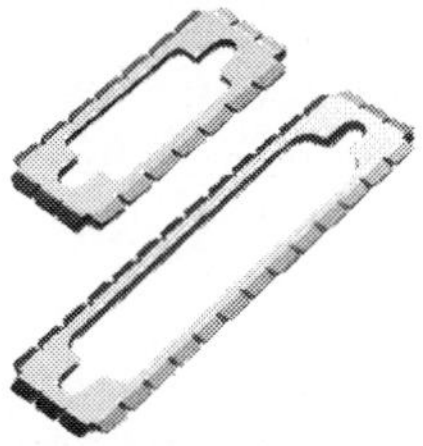

D-Sub Connector (Beryllium Copper)

Figure 6.36 Sample display of different gasket material.
(Courtesy-Tech-Etch)

6.9.3 Environmental Aspects of Gasket Use

Depending on where a product is intended to be used (environment), a gasket may be required to also be an environmental seal. This means protection from dust, moisture, vapors, salt spray and liquids. Selection of a sealing elastomer (non-conductive part of a two-part gasket) is as important as conductivity. To seal against dust and moisture, flat or strip EMI gaskets secured to a non-conductive sponge or solid elastomer may be adequate. Sponge elastomers, characterized by compressibility, are ideally suited for use in metal enclosures having uneven joints or warp with a required closure pressure that is generally low, between 5 and 15 psi. To avoid over-compressing sponge elastomers, physical compression stops are recommended. These stops can be designed into the enclosure or embedded in the material as a custom design feature.

The listing below presents important characteristics of the more commonly used elastomers:

- **Neoprene.** This poplar material will withstand temperatures ranging from -54°C to +100°C. Neoprene is lightly resistant to normal environmental conditions, moisture and some hydrocarbons. It is the least expensive of the synthetic rubber materials and best suited from a cost standpoint for commercial applications.
- **Silicone.** Silicone has outstanding physical characteristics and can operate continuously at temperatures ranging from -62°C to +260°C for solid versions, and -75°C to +205°C for closed cell sponge composition. Even under the severest of temperature extremes, silicone remains flexible and highly resistant to water and swelling in the presence of hydrocarbons.
- **Buna-n.** Butadiene-Acrylonitrile resists swelling in the presence of most oils, has moderate strength and heat resistance. It is not generally suited for low temperature applications.
- **Natural Rubber.** Natural rubber has good resistance to acids and alkali's when specially treated and can be used up to 160°C. It is resilient and impervious to water but will crack in a highly oxidizing (ozone) atmosphere and tends to swell in the presence of oils.
- **Fluorosilicone.** Fluorosilicone has the same characteristics of silicone with improved resistance to petroleum oils, fuels and silicone based oils. Since most environmental material used with EMI or conductive gaskets have elastomeric properties for stretch and compressibility, one must be very specific in dimensional tolerance during incorporation into a product.

6.9.4 Mechanical Problems When Using Gaskets

When using any type of gasket, it is far more efficient to have this material be initially designed into the product and not applied after the design has been frozen or is already in production. Once in production and the system fails EMC tests, one must find a quick fix by incorporating a gasket into an enclosure that was not designed to accommodate it, and in fact may never work in an optimal manner for the life of the product. It is sometimes considered [by management] to be a less costly solution to patch up a hole or seam than redesign a printed circuit board. The enclosure itself is an antenna no matter how well we seal gaps. Openings still exist for input and output signals (display, power cord, I/O cables, etc.) and a gasket will not minimize development of common-mode RF energy.

The main area related to mechanical problems that occur when choosing a gasket includes continuity and connectivity, compression, corrosion, joint unevenness and galvanic corrosion. It is critical for the mechanical designer to study vendor application notes carefully for conditions of use and incorporation!

6.9.4.1 Continuity and connectivity

- Knowing where gaps and slots for ventilation are located which are electrically large relative to the wavelength of signals present from internal printed circuit boards, is critical and do these openings require some type of shielding protection with a mesh type of gasket that allows for both air flow and EMI shielding simultaneously?
- Panels may not mate mechanically tight due to design errors or surface roughness between the panels leaving an air gap in the assembly, which in turn becomes a slot antenna (see *Joint Unevenness* below).
- The gasket may not be properly dimensioned for both compressed and non-compressed states. This is the cause for most gasket implementations to not be optimal after first use.
- Is there sufficient pressure (torque) to ensure full contact between gasket and metal without causing damage, and has too much force applied during installation that will degrade gasket performance.
- When a cover is removed for access to internal components, has "gasket set" occurred. This means the gasket is deformed permanently and does not return to its uncompressed state for optimal performance. When this happens, the gasket may now be ineffective for shielding purposes when the unit is reassembled by losing connectivity and continuity.
- Has any damage occurred to the gasket during installation and handling? Did hands that are dirty or contain oils touch the gasket and contaminate the surface coating?
- When using pressure sensitive adhesive, most of the time this double-stick tape is non-conductive, meaning the piece part is less expensive. When choosing a gasket family, if the physical dimension of the gasket is small, then the adhesive must be conductive (more expensive). If the gasket is several orders of magnitude wider than the width of the adhesive area (tape), then the tape can be non-conductive since under compression, the distance spacing from edge of the adhesive to edge of the gasket will ensure contact is being made.
- When using metal spring fingers (beryllium copper) on doors (e.g., anechoic chamber), periodic cleaning must occur with a solvent that removes dirt and oil yet does not provide an insulated barrier. Acetone is a widely used solvent however, this is a hazardous cleaning fluid that must be used with extreme care.
- Improper masking of paint or varnish on any metal surface will prevent a high quality, low impedance bond connection rendering the gasket useless. The metal surface must be clean of dirt, oil, contaminates or coatings (paint, anodization, varnishes, etc.).
- Non-conductive, anti-corrosion treatments (type of plating used on steel and aluminum) may prevent the gasket from making a solid connection between elements (must select gasket based on galvanic compatibilities).
- Environmental conditions can affect both electrical and mechanical performance.
- Oxidization is common with metal based gaskets, forming an insulating surface based on the environment where the system is located or operated within such as outdoor, indoor, wet, oily, hot or cold.
- Incompatibility between the material used for the gasket and contact area of a metal surface. Aluminum or zinc plated steel causes continuity and connectivity problems.

6.9.4.2 Gasket Compression (Compression Set)

Over compressing a gasket occurs frequently during manufacturing. Most gaskets have a mechanically designed recovery aspect designed to return to its non-compressed state. Too much compression may force the gasket material to not recover to original specifications, called compression set. This is especially true of gaskets incorporated with a foam or rubber material depending on type of material chosen for either internal or external applications. Once over compressed, the gasket may never recovery and therefore will no longer maintain an RF seal or barrier. This generally does not occur with elastomeric materials since elastomers (rubbers) are not compressible but deflect compression.

Selection of material for a seam or door which must be opened and closed frequently is, to a large extent, determined by the compression set characteristics of the gasket. Most of the resilient gasket materials will retain their original height after a sufficient length of time when subjected to moderate closing forces. As deflection pressure is increased, compression set increases.

Another consideration for pressure seals is the chemical permeability of the elastomer compound, if this material is chosen for a specific application. Chemical permeability is defined as the volume (cm^3) of gas that will permeate in one second through a specimen of one cubic centimeter. Finally, RF leakage can be reduced by using conductive grease if the application mandates use of this lubricant, generally found in highly corrosive environments. Compatibility of the grease with the seal elastomer and its application must be verified before selection.

6.9.4.3 Corrosion

One must select shielding material and finish which inhibits corrosion and are compatible with the enclosure material. A small amount of corrosion can render a gasket ineffective.

Equipment used in harsh environments, such as military applications (ships, planes and the battlefield), medical equipment, and transportation systems or industrial locations are often subject to high levels of corrosion. Corrosion is usually not a concern for commercial products since they are generally used in a fixed, office environment or one's home. Causes of corrosion include salt spray (ocean), air pollution (smog), ultraviolet light, high temperatures, ice and freezing temperatures below 0°C, metallic shavings floating in the air within a manufacturing environment, exposure to chemicals or similar environmental exposure.

To prevent corrosion, gaskets must be chosen to minimize the effects of dissimilar metals and electrochemical action. This is due to both metal contact and the environment. Another option to prevent corrosion from the environment is to seal out moisture at the gasket interface with an environmental seal (non-conductive gasket) secured to the conductive material (wire mesh or elastomer). When using a hybrid gasket, install the environmental seal on the *outer or outside portion* to protect the conductive portion from being damaged by the environment.

6.9.4.4 Joint Unevenness

Mechanical and electrical design considerations are interdependent yet engineers with different levels of expertise must work together. One important dependency is joint unevenness between metal surfaces. Joint unevenness refers to the degree of mismatch between metal surfaces and results when the mating portion makes contact at irregular intervals due to surface roughness or bowing of cover plates. Surface roughness is a result of the manufacturing process of the metal. At the microscopic level any metal surface is never smooth, illustrated in Fig. 6.37. There is nothing we can do about surface roughness since this is a manufacturing process at the factory that pours and forms sheet metal.

Joint unevenness also occurs due to:

- Improper material selection for environment of use and application.
- Thickness of the cover plate where weight is uneven causing off-centered loading effects.
- Too few fasteners to secure panels or excessive and/or uneven screw/bolt alignment.
- Improper sizing of gasket material under both compression and non-compression.
- Excessive heat warping metal or other material being used.
- Flexing of material (metal or gasket) either in the compressed or non-compressed stage.

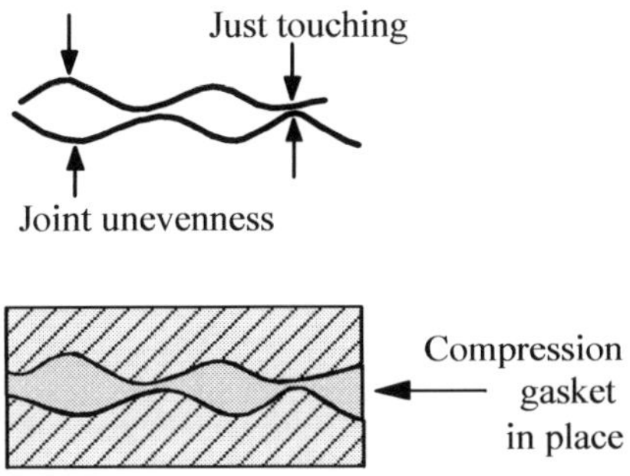

Figure 6.37 Surface roughness of metal panels requiring a gasket for full continuity.

Ideally gaskets should make even, continuous and uniform contact with all seam surfaces. Surfaces should be free of contaminates and insulating materials such as paint, varnishes or other decorative finishes. Joint unevenness and surface conditions can have adverse effects on the electrical performance of a gasket. An ideal gasket will remove irregularities in the metal without losing its properties of resiliency, stability or conductivity.

The primary function of a gasket is to maximize coupling efficiency of a seam between two boundary conditions. To provide for effective shielding, the seam design should incorporate the following:

- The mating surface should be as flat as possible, depending on type of metal chosen.
- The width of the flange of the mating surface should be at least five (5) times the maximum amount of expected joint unevenness of the material used.
- The mating surfaces requiring dissimilar materials should be selected based on electrochemical corrosion from the groupings of metals listed in Table 6.5. Metals at opposite ends of the table should be avoided with direct contact to other material.
- The mating surfaces should be cleaned first to remove all contaminates (dirt and oxide films) just prior to assembly.
- Dielectric protective/decorative coatings should be removed in the mating surface area before installation of the gasket. This mating surface should be

treated with chromate conversion coating for aluminum and plated with tin, nickel or zinc for steel.

- Fasteners such as screws should be tightened beginning from the middle of the longest seam toward the ends to minimize buckling and warping.

6.9.4.5 Gasket Design Errors with Elastomers

Figure 6.38 show various problems that can occur during the design cycle when using elastomers. The following guidelines should be followed.

- Minimum gasket width should not be less than one half of the thickness (height).
- Minimum distance from bolt hole (or compression stop) to nearest edge of sealing gasket should not be less than the thickness of the gasket material. When bolt holes must be closer, use U-shaped slots.
- Minimum hole diameter must not be less than gasket thickness.
- Tolerances should be conservative whenever possible.

DETAIL	WHY FAULTY	SUGGESTED REMEDY
Bolt holes close to edge	Causes breakage in stripping and assembly.	Projection or "ear" Notch instead of hole.
Metalworking tolerances applied to gasket thickness, diameters, length, width, etc.	Results in perfectly usable parts being rejected at incoming inspection. Requires time and correspondence to reach agreement on practical limits. Increases cost of parts and tooling. Delays deliveries.	Most gasket materials are compressible. Many are affected by humidity changes. Try standard or commercial tolerances before concluding that special accuracy is required.
Transference of fillets, radii, etc., from mating metal parts to gasket.	Unless part is molded, such features mean extra operations and higher cost.	Most gasket stocks will conform to mating parts without preshaping. Be sure radii, chamfers, etc., are functional, not merely copied from metal members
Thin walls, delicate cross section in relation to overall size.	High scrap loss; stretching or distortion in shipment or use. Restricts choice to high tensile strength materials.	Have the gasket in mind during early design stages.
Large gaskets mode in sections with beveled joints.	Extra operations to skive. Extra operations to glue. Difficult to obtain smooth, even joints without steps or traverse grooves.	Die-cut dovetail joint.

Figure 6.38 Gasket design problems when using elastomers. (Courtesy–Technit, a Division of Parker Hannifin Corp.)

6.9.4.6 Electrochemical Grouping

The primary function of a gasket is to ensure two mating surfaces make contact. In order to guarantee optimal contact, the surface finish of the metal must be compatible with the gasket material. By having optimal low impedance bonding, transfer impedance will also be low providing enhanced shielding performance. Galvanic corrosion is a major concern between dissimilar metals and gasket, and may introduce insulating surface corrosion.

Insulating surface corrosion occurs between dissimilar metals in the presence of an electrolyte. The rate of corrosion depends on the electrochemical potential between two dissimilar metals and conditions under which contact is made. Materials must be selected which provides the least corrosion activity due to galvanic action when in contact with each other for an extended period of time, along with an appropriate protective finish. Maximum galvanic activity occurs when dissimilar metals are exposed to salt atmosphere, fuels, chemicals and other liquids which may act as electrolytes.

To minimize corrosion, all surfaces must be free of moisture and containments. Therefore, EMI gaskets used in a corrosive atmosphere must be selected or treated to ensure compatibility. The design goal should be to use metals in the same group. When this is not feasible, a protective finish must be used to retard corrosion.

Electrochemical compatibility ensures corrosion does not occur. Select adjacent materials from a single group for best performance (Table 6.5). Material in adjacent groups can be used together, with appropriate protection. Items from separated groups should not be used under any condition. This means, for example, Group I is compatible with Group II but not Group III or Group IV.

Table 6.5 Electrochemical groupings to prevent corrosion between various material.

Group I	Group II	Group III	Group IV
Magnesium	Aluminum and alloys	Cadmium plating	Brass
Magnesium alloys	Aluminum alloys	Carbon steel	Stainless steel
Aluminum	Beryllium	Iron	Copper, copper alloys
Aluminum alloys	Zinc, zinc plating	Nickel, nickel plate	Nickel/copper alloys
Beryllium	Chromium plating	Tin, tin plate, solder	Monel
Zinc, zinc plating	Cadmium plating	Tin/lead solder	Silver
Chromium plate	Carbon steel	Lead	Graphite
Iron brass rhodium			
	Nickel, nickel plating	Stainless steel	Palladium
	Tin and tin plating	Copper and copper alloys	Titanium
	Tin/lead solder	Nickel/copper alloys	Platinum
	Lead	Monel	Gold

Tin, nickel and stainless steel are compatible with each other. There should be no galvanic problem when mated together in a joint or seam. These materials will retain their original electrical conductivity after aging occurs, when provided with a significant amount of mating pressure or compression. Aluminum, a common metal used for enclosures due to its light weight and strength, is only compatible only with a few other metals. If the protective plating on aluminum is clear or yellow chromate, nickel can be used however, the transfer impedance of chromates are high and may not provide optimal shielding protection.

Silver is not compatible with aluminum which is commonly found in conductive cloth gaskets (over foam) or elastomers. Galvanic corrosion will still develop over a greater period of time if exposed to a hostile environment that accelerates the corrosion process.

A photograph of galvanic corrosion between two dissimilar metals in the presence of an electrolyte is shown in Fig. 6.39.

Figure 6.39 Example of galvanic corrosion.

6.9.4.7 Closure Pressure

Shielding effectiveness and closure pressure have a direct relationship. The minimum closure force (P_{min}) is the recommended applied force to establish optimal shielding effectiveness and to minimize effects of minor pressure differences.

The maximum recommended closure force (P_{max}) is based on two criteria:

- Maximum compression set of 10% and/or;
- Avoidance of possible irreversible damage when pressure exceeds recommended maximum value. Higher closure pressures may be applied to most knitted wire mesh gaskets, however the gasket should be replaced when cover plates are removed or whenever the seam is opened.

6.9.4.8 Installation requirements

When incorporating a gasket, how it is installed determines total effectiveness. There are preferred ways to install a gasket, the correct way detailed by the manufacturer in their application notes, or a manner that is quick without regard to maximum effectiveness as a last minute fix to pass an EMC test. The following should be considered when choosing an EMI gasket, depending on application.

- How much shielding (in dB's) is required?
- How much compression is required and can the optimal amount be achieved?
- Will the gasket enter compression set and lose its shielding effectiveness after a period of time?
- What is the environment of use (wet, dry, salt spray, etc.)?
- Will the gasket be subjected to mechanical stress during maintenance or shearing?
- Can sufficient fasteners be provided at optimal distance spacing?
- What is the optimal gasket securement method?

Figure 6.40 shows techniques where a conductive gasket is installed on the inside of a screw or locking mechanism securing a top cover onto an enclosure [1]. This technique ensures electrical continuity under compression when the screw is tightened, as long as there is *"no"* paint, plating, varnish or insulating film present.

An example of plating material is gold versus clear anodyne commonly used on aluminum. A gold finish has significantly greater transfer impedance and may not permit use of gasket material without the ability to pierce through the coating, such as wire impregnated rubber or monel mesh.

What is not known by many engineers is that not all metal surfaces are truly conductive. This can be verified using an ohm meter in "buzzer" mode. Allow the probes to rest on the metal using gravity, not any finger pressure, and move them around. Do not use the tip to pierce the coating. One will quickly learn that the plating on the aluminum may *not* be at 2 mils thickness but 1 mil thick with a 1 mil thick varnish. Some unethical manufacturers plate the metal only 50% and then varnish it because it cost more money to plate a 2 mil thick finish instead of lower cost plate and varnish. If this type of metal is used, gaskets such as elastomers and cloth over foam are ineffective!

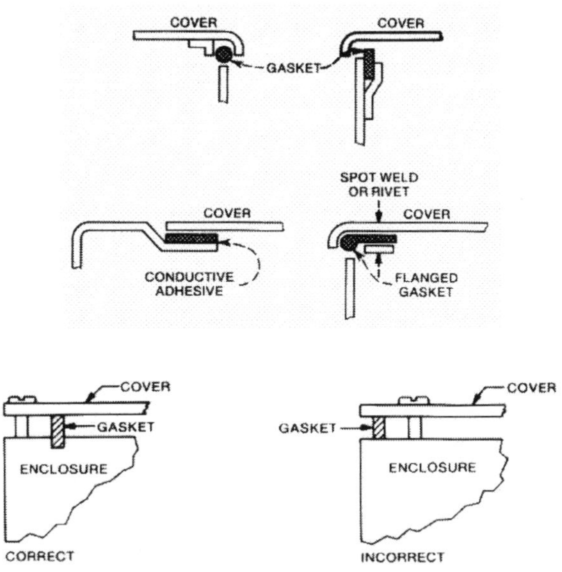

Figure 6.40 Correct method to install gaskets in sheet metal enclosures. (Source: Electromagnetic Compatibility Engineering–H. Ott)

An example on providing a gasket for a unique assembly is provided in Fig. 6.41. The panel meter has digital components that could radiate RF energy through an opening. In addition, some displays require a large cutout which can completely destroy the effectiveness of the shielded enclosure. Figure 6.41 is an example of a semi-permanent application. For optimum performance, the system must essentially be hermetically sealed against undesired field propagation through opening; slots and apertures.

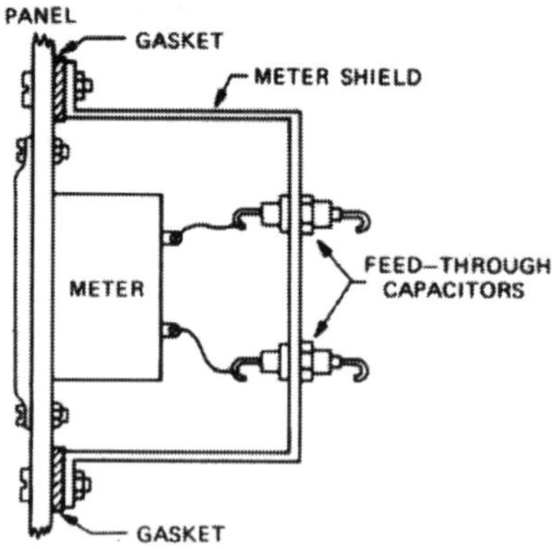

Figure 6.41 Method of shielding a meter hole in a panel.
(Source: Electromagnetic Compatibility Engineering–H. Ott)

6.10 Conductive Coatings

Commercial products are often packaged in nonmetallic enclosures made from plastic, composites or other non-conductive material to minimize weight and cost. These enclosures must achieve or improve EMC shielding and performance in some manner. There are various coating and metallizing techniques available to achieve this goal.

A frequently used coating is a conductive paint-like mixture containing metal particles. Conformal coatings are mixed with very fine conductive particles such as silver, nickel, copper and carbon. Other processes include conductive acrylic and polyurethane paint with silver particles embedded within. Surface resistance as low as 50 mΩ per square may be easily achieved. This is with a one-mil [0.001 inch] coating thickness, even though shielding effectiveness is compromised below approximately 100 MHz which is due to limited performance of the skin depth at that frequency. A lower surface resistivity of coating material provides enhanced shielding effectiveness. Shielding effectiveness of 60-100 dB are possible however, this can be significantly reduced by seams, holes and apertures. To be even marginally effective, conductive coatings must have a surface resistivity of a few ohms per square or less. For higher conductivity material to be used in a specialized application, silver, copper, zinc, nickel or aluminum should be used.

While the focus on conductive coatings is on maximum conductivity for shielding effectiveness, we must also consider another aspect of shielding effectiveness. If the system does not need a conductive coating to prevent radiated emissions from either causing harmful interference to electronic circuitry inside the enclosure, the printed circuit board and components may still be sensitive to the hazard of an ESD event. In this case we need a coating that is ESD dissipative with regard to charge energy. This means the surface resistivity of the coating will have to be high, generally a few hundred ohms per square. This is easily achieved with carbon or graphite based material. At issue however, is that when the ESD "spark" propagates within a lossy conductive enclosure, it may re-radiate a magnetic field, perhaps in close proximity to circuit devices and boards. If there are susceptible components to close-proximity radiated ESD spectra, operational perturbation may be exhibited.

The thickness of a conductive coating on plastic is usually very thin. As a result, absorption loss does not become significant until higher frequencies, therefore the choice of material used for the conductive filler must be chosen based on frequency range of interest and not just cost. The thicker the coating the greater the skin depth.

Common methods of producing conductive coatings on plastic enclosures include the following. Tables 6.6 and 6.7 provide details on each type.

- Conductive paint
- Electroless plating
- Metal foil linings
- Flame/arc spray
- Vacuum metalizing
- Metallic fillers molded into the plastic

- **Conductive paint:** Conductive paint consist of a binder (usually urethane or acrylic) and a conductive pigment (silver, copper, nickel or graphite)–ratio is typically 80% metal and 20% binder. This material provides good conductivity and is inexpensive to apply. A uniform coating is extremely difficult to achieve.
- **Electroless plating (chemical deposition):** This is a process of depositing a metallic coating (usually nickel) by a controlled chemical reaction that is catalyzed by the metal being deposited. Electroless plating produces a uniform film-thickness with good conductivity at reasonable cost.
- **Metal foil linings:** Metal foil linings are a pressure sensitive metallic foil (usually copper or aluminum) with an adhesive backing applied to the interior of the plastic part. The metal foil lining provides very good conductivity and is the preferred method for experimental work, not production due to intensive labor required.
- **Flame/arc spray:** A metal wire or powder, usually zinc, is melted in a special gun and deposited onto the plastic material producing a hard, dense coating of metal with excellent conductivity. This is an expensive process but popular among many manufacturers.
- **Vacuum metalizing:** A pure metal, usually aluminum, is boiled in a vacuum chamber and then deposited onto the surface of the plastic parts located in the chamber. Excellent adhesion and conductivity is achieved for complex designs. This is a very expensive process since specialized equipment is required to perform this method of metalizing.
- **Filler plastics:** Filler plastics are made by mixing conductive agents with plastic resin prior to molding the part. The part is then injection molded in the form of fibers, flakes or powders. Typical conductive fillers are carbon fiber, aluminum flakes, nickel-coated carbon or stainless steel fiber. Since the outer surface may not be conductive (the inside of the part is), a secondary machining operation may be necessary to expose the conductive material to the outside so that it may make contact with a conductive gasket, if required on a multi-piece assembly.

6.10.1 Concerns When Using Coatings

As with any EMI suppression process related to shielding there are concerns. These concerns differ from conductive gaskets. Manufacturer data sheets and application notes provide guidance on how to achieve optimal shielding effectiveness when using coatings.

1. **Will the coating peel or flake off into electrical circuitry?**
   This is more of a concern for sprayed on paints using a conductive binder in either aerosol form or molded plastic. Flaking occurs during handling and manufacturing. If the paint coating is damaged by scraping or other means, conductive fibers will circulate inside the enclosure and eventually land somewhere, generally between the power and ground pins of components causing a short, for example.

Table 6.6 Characteristics of common surface coatings.

Coating	Conductivity	Comments
Nickel paint	~ 1 ohm/sq.	At the upper level of conductivity for effective shielding.
Copper, copper alloy paint	~ ¼ – ½ ohm/sq.	Capable of very good performance if properly applied and well-connected together, if separate pieces are used.
Electroless Coating	Very high	Very high conductivity, excellent performance.
Zinc arc spray	Very high	Roughens the surface; must be carefully applied. Not a noble metal. Very conductive.
Vacuum deposition	Varies widely	May not provide reliable contact if deposition layer is too thin.

Table 6.7 Comparison of metallizing techniques.

Technique	Advantages	Disadvantages
Silver paint	Excellent conductivity	Expensive.
Copper paint	Good conductivity	Oxidizes and loses properties if not protected. Can flake off the material.
Nickel paint	Good conductivity. Resists flaking	Requires multiple coats.
Zinc arc spray	Excellent conductivity. Scratch resistant. Good adhesion.	Skilled labor required. Misses recesses. Special equipment required. May distort the plastic substrate. Expensive.
Cathode sputtering	Good conductivity and adhesion	Skilled labor and special equipment required. Expensive.
Vacuum metallization	Good adhesion and conductivity. Applicable to complex shapes.	Coating thickness is generally thin. Requires expensive equipment to apply. Requires surface treatment prior to metallization.
Electroless plating	Uniform thickness. Good conductivity.	Requires refinish on outer surface of enclosures to prevent remove of plating. Applicable to complex shapes and is flake resistant, plates both sides of the material.
Foil application	Good conductivity. Useful in development and experimental work.	Labor intensive. Unsuitable for complex parts or high-volume production.
Bulk fillers	No secondary plating or coating required.	Base properties of plastic may be altered. Seam bonding requires grinding of surface to get to the conductors. Questionable shielding performance.

2. **Will shielding effectiveness be consistent from part to part?**
As with any manufacturing process tolerances exist. If using electroless plating, flame/arc spray or paint applied with a spray gun, uniformity of the coating may differ between parts even though all parts are manufactured at the same time.

3. **Will the coating maintain shielding effectiveness over the life of the product?**
Products are used within many environments of use. Some gasket material is never physically touched during manufacturing, except for maintenance if required. Other products are abused by users such as hand phones and personal computers, where one takes off the cover to replace the battery or change configuration of printed circuit boards. Once covers are separated, it may be impossible to obtain factory level shielding effectiveness upon re-assembly. This can cause a problem for both emission and immunity levels.

4. **Will the coating affect flammability requirements on a particular grade of plastic used for shielding purposes?**
Product safety requirements mandate protection against the hazard of fire. If a fault condition occurs which in turn generates heat or fire, for example a short circuit in the printed circuit board, items within the enclosure can catch on fire. All printed circuit boards (core and prepreg) must be manufactured with a fire retardant having a flammability rating of V-1 or better. Plastic enclosures must be V-2 or better, but if used as a decorative barrier the flammability rating can be HB or better. Melted or molten plastics can fall onto the floor, desk, paper under the system, carpet or ignite flammable material that may be located in the vicinity of the system. All conductive coatings are therefore subject to flammability requirement if there is a potential for fire from systems contain high voltages or combustible components (e.g., batteries). Flammability ratings requirements in this paragraph are found in product safety standard such as IEC/EN 60950, UL 1950, CSA 1950, etc. A thorough description on ratings and testing are found in UL Standard 94 (Test for Flammability of Plastic Materials for Parts in Devices and Appliances).

5. **Will the coating reduce either the creepage and/or clearance distances required to minimize the potential hazard of electrical shock?**
As with the hazard of fire, electric shock could occur to a person or domestic animal when they touch a system containing hazardous voltages that has been energized. Hazardous voltage is defined as anything greater than 42.4 VAC or 60 VDC. As discussed in Chapter 5, Section 7, product safety standards mandate specific distance separation between components on a printed circuit board (creepage), and from components to the enclosure edge or other electrically active part through air (clearance). Conductive coatings can reduce this mandatory distance spacing requirement. An analysis must occur to determine if the coating will be located in the vicinity of electrical circuitry that may, under any abnormal or fault condition energize the enclosure case.

6. **Will the coating material be exposed to a user whom can physically touch the system (will a shock hazard be present should the enclosure become electrically live)?**
Along with reduction of creepage and clearance distances that may allow electrical circuitry to energy the conductive enclosure, the same is true when someone touches a metal or metalized enclosure, especially if they are wearing anything metallic such as a ring, watch, necklace, bracelet, earrings, etc. Will a shock hazard occur if a fault condition develops with internal circuitry shorting out?

6.11 Filters

Filters are used for both power line noise reduction and to minimize undesired EMI on transmission lines, both internal and external to a system. This Section examines what a filter is along with implementation. Filtering is used in conjunction with shielding to help achieve EMC requirements both emissions and immunity.

6.11.1 What is an EMI Filter?

An EMI filter is a passive device used to suppress interference present on both power and signal transmission lines. Filters not only suppress interference generated by the device itself, but also suppresses interference generated by other equipment to improve the immunity of a system from the electromagnetic environment present. Most filters include components to suppress both common- and differential-mode noise, and can also be designed with additional elements to provide transient voltage and surge protection.

All filters are a combination of lumped or distributed circuit elements arranged such that there is a frequency selective characteristic that passes some frequencies and blocks others, providing an effective means for reduction and suppression of EMI. Factors to consider when choosing a filter include the following.

- Insertion loss
- Impedance
- Power handling capability
- Signal distortion
- Tunability
- Cost
- Weight
- Size
- Rejection of undesired signals

6.11.2 Insertion loss

Filters have high reactive, discrete components relative to the impedance of a transmission line. This means the filter looks like a high value resistor generally for lower frequency signals, and an inductor for higher frequencies as well as their harmonic content. High impedance at a specific frequency, or range of frequencies, attenuates or reduces undesired electromagnetic fields within the transmission line (e.g., signal strength). By removing unwanted noise in the signal path, there is less coupling effect on other devices and circuits, or to comply with regulatory emission requirements.

A filter is simply a two-port device with a transfer function, $H(f)$ (6.20), illustrated in Fig. 6.42:

$$H(f) = \frac{V_L(f)}{V_{si}(f)} \tag{6.20}$$

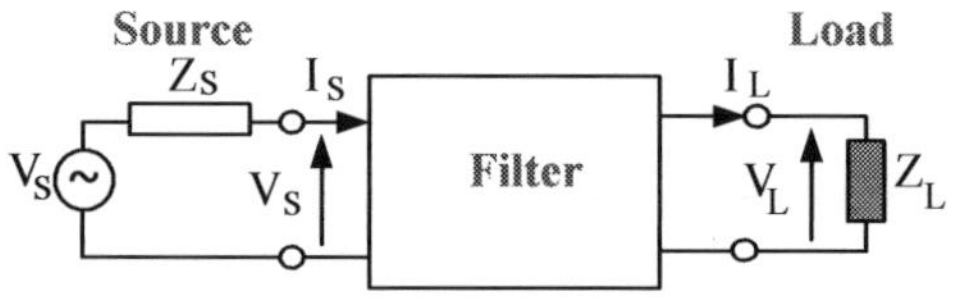

Figure 6.42 Transfer function of a typical filter.

Insertion loss (*IL*) is the reduction of signal power resulting from the insertion of a filter in a transmission line expressed in decibels (dB), or the effectiveness of the filter. If the power transmitted to the load before any insertion loss is P_t, and power received by the load after the loss is P_r, then insertion loss in dB is given by (6.21),

$$IL = 10 \log \left(\frac{P_t}{P_r} \right) \text{ (dB)} \tag{6.21}$$

This equation can also be described in a different form (6.22). For most filters, P_r will be smaller, meaning the insertion loss is positive, thus signal amplitude after the filter is less.

$$IL(f) = 20 \log \left(\frac{E_{L1}(f)}{E_{L2}(f)} \right) = 20 \log \frac{E_L(f) \text{ without filter inserted}}{E_L(f) \text{ with filter inserted}} \text{ (dB)} \tag{6.22}$$

Filter performance is dependent on both input and output impedance. Insertion loss data is given by manufacturers for use in a 50 Ohm system. While this provides a standard for comparison, it does *not* reflect actual performance when installed within a real product. Almost all manufacturers specify this 50 Ohm value when in reality, the impedance of transmission lines varies from several Ohms to hundreds of Ohms and are generally reactive, varying considerably over a large frequency range of interest. While there are methods for determining the actual impedances of source and loads, these values are usually unknown. Hence, selection of a filter using computational analysis is usually impractical or difficult due to lack of transmission line parametric values.

6.11.3 Basic passive filter elements

Two basic components create a filter; capacitors and inductors. Sometimes a resistor is included in AC mains line filters to discharge "X" capacitors that may be utilized. An "X" capacitor is a regulatory safety agency classification for capacitors placed across AC mains that ensures if they ever short out there can never be a risk of fire or arc flash. This resistor, usually 1 Mohm, is placed in parallel to dissipate stored energy quickly to minimize potential of electric shock hazard present should someone touch the AC mains terminals of the input receptacle while this large energy storage device remains fully charged.

Another capacitor used in AC mains filters are the "Y" capacitor configuration. A "Y" capacitor is connected between voltage and chassis ground. Thus, there must be two "Y" capacitors for a single phase system and three for a three-phase configuration. "Y" capacitors, like the "X" configuration must be safety agency approved. If they short-circuit, there is a risk of electric shock to the user should one touch an energized metal enclosure.

The simplest type (first-order filter) contains only a single reactive component. We can calculate impedance, X_c or X_L easily (6.23), which is the complex portion of the impedance equation ($Z=R+jX$).

$$X_C = \frac{1}{2\pi fC} \; ; \qquad X_L = 2\pi fL \tag{6.23}$$

Note that:

- Capacitive reactance, X_C, decrease with an increase in frequency up to its self-resonant value before becoming inductive.
- Inductive reactance, X_L, increases with an increase in frequency up to its self-resonant value before becoming capacitive.

Single component filter elements are not very useful as their attenuation changes at a rate of 6 dB/octave or 20 dB/decade. To achieve greater attenuation, should this be required, a second or higher-order filter containing more reactive components is required (capacitors or inductors). A two-stage filter thus provides 12 dB/octave or 40 dB/decade attenuation.

Large amounts of capacitance in a filter can cause the following problems:

- Leakage current in power supply circuits to chassis and earth ground, when used for AC mains filtering, may violate product safety requirements against the hazard of electric shock.
- Capacitance limits the desired signal bandwidth for data signals by rounding off signal edge transitions, which may be a serious signal integrity problem with higher frequency signals (e.g., high-definition video and GHz telecommunication networks).
- Capacitance, in addition to rounding the edges of a signal that requires fast edge transitions, will slow the velocity of propagation of the electromagnetic signal within any transmission line (Chapter 2).
- Capacitance in a transmission line will lower the characteristic impedance of the line which may have an effect on signal propagation, including the need for additional drive current. This presents another form of a signal integrity problem (Chapter 2).

Large inductor values must be carefully used for the following reasons.

- If there is any significant amount of DC current in a transmission line, this bias may saturate the magnetic core of the inductor in power line filters decreasing performance capability. In addition, large AC or DC bias will cause inductors to become hot, which then could be an ignition source to start a fire.
- Large inductance may yield little effect on limiting unwanted noise both above and below the desired bandwidth of a signal trace, depending on the distributed capacitance in the winding and the permeability of the core material at various frequencies.
- Large inductance will cause a voltage drop and thus establish a potential difference between two elements. This potential difference can establish common-mode current development.

Capacitive bypass filtering is used to shunt high frequency RF currents to an appropriate 0V reference or chassis ground. Inductive filters blocks undesired RF energy as a series element, or prevents incoming and/or outgoing RF noise from corrupting circuits and traces located on a printed circuit board or any cable assembly.

A series element can be a resistor or ferrite material (bead or core). Resistors can be used as a filter as long as the voltage drop in the transmission line can be tolerated across the resistor. Implementation of a resistor is similar to that used for series termination required for impedance control on a transmission line but in this case, it is not terminating a transmission line in its characteristic impedance but is used to dampen RF energy present. If a voltage drop cannot be tolerated across a resistor due to low-level voltage signaling required, ferrite material should be used instead which has a very low or minimal voltage drop, in comparison.

At lower frequencies, such as below the 10-30 MHz range, an inductor can be used with caution. At higher frequencies inductive reactance will increase ($X_L=2\pi fL$) and cause a different type of voltage drop to develop, which in turn creates a signal integrity problem. This different type of voltage drop is associated with the electromagnetic field present (V/m and A/m in the frequency domain), and not DC signal voltage levels commonly considered by design engineers (time domain). Above 30 MHz, ferrites are preferred for filtering due to having high AC impedance without affecting signal transmission quality (DC signal transfer).

Both capacitors and inductors are high-Q components and may create a problem at a particular resonant frequency. A very small resistor placed in series with the capacitor or parallel with an inductor will lower the Q and allow the filter to operate over a larger spectral bandwidth ($Q=L/R$). This low value resistor generally does not affect signal integrity, but is commonly used in power distribution networks for increasing the bandwidth of decoupling capacitors.

An alternative approach to understanding mismatching of impedance is that if a filter mismatches both source and load impedances, minimum transfer of *unwanted* signal energy (EMI) occurs. If the source driver's impedance of a component is high, the filter's input impedance should be low, provided by shunt capacitance to ground. If the source driver's output impedance is low, the filter input impedance should be high, or series reactive with an inductor. The same mismatch configuration should exist between the load's input impedance and the filter's output impedance.

Different applications require proper selection of passive components; one, the other or both depending on transmission line impedance. Most lower bandwidth circuits (≤ 10 MHz) benefit from use of a filter, which are not effective unless their placement is exactly adjacent to the interconnect or thc load side of a transmission line.

6.11.3.1 Capacitive Filtering

Chapter 4, *Power Distribution Networks*, discussed the behavior of a capacitor throughout the frequency spectrum and how they are used to decouple switching noise from power and return planes that affect component operation. When a capacitor is used as a filter element, the same behavioral characteristic exists with the only difference being application of use. A capacitor decreases impedance up to its self-resonant frequency before behaving as an inductor, which then increases with an increase in frequency. It is critical to select a filter capacitor, when used as a filter element, to have either the lowest possible impedance (power distribution network application which is to shunt undesired common-mode energy to a 0V-reference system), or high impedance as part of a multi-stage filter element to establish specific cutoff frequencies for optimal performance. This inductive aspect is due to loop and lead inductance present in the circuit layout. The Big-V curve in Fig. 6.43 illustrates actual performance of a typical capacitor. If there was no lead or loop inductance, impedance, impedance would go to 0 Ω and remain there. Figure 6.43 also provides insight into the application of a single discrete capacitor to shunt undesired RF energy from a signal line to the 0V reference or ground at a specific frequency of interest.

6.11.3.2 Inductive Filtering

Similar to capacitive filtering, inductors function identical except with 180 degree phase inversion (Fig 6.44). Capacitors at their self-resonant frequency will exhibit low impedance whereas inductors will have high impedance. Above resonance, the inductor then behaves as a capacitor and ceases to provide optimal filtering or performance. Like a capacitor, the Q of the resonant point can be made lower with small series resistance in parallel.

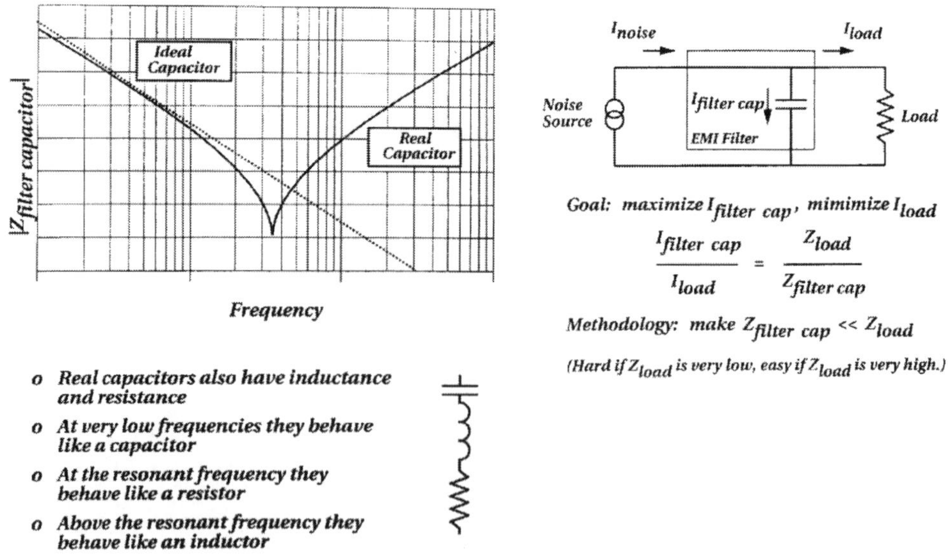
Ideal Capacitor
Real Capacitor
|Zfilter capacitor|
Frequency
Inoise
Iload
Noise Source
Ifilter cap
EMI Filter
Load
Goal: maximize Ifilter cap, mimimize Iload
Ifilter cap / Iload = Zload / Zfilter cap
Methodology: make Zfilter cap << Zload
(Hard if Zload is very low, easy if Zload is very high.)
o Real capacitors also have inductance and resistance
o At very low frequencies they behave like a capacitor
o At the resonant frequency they behave like a resistor
o Above the resonant frequency they behave like an inductor

Figure 6.43 Capacitor behavioral characteristic and shunting filtering capabilities.

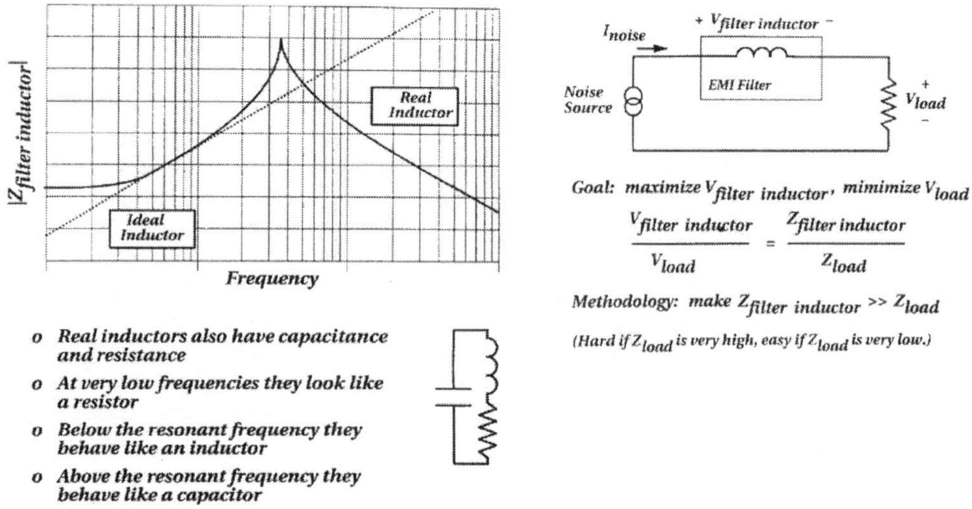
Real Inductor
Ideal Inductor
|Zfilter inductor|
Frequency
Inoise
+ Vfilter inductor −
Noise Source
EMI Filter
+ Vload −
Goal: maximize Vfilter inductor, mimimize Vload
Vfilter inductor / Vload = Zfilter inductor / Zload
Methodology: make Zfilter inductor >> Zload
(Hard if Zload is very high, easy if Zload is very low.)
o Real inductors also have capacitance and resistance
o At very low frequencies they look like a resistor
o Below the resonant frequency they behave like an inductor
o Above the resonant frequency they behave like a capacitor

Figure 6.44 Inductor behavioral characteristic and series filtering capabilities.

6.11.4 Parasitics related to filter components

Capacitors behave as an inductor above self-resonance and inductors behave as a capacitor. There is however another element that changes behavior characteristics, or when the phase changes due to parasitics. Figure 6.45 shows both series and shunt elements in their basic form with parasitics due to their physical implementation on a printed circuit board, along with the manufacturing process of the discrete device.

For a capacitor, lead and interconnect inductance is called *ESL* (equivalent series inductance). Most of this inductance comes from the connection between the capacitor terminal and the printed circuit board, which is generally a routed trace from the component, plus the via going directly to an internal plane. For capacitors, lead or loop inductance is generally magnitudes greater than the internal *ESL* within the device package (terminal to plate). There is little we can do to remove [parasitic] inductance due to manufacturing requirements and the how the capacitor is placed on the board. Placement and layout is thus critical for filtering a specific range of frequencies using capacitors without going inductive, or operating at a lower frequency due to loop inductance.

As for inductors, between each winding is a very small amount of parasitic capacitance (e.g., for a 1,000 winding inductor there are 999 capacitors in series between each winding). Also, between the two terminals is a second parasitic capacitor with free space acting as the dielectric between the two end. We now have many small value capacitors (999) in series, which are then in parallel with a larger value parasitic capacitor. One must take into consideration the equivalent Thevenin circuit to determine total parasitic capacitance present in an inductor, if one is concerned about parasitics affecting overall performance during computational analysis.

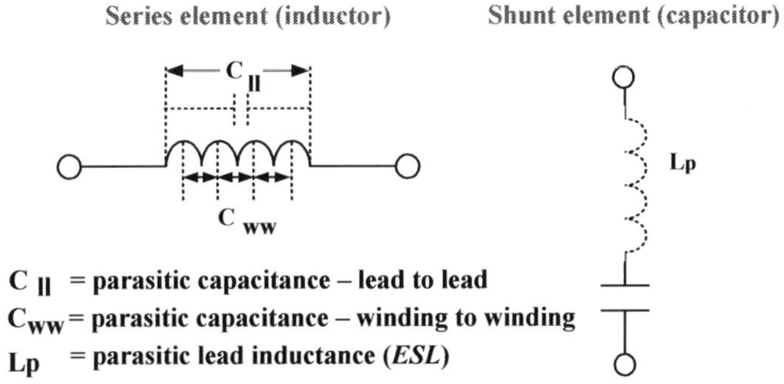

Fig. 6.45 Parasitic elements affecting filter element performance.

The equation that describes a self-resonant frequency due to parasitics is provided by (6.24). It is impossible to know what parasitics are going to exist during the development cycle, and designing a filter with a very specific frequency range of operation is difficult.

$$f = \frac{1}{\sqrt{2\pi(ESL)C}} \quad (MHz) \tag{6.24}$$

In essence, a low-pass filter may behave as a high-pass, and a high-pass may function as a low-pass. Trying to troubleshoot signal integrity problems with parasitics successfully is challenging, if not impossible. We do not want the single-element filter to change phase within any frequency range of interest. It is also important to remember that we must control parasitics if we are to obtain optimal performance for both series and shunt elements.

6.11.5 Basic filter configurations

For applications where additional attenuation of an unwanted signal is required, or a range of frequencies to be processed, use of multiple elements is required. Filters typically consist of reactive elements. The following configurations are used for loss reduction or attenuation of a signal.

- Discrete: Symmetrical filters (bi-directional)
- Series inductor: *T*-Filter
- Shunt capacitor: π (Pi)-Filter
- Asymmetrical: *L*-Filter

The advantage of using a discrete or single element filter is that only one physical device is required. Multi-element filters are more effective when a single element does not provide sufficient attenuation or covers a broadband enough range of frequencies that need to be addressed.

Series elements are usually inductors, except capacitors used in feed-through applications for implementing DC isolation between circuits or voltage level translation. For a series impedance filter element (inductor) to provide optimal performance, it must have an impedance larger than the sum of both the source and load impedance and thus, prefers low-impedance circuits and components.

Shunt elements are capacitors that divert undesired common-mode noise from a transmission line to chassis ground or another 0V reference and are therefore most effective in high-impedance circuits. For a shunt impedance filter element, its impedance must be less than the parallel combination of both source and load impedance.

Since impedance mismatches are required if the filter is to perform as desired, there are three basic configurations (Table 6.7).

- If both source and load impedance is low: Use a series element (inductive element)
- If both source and load impedance is high: Use a shunt element (capacitor)
- If either source or load has low impedance and the other side is high: Use both capacitors and inductive elements.

Table 6.7 Basic filter configurations.

Feed Through Filter	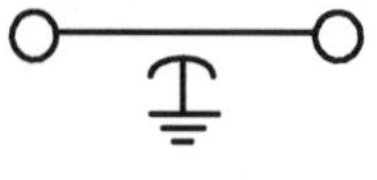	A low inductance device with a bypass capacitor to ground. Optimal when used with a high impedance source and load.
L Filter	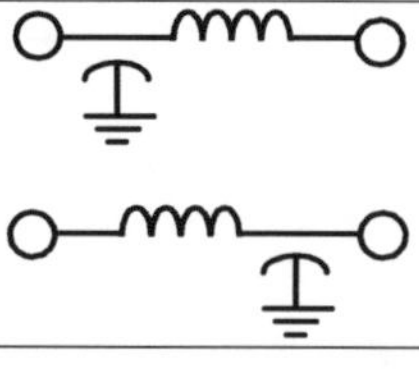	Used between a low-impedance source to a high impedance load, or vice-versa, with the capacitor facing the higher impedance circuit.
π Filter		Presents a low impedance to both the source and load and has a sharper roll-off than the feed-through or *L* section. Best suited for high impedance source and loads.
T Filter		Like the π filter, this has a sharper roll-off than the *L* configuration. Intended for applications with low impedance load and source

6.11.6 Common-mode and differential-mode filters

The reason we use filters is to remove undesired RF noise/energy present within transmission lines that includes AC mains, signals between functional sections, and component along with cable interconnects that either exit or enters a system, or which violates shielding integrity if the cable is not shielded and properly terminated. A filter must be designed to work with the mode that is most relevant to the interference present, either common-mode or differential-mode. A filter that attenuates the wrong mode is ineffective under all conditions of use.

6.11.6.1 Common-mode filters

Common-mode noise is undesired RF energy created as a result of an imbalance within a transmission line system. To filter common-mode noise, a shunt element or capacitor is required. This means the capacitor must be chosen for a specific frequency range (bandwidth) of interest and connected between both signal line and either the 0V return path or chassis ground.

A differential-mode filter does not affect common-mode noise since the shunt capacitor is located across two conductors carrying the same RF noise voltage and is thus invisible to the differential-mode interference present. The inductance in the series element is also shorted out by the lack of impedance in the opposite transmission line (return path).

When creating a common-mode "*L*" configuration, series impedance must be in *all* signal lines that require filtering. The capacitor(s) must be located across each transmission line and *earth or 0V reference.* When implementing an inductor, or resistor in this configuration, there are two ways to design the filter. One is to put the inductor or resistor in series with each and every line, which increases parts count. The other is to wrap all lines on a common core and wire them in opposite directions such that only common-mode current is attenuated. There are two advantages to doing this.

- Differential-mode current is a result of desired magnetic flux present in the circuit. Excessive or undesired common-mode flux cancels within the core. This configuration allow for greater amount of current to propagate in the transmission line without saturation that usually occurs when using discrete inductive elements only.
- The windings around a ferrite core do not affect signal integrity or differential-mode transmission, and is an excellent component for broadband signals.

The same is not true when using a capacitor located between a signal line and a reference. This capacitor will have a significant effect on wideband circuits and may degrade signal transition (signal integrity) of a digital signal by rounding the edge or making signal propagation slower. If a capacitor must be used for a signal line to shunt RF noise to 0V or chassis ground, it must be a very small value generally in the low pF range. However, for lower frequency or low impedance signal and power supply circuits, a shunt capacitor works as desired to remove common-mode noise with a higher capacitive value.

- When using capacitors for common-mode power line filtering, a critical aspect to success is having the capacitor connected to earth ground ("Y"-capacitor configuration). If an earth ground is not available, such as a battery operated system or an AC mains plug without a 3rd wire safety ground, then a series choke or resistor can be used to absorb undesired RF energy.

- If an earth ground connection is used, this filter element must be implemented with a very low impedance connection such as the bottom annular ring of a through-hole via on a printed circuit board to a metal standoff, or a flat braid to chassis ground. Round wire must never be used for bonding a capacitor or filter element to earth ground due to high lead inductance. Round wire has higher inductance than a flat braid (Chapter 3).

In addition to single element common-mode filters, there are additional configurations that provide a significant amount of attenuation as shown in Figure 6.46. It is critical to understand filter design before implementing either "*π*" or "*T*" topologies, since random selection of components based on prior design applications or knowledge will not work, mainly because of parasitics.

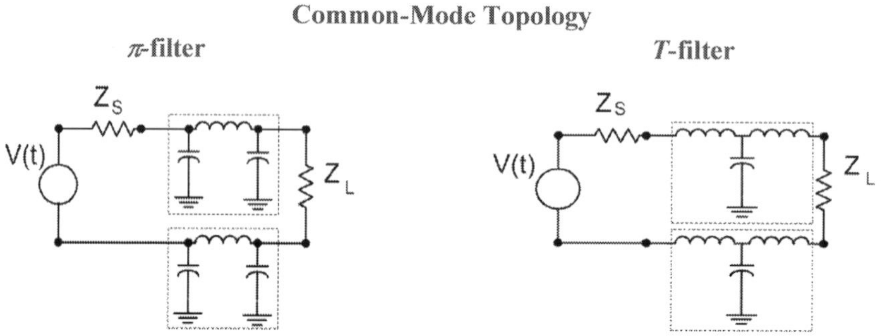

Figure 6.46 Common-mode filter topologies using multiple elements.

6.11.6.2 Differential-mode filters

Differential-mode filtering is used between a signal line and its return path *without* any connection to earth ground. Both lines must be in balance without any loss of signal performance (signal integrity). Magnetic flux from one line cancels with the flux in the other. Under this condition we have equal and opposite flux, thus optimal signal performance without EMI or crosstalk being established between the two lines.

When differential-mode filtering is required, line-to-line capacitors ("X"-configuration) and a discrete series inductor can be used (Fig. 6.47). As with any filter that needs to cope with high levels of low-frequency differential-mode noise (e.g., from switch-mode converters, phase angle power controller and motor drives), we may need more differential-mode attenuation than can be achieved by using "X"-capacitors alone. One will need to use a differential-mode choke. This choke can be two windings on the same core with both wires wound in the opposite direction to each other (similar to the common-mode choke configuration). It is difficult to get significant differential-mode inductance into a small package due to core saturation, which means a larger size component and more expense. Adding additional stages to a simple single-stage filter is usually the most cost effective means of enhancing a filter's performance. Parasitics limit attenuation capabilities. When using multi-stage filters, parasitics also plays a smaller role in degrading filter performance. In addition, a multi-stage filter is less affected by the extremes of both either source or load impedance.

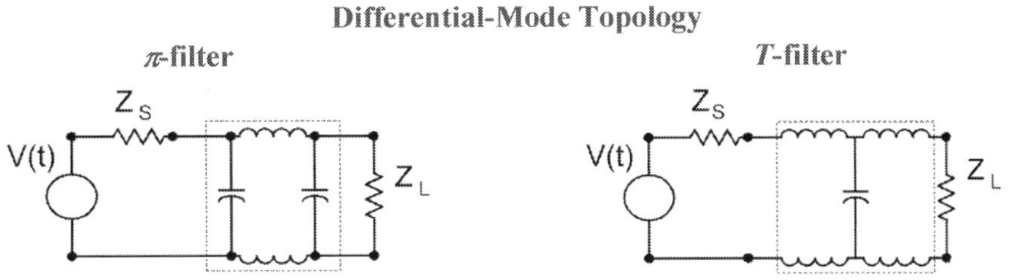

Figure 6.47 Differential-mode filter topologies using multiple elements.

6.11.7 Signal line filter configurations

Filters, when used for signal line applications, are generally classified with a cutoff frequency either upper and/or lower boundary. This cutoff frequency applies to a specific point in the spectrum where the filter performs in an optimal manner.

When defining a cutoff frequency, this is usually defined as the 3 dB corner, a frequency for which the output of the circuit is −3 dB of the nominal band-pass value, which corresponds approximately to half the electromagnetic field power in the transmission line. As a voltage ratio, this is a reduction to $\sqrt{0.5} = 0.707$ of the pass-band voltage level.

Alternatively, a band-reject corner frequency may be specified as a point where both transition and stop-band meet. This is a frequency for which the attenuation is larger than the required stop-band attenuation, which for example may be 30 dB or 100 dB.

There four basic types of filter configurations for signal lines illustrated in Fig. 6.48:

- **Low pass** – Rejects undesired RF energy above a desired set point, passing frequencies below this point with little or no attenuation, and are commonly used for applications such as AC mains and low frequency signal lines.
- **High pass** – Rejects undesired RF energy below a desired set point, passing frequencies above this point with little or no attenuation.
- **Band pass** – Passes a range of desired frequencies with little or no attenuation, rejecting frequencies outside a specific range of interest.
- **Band reject** – Rejects a range of frequencies within a particular frequency band of operation while passing all other frequencies outside this band.

For both band-pass and band-reject, the 3 dB point is shown in the insertion loss curve (Fig. 6.48), or the value which corresponds approximately to one-half power in the transmission line being attenuated. Many filter designs specify a 3 dB point. However, some filters are designed to have a steeper slope for both rising and falling edges to establish instead a 6 dB cutoff point at where the 3 dB point normally would have been. This means the filter has a higher Q and a much more narrow bandwidth of operation.

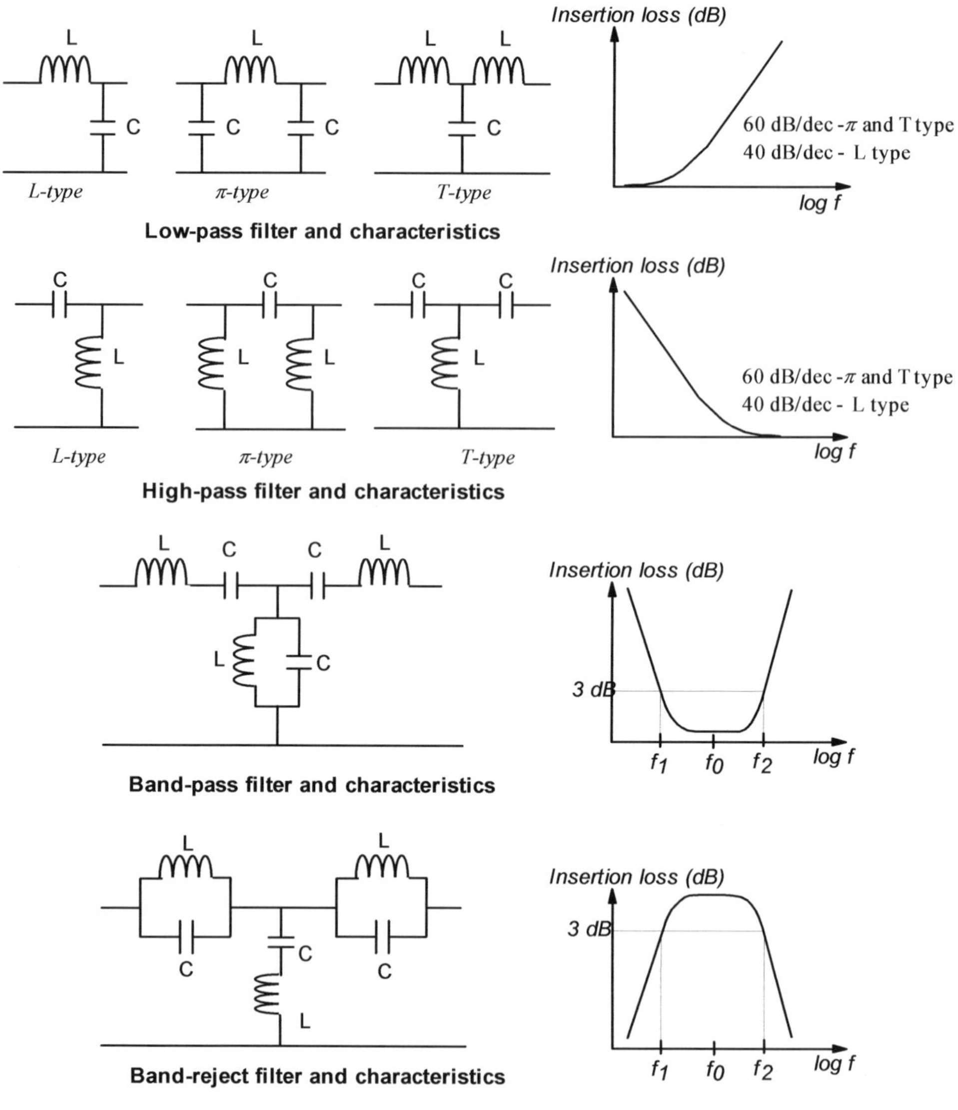

Figure 6.48 Basic filter configurations and behavior curves.

6.11.7.1 Common-Mode Rejection

Balanced circuits use differential signaling or two transmission lines referenced to each other. Drive current in both lines should be equal and preferably in balance. If there is any imbalance, the magnitude of this imbalance is the magnitude of common-mode noise that gets created. This imbalance can be the result of different line lengths or impedance discontinuities in one of the two transmission lines.

A typical differential pair transmission line is shown in Fig. 6.49, commonly found in op-amp configurations. All differential signals must have two transmission lines with equal and opposite paths in balance. To illustrate the effect of being out of balance, assume common-mode noise is present on one or both of the signal and return paths with respect to 0V reference (a third transmission line path). Differential-mode signals are thus referenced to each other. If there is perfect balance between two signal lines, no differential-mode voltage *Vdm* would appear across the input of the receiver. If there is any imbalance in the network regardless of magnitude due to excessive impedance somewhere in the trace routing, a small amount of differential-mode noise voltage *Vdm* will appear across the input of the amplifier. As a result, common-mode voltage *Vcm* will be established. Common-mode noise is the primary cause of EMI.

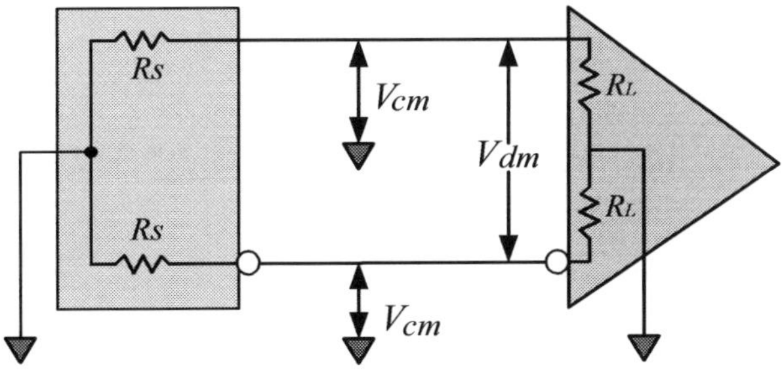

Figure 6.49 Differential pair signaling with regard to CMMR definition.

Differential-mode components (op-amps) are provided with a Common-Mode Rejection Ratio (CMRR) number by the manufacturer. This number is a metric used to quantify the degree of balance, or effectiveness of a balanced circuit in rejecting common-mode noise voltages established in the signaling system, passing desired differential-mode signals without degradation.

CMRR is defined as the ratio between common-mode and differential-mode noise voltage (6.25):

$$CMMR = 20\, log\left(\frac{V_{cm}}{V_{dm}}\right) \quad (dB) \tag{6.25}$$

CMRR identifies how much undesired common-mode noise will be rejected from entering the device. The better the balance between the differential pairs a greater the amount of common-mode rejection will occur. At higher frequencies, achieving a large CMRR value may be difficult to accomplish. Typically 40-80 dB of CMMR is found in well-designed components. Additional components can be provided to increase this rejection level. Common-mode chokes are generally used on cable interconnect circuitry since the impedance of external cables will generally be different that the impedance of the transmission lines on a particular routing layer within a printed circuit board (a.k.a. traces), or impedance imbalance created within the interconnect connector due to pin assignment and configuration.

6.11.7.2 Common-mode and differential-mode filters and chokes

Common-mode filters are used in differential-mode circuits to remove unwanted common-noise from differential pair transmission lines, AC mains connection to building power distribution networks (to prevent LCI-Line Conducted Interference) and high-speed communication circuits. Common-mode chokes do not work on single-ended signal transmission lines or a single routed trace between components. Many traces routed on a printed circuit board are routed as differential-pair signaling (e.g., Ethernet, USB, Firewire, etc.) which benefits significantly from use of common-mode filters and chokes.

Common-mode filters are more difficult to design than differential-mode for basically two reasons:

- We generally do not know ahead of time the differential-mode signal characteristics of the transmission line pair and how common-mode current will get created, unless extensive computational analysis occurs ahead of time.
- Both source and load impedance is usually unknown, yet manufacturers provide specifications in a theoretically perfect 50 Ohm test configuration within their datasheets.

The effectiveness of a common-mode filter depends on knowing both source and load impedance. Differential-mode or single ended signals are easily filtered with a ferrite device since we generally know both output impedance of the driver and load impedance, which is always several magnitudes greater. Common-mode noise is generated by parasitics of the circuit layout and if interfaced to a cable assembly, the load impedance may be unknown. Everything is frequency and impedance dependent. Any common-mode noise present at interconnects will drive cables which are efficient radiating antennas.

Differential-mode filters should be located as close to the source driver as possible to prevent undesired RF current from propagating along the transmission line. Sometime a small series resistor will act as a differential-mode filter. One can however still use a ferrite bead in series with the resistor for additional performance.

Common-mode filters on the other hand must be positioned as close to where the load, or external cable assembly interconnect is located to prevent driving an external cable which may be an efficient radiating antenna. These common-mode filters are usually of the low-pass configuration. Common-mode filters consist of either a single-element (in series or shunt mode; inductance or capacitance) or multi-element (L, T, or π configuration). The single-element component is simple to utilized where multiple-element will often be effective only when single-element components does not provide sufficient attenuation of an unwanted signal.

To use one component to remove common-mode noise, a specially designed filter can be used. This specially designed filter is a toroidal core with two sets of wires, one for each of the positive and negative signal line of a differential-mode transmission line. The wires on the toroid coil are wrapped in opposite directions to cancel undesired magnetic field flux. On each transmission line we propagate an electromagnetic field from source to load with a magnetic field present in the dielectric surrounding the transmission line.

We are dealing in reality with V/m and A/m, or P/m^2 (total electromagnetic power). Engineers generally think in the time domain using voltage and current as metric units of measurements as this is an easier way to perform design engineering and understand circuit theory. Faraday's Law, or the Right Hand Rule states that when there is time variant current in a transmission line a time variant magnetic field is created propagating counterclockwise with respect to the direction of current flow. If we take the two transmission lines of differential-mode signaling and wrap them in opposite directions around a core, undesired magnetic flux (common-mode RF noise) on one differential signal line will cancel out with the second differential line leaving a pure differential-mode signal without any common-mode current. A common-mode choke is shown in Fig. 6.50. The lines wrapping the core traveling in opposite directions are the different-mode currents.

A common-mode multi-winding choke and low value capacitors, when provided in AC mains line filters, can be designed to have negligible effects on differential-mode current propagation. A higher value parallel capacitor to earth ground will contribute to differential-mode attenuation. The leakage inductance of the common-mode choke will appear in series with the differential-mode circuit and which must be minimized with careful design. This is especially true if doing a filter design using discrete components.

The filter must provide a desired level of common-mode attenuation without contributing to choke saturation or minimizing desired different-mode current (signal).

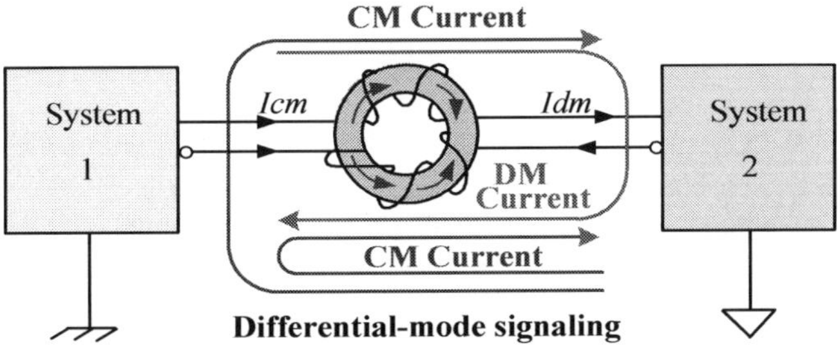

Figure 6.50 Common-mode choke implementation in differential-mode configuration.

6.11.8 Criteria in selecting an EMI filter for AC mains applications

The proper selection of an EMI filter involves not only electrical characteristics and electromechanical configuration, but also how it is to be utilized such as a purchased AC mains line filter installed on a printed circuit board or at the AC main input connector. The filter's mechanical footprint including mounting and terminations may impact effectiveness. Primary considerations also include leakage current, insertion loss, rated voltage and current handling capabilities, and agency approvals required. Regulatory safety approvals apply to line filters that connect to AC mains voltages or other levels of high voltage to prevent harm from electric shock or fire.

Concerns with purchasing off-the-shelf line filter for AC mains applications include:

- Choosing one with proper insertion loss.
- What is the steady-state voltage and current rating and does the filter have the same performance under both minimum and maximum load conditions?
- What is the transient voltage rating for the front-end "X" capacitor(s), if used?
- How much transient energy can the filter handle before damage occurs such as surge, fast transient or ESD?
- How much non-linear distorted AC mains current does it take to cause to cause the common mode choke to go into saturation?
- How much non-linear distorted current from the system is there to affect filter performance?
- Will the filter work when different operating modes occurs?
- Will performance vary with changes in input and output impedance based on end use application, if provided as a module sold for implementation into another company's product?

Virtually all off-the-shelf power line filters are designed to handle differential and/or common mode noise. It is difficult to determine the interference mode of the equipment and power mains cable impedance ahead of time, thus the type of filter required may not be determined until after the system has been designed. An illustration of a common AC mains filter configuration is shown in Fig. 6.51. For filters used in medical applications where isolation is required to prevent a shock hazard when invasive contact is made with a patient, or where a 2-wire plug configuration without a 3rd-wire protected earth ground is required, the filter will be isolated from chassis ground.

It is also a common oversight by engineers to route a 3rd-wire protective ground wire from a *metal* line filter to a *metal* chassis because the filter contains a terminal lug for such wire. This wire provides no functional use and is an added expense in material and labor to install. This lug on the filter is required however if the unit is installed within a plastic enclosure.

If the metal case of the line filter is already bonded to a metal enclosure or metal mounting panel, then why provided an additional ground wire- this is a redundant connection!

The common-mode choke (L_{cm}) within an AC line filter contains two identical windings on a single high permeability core wound so that differential (line-to-neutral or hot-to-hot) currents cancel each other out. This configuration has high inductance, typically 1-10 mH in a package size that does not allow for choke saturation caused by high levels of mains supply current. The full inductance of each winding attenuates common-mode currents with respect to earth ground; inherent leakage inductance also attenuates differential-mode interference.

The differential-mode choke (L_{dm}) is a single element device with one power line going through the it. This choke can also be a single component containing both wires (transmission lines) wound in opposite directions, similar to the common-mode choke configuration. This differential-mode choke provides series impedance to remove excessive RF noise in the differential-mode from passing through and does not affect common-mode noise.

Common-mode capacitors (C_{y1}/C_{y2}) attenuate common-mode interference. If the differential-mode capacitor C_{x2} is large, then there is no significant effect on differential-mode noise. Effectiveness of the common-mode capacitors (C_y) depends significantly on the common-mode source impedance of the circuit. The attenuation provided by the two *Y* capacitors is generally 15-20 dB. The common-mode choke is thus the more effective component in this filter assembly.

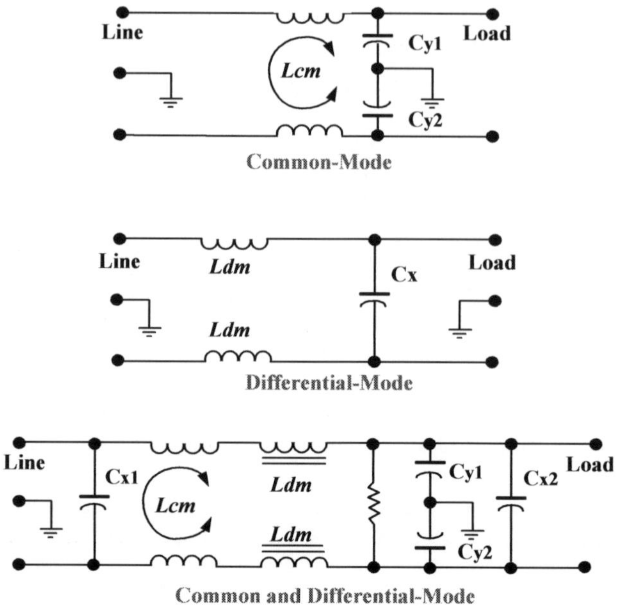

Figure 6.51 AC mains line filter configurations.

There will be situations where C_y is severely limited in performance due to leakage current that may be present that would violate product safety shock hazard, in which case more than one common-mode choke may be required to enhance overall performace.

The differential-mode capacitor (C_x) attenuates differential-mode noise by balancing out RF energy on both power lines and can have a fairly high value of capacitance, generally 100-470 nF (0.1-0.47 μF). Depending on filter performance, and that the source and load impedance may be too low for the capacitor to be useful, this element may be removed from the filter assembly. A 100 nF (0.1 μF) capacitor has an impedance of 10.6 Ω at 150 kHz [$Z=1/(2\pi fC)$]. The differential-mode source impedance observed may be considerably less than this for a power supply in the hundreds of watts range, meaning that this capacitor would have no effect at the lower end of the frequency range where attenuation is required to be at a maximum level.

When using *Y* capacitors, a serious product safety concern exists which is leakage current. Capacitors C_{y1}/C_{y2} must be limited in their capacitive value to minimize continuous leakage current flow to earth ground. Earth ground is usually accessible to users and if a metallic enclosure is touched with too much leakage current, hazard of electric shock could occur that may include death. Values for this current related to product safety must be less than 0.25mA, however under certain installation conditions this value can go to 5 mA. Medical equipment requirement have a maximum leakage requirement of 0.1 mA. This applies to the total leakage current of the system that includes transient suppression devices. Both *X* and *Y* capacitors must be safety agency approved against shorting out, or if a failure occurs they will always open up and never short out.

6.11.9 Using ferrite material for filtering

Within every transmission line is time variant current that creates an electromagnetic field (Faraday's Law). For purpose of filtering unwanted levels of a magnetic field, we focus herein on transmission lines (wires, cables and printed circuit board traces) and not free space to minimize radiated EMI. Free space minimization is through the use of shielding whereas filters are used with transmission lines.

When a magnetic field is present in a non-permeable material, such as air or plastic, magnetic field strength and flux density are equal. Using a magnetically permeable material in or around transmission lines will increase the flux density for a given field strength and therefore will increase the inductance of the transmission line. The inductance of a straight wire is typically 20 nH/inch (51 nH/cm). With inductance in transmission lines we have a loss of energy transfer throughout the entire length of the wire.

Ferrite is a material that changes magnetic flux density through the property of permeability (μ). The permeability value of ferrite material is dependent on the composition of different oxides such as nickel, zinc and manganese. All of these materials affect magnetic field propagation and are frequency dependent, therefore a different composition of oxides and their ratios determine if operation is to occur at low or high frequencies.

The permeability of ferrite material is complex with both real and imaginary parts, which translates into both resistive and inductive elements described by (6.26). It is these two parameters that determine total impedance presented to the circuit as essentially a high-frequency resistor without any voltage drop to affect signal integrity.

$$Z(f) = R + j\omega L(f) \tag{6.26}$$

where:
$Z(f)$ = Impedance at a particular frequency
R = Resistance inherent in the ferrite material
ω = $2\pi f$
$L(f)$ = Inductance of the ferrite material which is frequency dependent.

The ratio of the metallurgy varies with frequency. At higher frequencies the resistive element dominates. The ferrite material can thus be viewed as a frequency dependent resistor and the device becomes a lossy element to the magnetic field surrounding the transmission line. This means that undesired RF energy (magnetic field) is dissipated as heat or reflected back to its source. There is no interaction with parasitic capacitance therefore resonances that may occur with a standard inductor are either avoided or damped, which is a significant advantage over using inductors as an in-line filter for transmission line data signals.

The selection of a particular ferrite material must be based on the impedance the device presents to the circuit according to the permeability (μ) of the oxides used. This impedance is in reality a series combination of inductive reactance ($j\omega L$) and loss resistance (R), upon which the inductive reactance portion is frequency dependent ($j\omega L$) per (6.26). The real component represents the reactive portion and the imaginary represents losses. This means that interference energy tends to be absorbed rather than reflected. This reduces the Q of the circuit and minimizes resonance problems.

Ferrites are designed to have high loss in contrast to lower frequency or power inductors which are required to have minimum loss. Since a ferrite choke operates similar to a lossy inductor, it is useful only between low impedance circuits. A ferrite provided in a high-impedance transmission line thus offers little or no attenuation. Most circuits and cables have impedance in the range of 10-150 ohms. A single ferrite element can provide significant attenuation if chosen for use with a high impedance at a particular frequency of interest. An equivalent circuit for ferrite when used in a typical application is shown in Fig. 6.52.

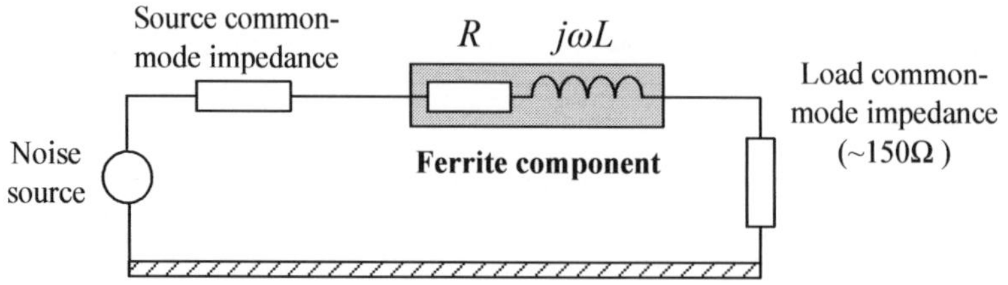

Figure 6.52 Equivalent circuit for a ferrite device.

Figure 6.53 is an example of a typical attenuation curve for a ferrite bead with respect to frequency. The manufacturer's databook value will identify this as a 200 Ωs at 100 MHz, and sometimes they specify impedance at 25 MHz. If we are interested in filtering a 500 MHz signal, the impedance will be approximately 220Ω and at 50 MHz impedance will be 175Ω . Ferrites generally have impedance from 10Ω to 2000Ω, all with different performance curves and bandwidth of operation. It is important to select a bead for maximum impedance for a desired range of frequencies that require filtering. In Fig. 6.53, both resistive and inductive reactance is also shown in addition to total impedance, which is useful for signal integrity purposes and to make sure that we do not attenuate a desired frequency below operational margins or introduce a significant voltage drop (ensure signal integrity). The value is Z in the plot is based on the equation of (6.26).

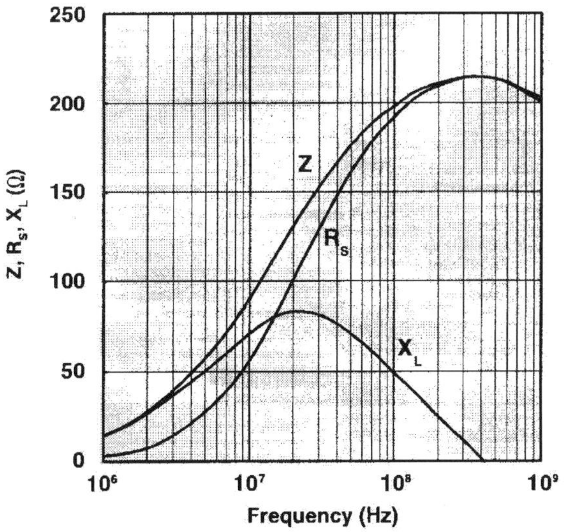

Figure 6.53 Typical impedance curves for a ferrite device.
(Plot provided courtesy Fair-Rite Products Corp.)

Permeability is the degree of magnetization of a material in response to a magnetic field. It is the measure or ability of oxides to support formation of a magnetic field within itself, or the degree of magnetization in response to an applied magnetic field. Magnetic permeability is represented by the Greek letter "μ" and coined in September, 1885 by Oliver Heaviside. The greater the permeability of the oxides used to create the ferrite device, a lower operating range of frequency will occur. An example of permeability and typical range of frequencies is shown in Table 6.8.

Table 6.8 Permeability of ferrite material and frequency range supported.

Permeability	Frequencies Suppressed
2500	30 MHz or less
850	25 to 250 MHz
125	200 MHz and above

Values provided courtesy Fair-Rite Corp.

6.11.10 Use of ferrite material on cables

When using ferrites on cable assemblies, cores could be configured to be a snap-on clamp or a solid core molded onto the assembly with a protective molded cover. Regardless of the manner of implementation, performance for both configurations is identical however, they must be used appropriately to remove common-mode currents.

6.11.10.1 Differential-mode currents on a cable

Differential signaling uses a two-wire transmission line, sending a signal between source and load and having an RF return path without reference to a 0V reference or chassis ground. Twisted pair cable used in data communication networks are referenced off of each other thus no ground loop current can flow to create common-mode noise. If the transmission lines are routed adjacent to each other (differential-pair signaling) or closely coupled, magnetic flux cancellation occurs however, there may still be common-mode current due to an unbalanced network or changes in source and load impedance due to transmission line routing.

If a ferrite core or clamp is placed over a cable assembly, this component will remove radiated common-mode current (magnetic flux) that surrounds the entire assembly (electromagnetic field propagation within the dielectric surrounding the transmission line path), and will have no effect on differential-mode signal characteristics. One can put many wires through a core for optimal performance as long as the total sum of the differential-mode current in the cable assembly is zero.

Only a series device in the transmission line will filter differential-mode current, should any be present. What is present within transmission lines that can be considered differential-mode noise is ringing with overshoot in both the positive and negative polarity. This ferrite device will provide a high-impedance to the propagating magnetic field and provide loss to overshoots without affecting the DC bias level of the signal.

6.11.10.2 Common-mode currents on a cable

Cables carry common-mode current caused by numerous losses. This means there is minimal or no flux cancellation between differential signals. As a result, residual undesired common-current will flow in the same direction and will phase add with all other common-mode current present. This is generally cause by either:

- Earth-reference noise due to capacitive coupling, generally within the interconnect which is usually referenced to the system's main 0V reference or earth ground.
- An imbalance in impedance of the signal and return to earth ground or 0V reference, thus allowing a certain percentage of the RF return current to travel back to its source in a manner that has no electromagnetic relationship to the cable assembly, such as free space.

Shielded cables can also carry common-mode noise if the shield is not properly terminated. This was discussed in Chapter 5. Although the magnitude of the current may be small, the propensity of having a much greater interfering potential will exist because the RF return path is essentially uncontrolled. In addition, transient noise or coupled RF energy (e.g., crosstalk) generally creates common-mode RF. This is called mode conversion.

Common-mode current may therefore be present on cable assemblies, even shields. This current is a magnetic field and a ferrite placed over the cable will increase the cable's RF impedance, preventing the flow of undesired current, which may cause potential disruption to circuits and components or non-compliance to regulatory emission standards.

6.11.11 How to select a ferrite device for use on printed circuit board traces

Before choosing a particular oxide mixture or physical size and configuration, the first step in achieving success is to know the range of frequencies of unwanted noise that is to be suppressed. During the preliminary design process, this information need not be accurate but approximate. After the printed circuit board has been manufactured, transmission line parameters will then be known as well as the signature profile of the noise that may be present. Mounting pads should be placed in the layout which makes it easy to experiment when selecting the right device for optimal performance, if experimentation is necessary. Be aware that environmental conditions will need to be addressed before selection becomes final. These environmental conditions are discussed below in this section.

The attenuation of a ferrite device depends on both the source and load impedance of the transmission line. To be effective, the ferrite must have impedance greater than the sum of both source and load impedance at the frequency of interest, even broadband if necessary. Ferrites having an impedance of a few hundred ohms or less are more effective in low-impedance circuits. If a single ferrite does not provide sufficient attenuation of undesired RF noise, a second ferrite can be used in series, or if this is a bead-on-lead with a wire wrapping through the beads, additional turns can be added but at the disadvantage of having the ferrite work within a smaller bandwidth of frequencies due to capacitive coupling between windings.

First, determine attenuation desired for a particular application without affecting the signal integrity of the transmission line (6.27).

$$Attenuation = 20\log\frac{|Z_s + Z_L + Z_{sc}|}{|Z_s + Z_L|}\ (dB) \tag{6.27}$$

where:

Z_s = source impedance
Z_L = load impedance
Z_{sc} = Suppressor core impedance

There are three common applications for ferrite material, listed from least to most used.

1. A shield to isolate a facility, transmission lines, electrical components or circuits from stray electromagnetic fields and high-power low frequency transmitting systems.

   This is one application of ferrite material for use in, for example, a shielded enclosure to keep RF energy from either entering or leaving a zone containing electrical elements. An example is the room that houses a MRI (magnetic resonant imaging) scanner used in hospitals, or anechoic chamber construction (tile and cone hybrid). The magnetic field strength is very large inside this zone and can cause serious disruption to electrical products located a considerable distance away. The walls of the room must contain both ferrite material (*H*-field) and steel (*E*-field) to contain the electromagnetic field. With high power, low frequency communication systems or sensitive scanners using transducers to measure a desired image, magnetic field disturbance can occur to not only itself but other systems in the near vicinity if a shield barrier is not provided using ferrite material to reduce magnetic field propagation.

2. When used with a capacitor, a low or high pass filter can be created (*LC*).

   This is an application when used with other discrete components such a capacitor, resistor or inductor creating either a high pass or low pass filter. This application is generally found in interconnects to prevent common-mode RF energy from either entering (immunity protection) or leaving the system (emissions). Ferrite cores placed over cable assemblies perform as a common-mode choke, and also prevent high-frequency noise from being conducted both into and out of a circuit or even between systems.

3. To prevent parasitic oscillation or attenuate unwanted signal coupling traveling along component leads, interconnecting wires, transmission lines (printed circuit board traces) or internal cables.

   This is an application where a ferrite bead is used on a printed circuit board trace to dampen common-mode noise created as a result of undesired switching energy present. Typical implementation includes: suppressing oscillations, as an RF absorber tile secured on top of a processor to dampen switching energy from clock generating circuits, or any other application where a ferrite can suppress unwanted noise including series termination of a transmission line. Small ferrite beads are effective in damping high-frequency oscillations generated from switching transients or parasitic resonances within the circuit or transmission line.

When using ferrite material for mainly preventing oscillations or to attenuate unwanted signal coupling, there are limitations to their use, both electrical performance and environment. These are:

- Core size or shape determines impedance.

  The impedance of a ferrite device, for most configurations of use such as beads (surface mount or bead-on-lead; single or multi-turn), absorber sheets or cores/clamps is directly related to the physical size of the device. In other words, the more ferrite material provided the larger the impedance, under most conditions of use. To achieve greater impedance, select a core with a longer length (circular, flat, or toroidal) either width or thickness. In addition, the permeability of the ferrite material will determine the frequency range of operation.

  The impedance of a core is a combination of the intrinsic material characteristic, mostly permeability. Permeability, as discussed above, consists of a real and imaginary part. The real component represents the reactive part and the imaginary element loss. These two parameters are expressed as a series component (μ_s' and μ_s'') or parallel component (μ_p' and μ_p''). At lower frequencies μ_s' is high, however as the frequency increases, μ_s'' becomes dominant and is the biggest contributor to overall impedance since the material is now mainly inductive. At higher frequencies, μ_s'' becomes more significant and impedance more resistive, absorbing the unwanted RF field.

  In other words, ferrites are represented by a parallel combination of a resistor and inductor (Fig. 6.54). At lower frequencies the resistor is shorted out by the inductor, whereas at higher frequencies the inductive impedance is so high that it forces the current through the resistive element. Ferrites are "dissipative devices" where they dissipate high frequency energy as heat. This can only be explained by the resistive, not inductive effects.

While the impedance of a ferrite material will vary as a function of frequency, environmental conditions of use all also affect performance. The most significant ones are:

- Environmental conditions

  The environment affects magnetic parameters of the material and is independent of physical geometry. Impedance increases as a function of temperature. Manufacturers publish impedance plots that show the magnitude of change with respect to ambient temperature.

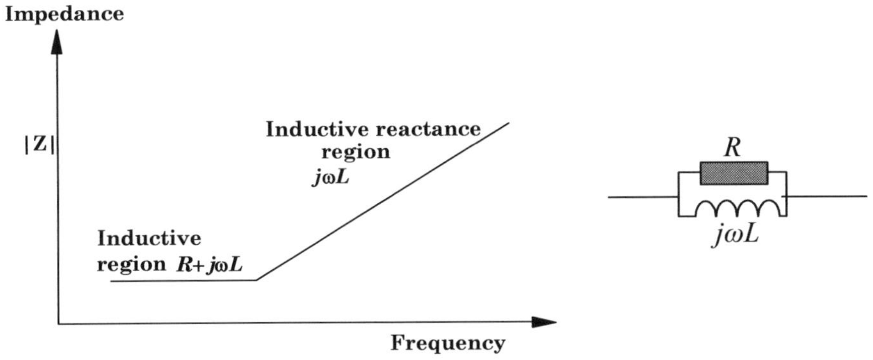

Figure 6.54 Performance characteristic of ferrite material both resistive and inductive.

- **DC bias**
  DC bias refers to the amount of DC current passed through the ferrite core. We are not concerned with DC current since the device is designed to be an attenuator for RF energy. An increase in the DC voltage level within the transmission line for this series element will decrease total impedance of the circuit more than any other parameter. Most surface mount ferrite beads are capable of handling 100-200 mA of current, which is typical for most signal transmission lines on a printed circuit board. If current is greater than 200 mA, the device may not work related to filtering but could be destroyed by excessive internally generated thermal heat. One should never exceed the current rating value of an in-line ferrite bead above 50% of its rated value to prevent permanent damage.
- **Resistivity and Curie temperature**
  Both resistivity and Curie temperature varies depending upon the value of DC current present. The Curie temperature is the point or transition temperature upon which the ferrite material loses magnetic properties. Once this temperature is reached, the component no longer performs its intended function but will return to normal operating conditions when the temperature returns to ambient. If the ferrite material is too resistive then a voltage drop can occur which would affect the signal integrity of the transmitted signal.
- **Multi-turn beads**
  To increase impedance, additional turns of wire may be added. Adding turns of wire through a ferrite core will increase impedance in direct proportion to the turns squared; however, the frequency at which maximum impedance is reached is lowered as well as bandwidth of operation. To illustrate this further, the bead has maximum impedance over a range of 400 MHz. With an extra turn, the bandwidth of operation may be decreased to 300 MHz, and with 6 turns the bandwidth go down to 50 MHz. However, the total impedance of the bead would increase from 100Ω to 1000Ω, excellent if high attenuation is required in a small package device that can handle amps of DC current, but now only for a very narrow bandwidth range of frequencies.

A ferrite core or bead can be added to a physical inductor to improve, in two ways, its ability to block unwanted higher frequencies RF noise. First, the ferrite will concentrate the magnetic field in one tight location and increase inductance of the transmission line. This reactance will impede or filter out undesired RF noise. Second, if the ferrite is properly chosen it can also add an additional loss in the form of resistance that occurs because of loss in the material itself. The ferrite creates an inductor with a very low Q factor ($Q=L/R$). This loss heats the ferrite but normally it is an infinitesimal amount. Although this loss may be large enough to cause interference, or undesirable effects in sensitive circuits, the energy absorbed is typically quite small. Depending on application, resistive loss characteristic of the ferrite may or may not be of concern. When a ferrite bead is to be used to improve noise filtering, specific circuit characteristics and frequency range of operation must be considered to ensure optimal signal integrity.

A ferrite choke is also especially effective to slow down the fast edge-rate transition of an ESD (electrostatic discharge) event induced on cable assemblies. This transient energy is absorbed in the ferrite material instead of being propagated or reflected to circuits and components that can be disrupted.

6.11.12 Feedththrough capacitor filter

Feedthrough capacitors were originally designed for DC power lines or use between RF assemblies installed in a single-box system or between two areas such as an anechoic chamber and control room. The capacitor passes lower frequency signals but blocks higher frequencies energy from corrupting secondary circuits using a hole drilled into the metal bulkhead. One side of the capacitor is bonded to the enclosure while the other terminals connects to the signal line.

A feedthrough capacitor is essentially a leaded capacitor (axial configuration) that contains a composite dielectric element made of resin material and a dielectric powder that forms one side of the capacitor plate. An axial lead enters from one side and exits to the other. There is an outer electrode terminal surrounding the internal plate design designed to be at ground potential (a polarized capacitor). Thus, capacitance exists between the signal line and outer electrode terminal bonded to chassis ground.

Inductance in the leads of a capacitor limits its effectiveness. This is due to total inductance (*ESL*) both internal to the capacitor package along with the interconnect leads and interface to circuitry. For enhanced performance, especially where penetration of a metal enclosure must be protected at UHF frequencies and above, feedthrough construction is required. In this application, the ground connection is made by screwing or soldering the outer body of the capacitor directly to the meal enclosure. Because RF current to ground can spread out in a 360° manner around the center conductor, there is effectively no inductance associated with this capacitor configuration, and performance can be achieved well into the GHz region. This performance however is compromised if a 360° connection is not made correctly or the bulkhead is limited in extent by a plating or finish with high transfer impedance or paint.

The inductance of the through-leads can be increased to create a π-section filter by separating the plates inside the package and incorporating a ferrite bead or element internally. Feedthrough capacitors are available in a wide range of voltage and capacitance ratings along with configuration, all at additional cost.

An example of a feedthrough capacitor and its construction is shown in Fig. 6.55.

6.11.13 Three terminal capacitor filter

All filter combinations, except the single element inductor uses a capacitor. A theoretically perfect capacitor provides an increase in attenuation at a constant 20 dB per decade as frequency increases however, stray inductance will limit this attenuation and increases impedance as the frequency goes higher. This stray inductance comes from the connection between capacitor plates and its terminals as well as the connection to a transmission line. This is illustrated in Fig. 6.56. Impedance decreases until a resonant frequency occurs upon which impedance will then increase.

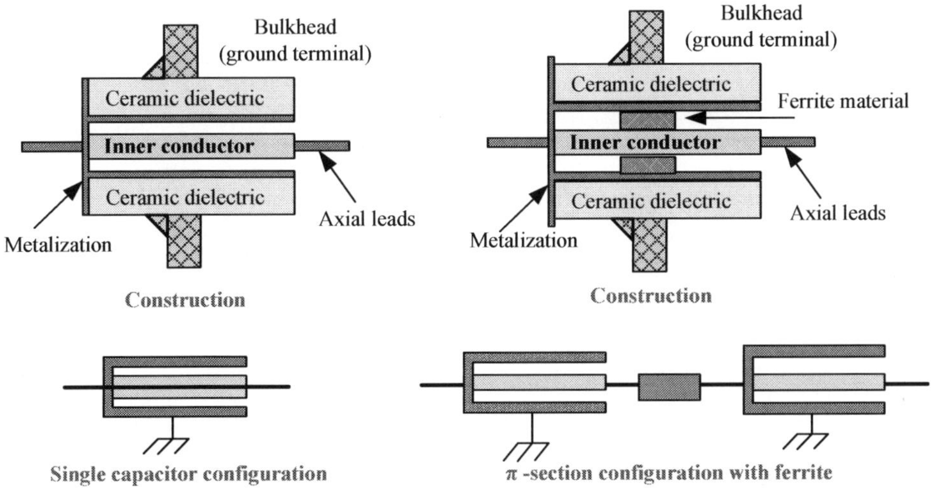

Figure 6.55 Feedthrough capacitor construction and configurations.

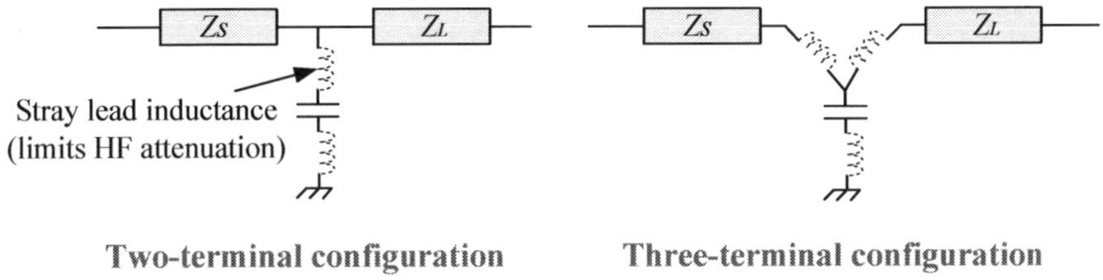

Figure 6.56 Three terminal capacitor configuration.

Lead inductance can be useful for filtering if designed into a package that contains a capacitor. In addition, a ferrite bead can be included in the manufacturing of the filter to create a "*T*" or 3-terminal unit with a sharp *Q* resonance. For the two-terminal configuration, the lead inductance forms the *T*-filter with the capacitor. This improves higher-frequency performance. When a ferrite bead is included, this further enhances lead inductance and increases effectiveness when used in a relatively low impedance network.

A three-terminal capacitor can extend the frequency range of operation, which is useful for interference in the VHF band. To achieve optimal performance, the middle terminal or ground wire must be connected to an earth ground or 0V potential using a low inductive bonding technique. If this low inductance connection is not optimal, the inductance in this device will defeat the capacitor's purpose or functional use.

6.11.14 Installation Guidelines for Filters

The only way to be sure that a filter will reduce EMI to compliant levels is to test the equipment for both radiated and conducted emissions first, and then be prepared to try several different configurations having anticipated the need ahead of time. This trial-and error approach may be unscientific and take time, but in most situations proves to be the fastest, most cost effective approach at minimum risk since it is nearly impossible to simulate the existence of common-mode current using computational analysis as many unknown or hidden parasitics exist with transmission line routing. As a result, incorporation and installation of a filter is extremely critical for optimal performance. This includes minimizing excessive inductance in transmission lines, or having an optimal low impedance bond connection to 0V or earth ground for data signals.

6.11.14.1 AC mains filtering

An AC mains filter case-to-frame ground connection must have low impedance over the frequency range of interest. The input-to-output leads must have maximum physical isolation. In the case of power and I/O line filters, the filtered lines must be as close as possible to the enclosure entry point (Fig. 6.57).

6.11.14.2 Signal line filtering

There are several rules or guidance in using a ferrite bead on signal transmission lines.

- It is best to filter at the source and not let unwanted RF energy propagate down a transmission line.
- Suppress all spurious signals, including harmonics. This means using a broadband filter that ensures those signals that need to be suppressed are covered.
- Determine if filtering is to solve a radiated field or conducted current problem. There are significant differences between these two modes of emissions as well as choice of filter configuration.
- Design susceptible circuits that may be affected by radiated fields and/or conducted currents by using proper system design and layout design techniques [7] to minimize the generation and propagation of EMI, detailed in earlier Chapters.
- Ensure all filter elements interface properly with other EMC components; i.e., proper mounting of a filter in a shielded enclosure and that the impedance of the filter does not cause signal integrity disruption.
- Do not select a filter based on databook values of insertion loss or attenuation. These numbers are based on a perfect 50-ohm system, which in real products do not exist.
- Determine if the noise to be filtered is common-mode, differential-mode or both. Sometimes two filters in series or a multi-element device is required.

T, L & π filters are reflective elements

- Signal “loss” is due to the mismatch between the noise source and filter’s input impedance. If the filter does not have proper impedance relative to the transmission line, performance will be affected.
- Undesired noise sources are shunted to ground with a capacitor, or reflected back to the source with a series inductor or ferrite material with hopefully little effect on desired signal propagation. If a capacitor is used to shunt energy to ground for a signal trace, capacitive loading may round off the edge transition or slow down signal propagation and cause a functionality or signal integrity problem.
- Install the leads of capacitors with the shortest physical length possible to minimize loop inductance.
- Install resistors and inductors with their leads as far apart as possible to minimize [parasitic] capacitive coupling.
- Filters should be placed as close to the physical entry point of enclosures or interconnects.

- Locate inductors as far from metal panels and the printed circuit board return planes as possible to reduce the presence of stray or parasitic capacitance.

Use pot core ferrites rather than toroids to reduce stray magnetic emissions in power supplies.

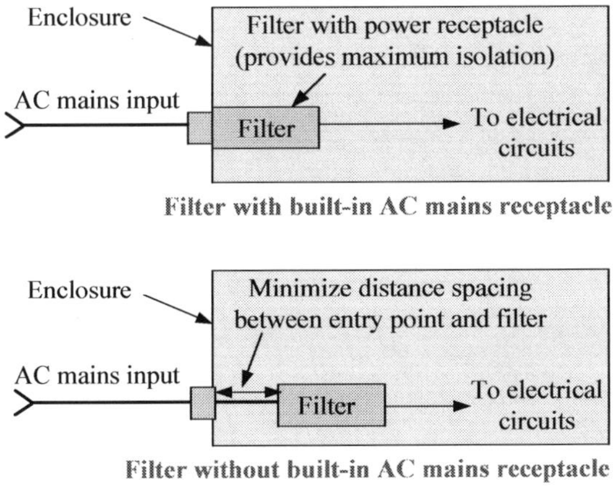

Figure 6.57 Installing a filter for AC mains protection.

REFERENCES

1. Ott, H. 2009. *Electromagnetic Compatibility Engineering.* Hoboken, NJ: John Wiley & Sons.
2. Paul, C. R. 2006. *Introduction to Electromagnetic Compatibility*, 2nd ed. Hoboken, NJ: John Wiley & Sons.
3. Chomerics. 2000. *EMI Shielding Engineering Handbook.* Wobun, MA.
4. Fair-Rite Corp. Product Catalog, 17th edition.
5. Joffe, E. & Lock, K. S. 2010. *Grounds for Grounding-A Circuit-to-System Handbook.* Hoboken, NJ: John Wiley & Sons/IEEE Press.
6. Kraus, J. D. and Marhefka, R. J. 2002. *Antennas.* 3rd ed. New York, NY: McGraw-Hill.
7. Montrose, M. I. 2000. *Printed Circuit Board Techniques for EMC Compliance-A Handbook for Designers.* Hoboken, NJ: John Wiley & Sons.
8. Quine, J. P. 1957. "Theoretical Formulas for Calculation of the Shielding Effectiveness of Perforated Sheets and Wire Mesh Screens." *Proceedings of the Third Conference on Radio Interference Reduction.* Armour Research Foundation, Vol. (2), pp.315-329.
9. Schelkunoff, S. A. 1943. *Electromagnetic Waves.* New York, NY: Van Nostrand Reinhold.
10. Tsaliovich, A. 1995. *Cable Shielding for Electromagnetic Compatibility.* New York, NY: Van Nostrand Rheinhold.
11. Williams, T. 2007. *EMC for Product Designer.* 4th ed. Oxford, UK: Newnes.
12. Williams, T. & Armstrong, K. 2000. *EMC for Systems and Installation.* Oxford, UK: Newnes.

Appendix A – Maxwell Made Simple (The Five Equations)

This book is about *EMC Made Simple®*, or essentially *Maxwell Made Simple®*. Both are internationally registered trademarks of Montrose Compliance Services, Inc. What is presented in Chapter 1 through 6 is a simplified explanation of complex electromagnetics as published by James Clerk Maxwell (1873) with a series of 20 equations and 20 variables. Oliver Heavyside reformulate 12 of these 20 equations into four equations with four variables (**B**, **E**, **J**, and ρ), the form by which they have been known ever since (1881). "Maxwell developed the science and deserves full credit," according to Heavyside.

Maxwell's equations deal with complex field theory in the frequency domain using vector calculus. Most design engineers tend to shy away from solving equations based on field propagation but instead prefer to simplify engineering analysis using easy algebra. This however does not diminish the important of field theory as the field of electrical engineering is based on Maxwell's equation. If there is a lack of understanding or confusion on how field theory applies to circuit development, how can one design an optimal printed circuit board or system?

In order to present Maxwell's equations for ease of understanding, there are five algebraic equations that *overly simplify or describe applied field theory in a manner that every engineer can understand, especially those who never studied electromagnetic in a university environment.* Most design engineers prefer to think in the time domain whereas those doing EMC prefer to analyze signals in the frequency domain. One must be comfortable working in both domains, or using spectrum analyzers/receivers and oscilloscopes at the same time to observe different aspects of signal propagation, which in reality are identical. A signal measured on an oscilloscope may be acceptable for functional performance yet the signal being propagated may create undesired common-mode current that will radiate as a propagating field due to losses in the transmission path loop, which can only be measured with a spectrum analyzer or receiver, not oscilloscope. So which instrument do we use? The answer is: *It depends.*

It is important to highlight, in this Appendix although stated slightly differently in prior Chapters to illustrate or discuss a complex concept making field theory simple to understand, that time and/or frequency domain analysis have in reality nothing to do with field or circuit representation. It is implied in earlier chapters that the time domain (DC analysis) is based exclusive on Ohm's law or $V=IR$. In reality, DC circuit representation also contains AC elements ($V_{RF}=I_{RF}Z$), which is a different manner of observing the same signal transmission depending on frequencies of concern. We consider this comparison between time and frequency domains to be, in other words, Kirchhoff's equation which are in fact a direct simplification of Maxwell with some assumption regarding the time domain ($dE/dt=0$, $dH/dt=0$), as well as electrically small circuits which is equivalent to the former. Also, Kirchhoff assumes that there are no rotational fields in the environment which is represented by the time domain version of Ohms law using R instead of Z.

With this said, in reality, circuits effects such as $Z=Z(t)$, $I=I(t)$ and $V=V(t)$ can be described in either the time or frequency domain, yet operation performance may differ based on functional requirements, either enhanced signal integrity or EMC compliance.

The field of electrical engineering is based exclusively on transmission line theory. How the transmission line is defined is irrelevant for this discussion. In the frequency domain, we generally deal with propagating electric and magnetic fields or power density (Maxwell's equation). In the time domain we usually work with voltage and current amplitude with little thought about the power level in the transmission line (Ohm's law).

It is critical to note that frequency domain components also include voltage and current, and time domain analysis includes propagating fields. So what do we have in reality?

What we have in reality is electromagnetic energy or an electromagnetic field traveling between circuits somehow in a transmission line. This EM field is generated by digital components switching logic states. It is the *rise* and *fall* time (edge transition) of a digital signal that creates RF energy using Fourier Analysis (Appendix C), not the DC or steady state value such as 0V (ground) or Voltage that define the operation of digital circuits in the time domain. Both are described per Maxwell but better visualized and easier to manipulate using Ohm's law.

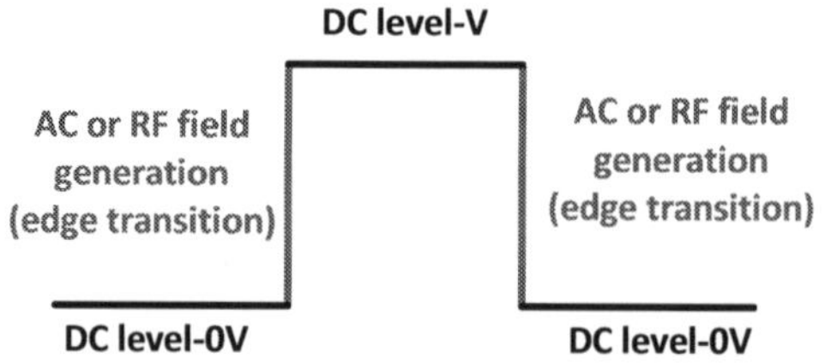

When mentioning frequency domain aspects of signal propagation, we also include electrodynamics, quasi-static fields and electrostatics. In circuits, we deal with voltages and currents therefore we have both AC and DC current, voltages or fields. Also, Z could be dynamic based on resistance, inductance and capacitance within every transmission line. Remember, field theory also applies to circuit theory, except with a different viewpoint on how it is implemented within a transmission line structure on a printed circuit board.

Electric and magnetic fields are measured in units of V/m and A/m, respectively. This corresponds to Power as Watts/m^2. If we remove the "per meter" notation from this unit of measurement (frequency domain), we are essentially converting complex field theory into simple algebra (time domain) that is easily understood using Ohm's law. Although *not mathematically or technically correct*, if we "*visualize*" what is occurring in transmission lines, our job as designers become easier.

When incorporating component on a printed circuit board schematic during the design period, a line is drawn between elements. Do we identify for each and every trace the signature profile of the propagated signal, for example 3.3 V/m at 2 mA/m (6.6 mW/m^2) [frequency domain-Maxwell] or 3.3 V at 2 mA (6.6 mW) [time domain-Ohms law], or just assume the signal has a typical digital or analog waveform and only issue a net name?

Power density is rarely taken into consideration by most *digital circuit designers* who are not working in the field of intentionally transmitted RF energy. So what unit of measurement do we put on documentation or schematics, since both units are technically the same, with one easier to comprehend and work with?

Once conversion to Ohm's law occur *conceptually*, simple numerical understanding and analysis on EM field propagation is possible, realizing that all transmission lines are based on complex field theory, not digital logic levels operating at a specific frequency with voltage and current amplitudes which one generally thinks of, especially digital design engineers.

Five equations that describe the field of electrical engineering in the time domain:

(OVERLY SIMPLIFIED)

$V = I\,Z$ (Ohm's law-DC analysis) (A.1)

$V_{RF} = I_{RF}\,Z$ (Maxwell's equations-AC analysis)

$$Z = R + jX \quad (A.2)$$

where:

V = voltage potential (Ohm) or electric field propagation (Maxwell) in the transmission line

I = current in the transmission line (Ohm) or magnetic field propagation (Maxwell) in the transmission line

R = DC resistance in the lower frequency range (usually <100 kHz)

$jX = 2\pi fL$ (inductive reactance of the transmission line > 100 kHz).

One item to note is that for DC energy transfer, which is at a steady state condition, $V=IR$ must be addressed in the time domain as the function is time invariant. Low frequency phenomena using quasi-static assumptions can also use time domain analysis which Kirchhoff describes such as lightning effects. However, when frequencies are higher, $V=IZ$ [$V=I(2\pi fL)$] the complex portion of the equation now becomes dominant, not the resistive portion ($V=IR$).

When the operating frequency of the EM field in the transmission line starts to increase, generally above 100 kHz, the value of jX_L will become much larger than R due the variable f in the equation, thus we are now in the AC (RF) environment and not DC or steady state. *Ideal* inductors do not create loss in energy although this was implied in earlier Chapters to highlight a concept of transmission line physical characteristics; *ideal* inductors store and return energy. However, *non-idea* inductors will lose some energy.

Differential-mode to common-mode conversion occurs within *non-ideal* inductors, and vice-versa, which are not strictly speaking "losses" per se but rather conversion. Sometimes they are called "conversion losses" when in reality they function completely differently.

At higher frequencies, analysis is typically done in the frequency domain because it is far more intuitive. This is not always the correct way to do analysis, it really depends on the nature of the problem. For example, nuclear electromagnetic pulses are typically addressed in the time domain although their spectrum extends to about 100 MHz simply because of the nature of the interaction of the electromagnetic fields present.

We break down or analyze Equations (A1 and A.2), for *visualization* purposes only, realizing that there are many other factors and variables involved (such as X_c), but to simplify *EMC Made Simple®* at this time, we consider the following to highlight the difference between the time and frequency domains aspects of signal propagation using the following notation, remembering that both are identical; *it is the manner of which one wishes to work in*:

Field of electrical engineering = *Real* + *jComplex*

Real component	Complex component
Time domain analysis;	*Frequency domain* analysis
DC signal propagation;	*RF signal propagation* (AC currents)
Ohm's law	*Maxwell's equations*

To further examine (A.2) and its relationship to (A.1), if we have any loss in a transmission line (*Z*) with fixed drive voltage, propagating current will increase (or decrease) per Ohm's law. To again, overly simplify a complex topic in the time domain, switching noise in the power distribution network may cause power and/or return planes to bounce, which in turn may create a signal integrity problem. In the frequency domain, bouncing planes is one form of creating undesired common-mode current, or an EMI event.

Now that we have examined basic transmission line theory made simple (A.1, A.2), we consider the magnitude of RF energy developed and how undesired common-mode energy is generated (A.3).

$$E = -L\,(dI/dt) \tag{A.3}$$

where:
E = magnitude of common-mode current that is developed
L = inductance in the transmission line
dI/dt = total amount of switching current with respect to time.

Equation (A.3) contains three variables. The greater the inductance in a transmission line, the greater the amount of *loss* that occurs to the propagating EM field, which in turn creates common-mode current (*E*) since that loss must be made up somehow to satisfy Kirchhoff's law. This is the loss referred to in (A.2) as "*Z*". Per (A.3), either a greater amount of switching current and/or a faster edge transition will also create common-mode current in direct proportion to inductance based on the loss in the transmission line. Inductance in a transmission line is discussed in Chapter 3. Substituting *E* into (A.1 and A.2) with variable "*V*", we now have an EMI event with RF current established across the impedance in the transmission line.

Propagating fields needs an efficient antenna structure to propagate RF energy. There are two basic equations related to antennas.

$$f = \frac{1}{2\pi\sqrt{LC}} \tag{A.4}$$

$$\lambda = c/f \tag{A.5}$$

where:
f = frequency of the signal (Hertz)
L = inductance in the transmission line (Henrys)
C = capacitance between transmission line and return path (Farads)
λ = the wavelength of the propagating signal
c = speed of light ($3*10^8$ m/seconds)
f = frequency (Hertz).

What these two equations tell us is that with a fixed value of inductance and capacitance in a transmission line (A.4), a resonant RF frequency will exist, regardless of the magnitude of voltage or current present; voltage and current does however determines the amplitude of signal propagation. With reference to the speed of light, a propagating field at frequency (*f*) will occur at a specific transmission line wavelength (λ).

When troubleshooting a radiated field problem, it is easy to locate the propagating transmission line based on physical dimensions (wavelength) should a radiated emissions problem develop. We use a ruler or tape measurer to achieve this goal. For example, if we have a radiated signal at 30 MHz, we need to look for a transmission line length of 10 meters (from A.5). From here we subdivide this length into smaller segments. This means $\lambda/2$ = 5 meter, $\lambda/4$= 2.5 meters, $\lambda/8$ = 1.25 meters, etc. For signals that are at a higher frequency, the length of an efficient antenna need not be large, especially in the GHz range and if we take into consideration the 10th harmonic of a switching signal almost any transmission line length becomes an efficient antenna.

To summarize Maxwell's Equations Made Simple

1. Based on the edge rate transition of digital components (Fourier analysis), RF switching energy is created based on total transmission line loop inductance and the time rate of change of the current [magnetic field] (Faraday's Law)–Equation (A.3).
2. This RF field propagates within a transmission line (A.1 and A.2) and depending on the frequency of the signal (A.4), with regard to inductance in that line (A.3), a voltage drop is created or mode conversion from differential-mode to common-mode occurs. To ensure conservation of energy (power), conceptually based on Kirchhoff's law, an increase in RF current must occur since components operate on fixed voltage levels described by Ohm's law (A1). This increase in RF current is unwanted common-mode energy.
3. Since we have time variant current in a transmission line created by inductance (loss) in that line (A.3), this energy will find an antenna structure (A.5) to propagate from based on the self-resonant frequency of the transmission line (A.4) due to both inductance relative to a return path (loop inductance) and the capacitance between the two transmission line paths-source and return.

Appendix B – The Decibel

In the field of engineering, a common unit of measurement or reference is required. This often misunderstood unit, a logarithmic function, is the decibel (dB). We use the decibel because of the scaling range of units involved. Most ratios are dimensionless whereas some are magnitudes expressed in dB with a reference measurement value.

The basic form of the dB is the logarithmic ratio of two products. Absolute power, voltage or current levels are expressed in dB by giving their value *above* or *referenced* to some *base* quantity. The following describes power gain ($P2 > P1$) or loss ($P2 < P1$) in a system.

$$\text{Power Gain: } dB = 10 \log\left(\frac{P_{out}}{P_{in}}\right)$$

In many situations, reference must be made for voltage, current, field strength, and the like instead of power. The following describes formulas for voltage and current gain ratios. The unit dB is dimensionless.

$$\text{Voltage gain: } dB = 10 \log\left(\frac{V_{out}^2 / R}{V_{in}^2 / R}\right) = 10 \log\left(\frac{V_{out}}{V_{in}}\right)^2 = 20 \log\left(\frac{V_{out}}{V_{in}}\right)$$

$$\text{Current gain: } dB = 10 \log\left(\frac{I_{out}^2 R}{I_{in}^2 R}\right) = 10 \log\left(\frac{I_{out}}{I_{in}}\right)^2 = 20 \log\left(\frac{I_{out}}{I_{in}}\right)$$

A pattern follows for voltage and current. In the field of electrical engineering, everything is based on a common reference of a single dB *above or below one milliwatt of power, denoted as 0 dBm.*

Radiated electromagnetic fields are described in terms of field intensity. These units are V/m (volts per meter) for electric field strength or A/m (amperes per meter) for magnetic field strength. The common unit of measurement for the following voltage and current field strength intensity are

1µV/m	= 0 dBµV/m
1mV/m	= 0 dBmV/m
1µA/m	= 0 dBµA/m
1mA/m	= 0 dBmA/m
1mW	= 0 dBm (dBm represents power, not voltage or current)

Most regulatory limits are described in µV/m. For example, 100 µV/m limit translates to 40 dBµV/m. The equations that describe this conversion are

$$dB\mu V/m = 20\log\left(\frac{V/m}{1\mu V/m}\right)$$

$$dB\mu A/m = 20\log\left(\frac{A/m}{1\mu A/m}\right)$$

Conversions between units are easy. For example:

1 μV = 0 dBμV = dBm - 107 dBm	For a 50 Ω system
V(dBμV) = 90 + 10 log (Z) + P (dBm)	For a given impedance Z in ohms

The proof for this conversion is

$$0\ dBm = 1\ mW = 0.001\ Watts$$

$$E = \sqrt{mW * R} \qquad Assume\ R = 50\ ohms\ (Note - for\ dBm, the\ reference\ is\ 1\ mW)$$

$$E = \sqrt{1mW * 50\,\Omega} = 0.224\ V$$

$$dB\mu V = 20 log_{10}\left(0.224\ V / 1\,\mu V\right) = 107\ \ dB\mu V$$

$$Therefore,\ 107\ dB\mu V = 0\ dBm\ in\ a\ 50\,\Omega\ system$$

$$The\ scale\ factor\ is\ thus:\ dBm + 107\ dB = dB\mu V$$

Five commonly used variations exist for the decibel. An example of this variation follows to present the concept of *dBs* using different units.

$$dBm = 10\ log\left(\frac{P}{0.001W}\right)$$

$$dB\mu V = 20\ log\left(\frac{V}{1\,\mu Volt}\right)$$

$$dB\mu A = 20\ log\left(\frac{A}{1\,\mu Amp}\right)$$

$$dB\mu V\ /\ m = 20\ log\left(\frac{V}{1\,\mu Volt/meter}\right)$$

$$dB\mu A\ /\ m = 20\ log\left(\frac{V}{1\,\mu Amp/meter}\right)$$

$$dB\mu V\ /\ m\ /\ 120KHz = 20\ log\left(\frac{A}{1\,\mu Volt/meter}\right) at\ a\ 120\ kHz\ bandwidth$$

Several pitfalls are related to use of the decibel, owing to the impedance of the transmission line. Because not all transmission lines have the same impedance, different values will be obtained.

- dBm = 10 log (P1/0.001 watts)
- 1 volt in a 50 ohm system is equal to:

$$dBm = 10\,log\left(\frac{1\,volt^2 / 50\,ohms}{0.001\,watts}\right) = 10\,log\,(20) = 13\,dBm$$

- 1 volt in a 600 ohm system is equal to:

$$dBm = 10\,log\left(\frac{1\,volt^2 / 600\,ohms}{0.001\,watts}\right) = 10\,log\,(1.67) = 2\,dBm$$

Most engineers make a common mistake when performing decibel (logarithmic) math. This is known as the 6 dB problem. We must ask ourselves, "When does 6 dB not equal 6 dB"?

Examples of this mistake follow. If the reference level is doubled, the logarithmic function increases by 6 dB (50%).

If we decrease RF energy by 6 dB, we essentially remove 50% of the power present in the transmission line, or increase by 50% if an implemented change increased the energy level by 6 dB.

A three times increase in the reference is a 9.5 dB increase over the base value.

- 1000 µVolts = 60 dBµV

- <u>1000 µVolts = 60 dBµV</u>
2000 µVolts = 66 dBµV

- <u>1000 µVolts = 60 dBµV</u>
3000 µVolts = 69.5 dBµV

Appendix C - Fourier Analysis

Every signal exists in both the time and frequency domain. Conversion between domains is accomplished through use of Fourier analysis. Digital circuits are always discussed in terms of operating frequency. Although this frequency value is important for speed of operation, it is the physical edge rate of the periodic signal (rise and fall time) that determines total RF spectral distribution of energy created. RF energy will find a propagation path either through free space or within metallic interconnects.

There are four primary aspects associated with Fourier analysis; frequency, amplitude of the signal, duty cycle and edge rate transition time. To summarize these four aspects:

- Frequency: Where in the frequency spectrum can we expect to observe the signal and at this frequency will it cause harmful interference to other electrical devices?
- Amplitude: How strong is the signal and its propensity to cause harmful disruption?
- Duty cycle: At what frequency will the amplitude of the RF energy start to decrease?
- Edge rate: How high up in the frequency spectrum do we need to go to when measuring an RF signal before the amplitude falls into the non-disruptive region?

The mathematical operations in performing Fourier analysis are time consuming but not necessary for the working engineer to solve. Understanding what the math tell us is however critical to successful system design (*EMC Made Simple®*). As presented in Chapter 1, the only signal that exists in a transmission line is time variant propagation of an electromagnetic field described by Faraday's law. All time domain signals (a.k.a., digital pulse in visual appearance, or a sine wave with an infinitely fast rise time) creates a continuous waveform (electromagnetic field) in the frequency domain.

Signals are easily measured on both oscilloscopes (time domain) and spectrum analyzer or receiver (frequency domain).

Periodic signals are represented by a series of sine and cosine functions.

$$f(t) = \frac{A_o}{2} + \sum_{n=1}^{\infty} \left(A_n \cos(n\omega_o t) + B_n \sin(n\omega_o t) \right)$$

where

$$\omega_0 = \frac{2\pi}{T} = natural\ fundamental\ frequency$$

$$A_o = \frac{2}{T} \int_{to}^{to+T} f(t)dt,$$

$$A_n = \frac{2}{T} \int_{to}^{to+T} f(t)\cos(n\omega_o t)dt,$$

$$B_n = \frac{2}{T} \int_{to}^{to+T} f(t)\sin(n\omega_o t)dt,$$

These equations illustrate that a periodic signal is a summation of sinusoidal signals of multiple frequencies and amplitudes. Therefore, a periodic signal corresponds to a particular frequency range. Fourier transform converts time-based signals (digital) to the frequency domain (analog). The actual Fourier transform is shown below.

$$F(\omega) = \int_{-\infty}^{\infty} f(t) e^{-j\omega t} dt$$

The Fourier envelope is used to quickly calculate the worst-case frequency spectrum envelope. For a given periodic square signal with finite rise (t_r) and fall time (t_f), the frequency spectrum envelope is shown below with numerical analysis on what we observe based on the following parametric values. These values are easily calculated by the equations below:

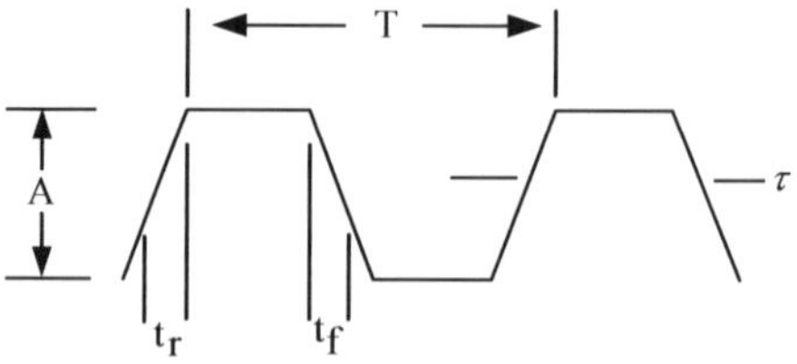

where:
A = peak amplitude (volts or amperes) in the time domain
τ = pulse width or duty cycle (measured at half-maximum)
T = pulse period (time it take for one digital pulse to occur)
t_r = rise time – assume 10–90% of the edge transition
t_f = fall time – assume 90–10% of the edge transition

If $t_r \neq t_f$, the smallest (or fastest value) must be used.

The amplitude of the signal in the frequency domain, A_f, is calculated using

$$A_f = 2A\frac{\tau}{T}$$

The corner frequencies f_1 (lower) and f_2 (upper) are calculated using the following simple equation. Duty cycle, δ, is also easily calculated.

$$f_1 = \frac{1}{\pi \tau_r} \qquad f_2 = \frac{1}{\pi \tau_f} \qquad \delta = \frac{\tau_r + \tau_f}{T}$$

Examining both corner frequencies, we note that the rising and falling edge of the periodic signal may not be the same and is often very different due to the manufacturing process of the silicon die or semiconductor device. With this situation, use the faster of the two corner frequencies when doing numerical analysis; generally the transition from high-to-low (falling edge) is faster for most semiconductor manufacturing processes.

Example: the rising edge may be 1 ns from 0-100% whereas the falling edge could be 600 ps. When performing Fourier analysis, use the 0-100% portion of the waveform whereas digital design engineers generally concern themselves with either the 80/20 or 90/10 percentage of the edge transition, which is where V_{ih} (voltage high input level to a digital device to switch to logic level 1) and V_{il} (voltage low input level to switch to logic level 0). Digital designers usually do not consider the full waveform profile (0-100%), which is how the signal should be analyzed related to frequency domain concerns and EMI compliance.

The amplitude or radiated field strength of the time variant RF signal (frequency domain) decreases at –20 dB per decade up to corner frequency f_2 (duty cycle value). Above the f_2 breakpoint (edge rate transition value, t_r), the signal amplitude then begins to fall off at –40 dB per decade. Thus, the spectrum amplitude above f_2 is defined by:

$$A_f = \frac{2A}{f^2 \pi^2 t_r T}$$

The RF current for the nth harmonic is thus calculated by:

$$I_n = 2I_d \frac{\sin(n\pi d)}{n\pi d} \frac{\sin(n\pi t_r / T)}{n\pi t_r / T}$$

where:
I = peak-peak amplitude of the wave
d = duty cycle
t_r = edge rate rise time
T = period of the signal
n = harmonic number.

The unit of I_n is the same as I_d, thus the equation is dimensionless. For harmonic calculations, let's assume the rise and fall time edges are the same. If the two edges are different, the smaller of the two must be used for worst-case analysis. For a 50% duty cycle ($d = 0.5$), the first harmonic (fundamental) contains an amplitude of $I_1 = 0.64I$ with only odd harmonics present. This is for the case where the rise time (t_r) is much less than period (T).

The following figure illustrates the envelope of harmonics for a symmetrical square wave. The amplitude of all harmonics decreases with frequency at -20 dB per decade up to the frequency of $1/\pi t_r$. Beyond this corner frequency, harmonics fall off at -40 dB per decade. As the rise time increases (becomes slower or longer), RF energy in the higher order harmonics decreases.

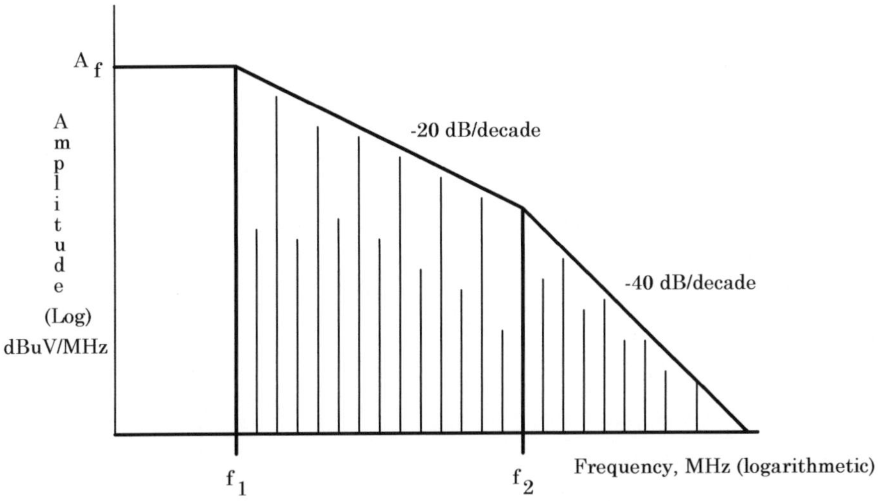

Any change in the duty cycle and transition time will either increase or reduce the frequency spectrum envelope. As a frequency is doubled, electromagnetic radiation increases by 6 dB if all parameters remain the same. If the frequency is doubled, we must reduce the edge rate transition time (50%) to accommodate for this faster signal. A decrease in edge rate transition time (i.e., faster signal) will thus increase the amplitude of the radiated RF signal by 12 dB and be observed at higher frequency points in the spectrum.

In looking at the above figure for a 50% duty cycle, only odd harmonics are shown. For smaller duty cycles, that is when the period becomes significantly long compared to the pulse duration, only a few harmonics within the envelope will reach the maximum amplitude level. There harmonics are thus observed as radiated EMI because of this larger amplitude of electromagnetic energy.

The spectral profile envelope of a signal does not show phase or polarity. At every multiple of $1/\tau$ there is a 180 degree reversal due to the fact that harmonics follow a sine or cosine function of frequency. Measurement equipment, such as spectrum analyzers or receivers, are insensitive to phase and will only display the absolute value.

When harmonics are spaced close to each other (many harmonics) they will not add together in phase. When measurements are performed using a spectrum analyzer, we generally tune the analyzer with a specific resolution bandwidth. The resolution bandwidth of a spectrum analyzer is the ability to display discrete frequency components within a specific frequency span (beginning and ending frequency range). The noise that is displayed from a periodic signal will thus appear as a narrowband signal, unless there are many signals close together and the resolution bandwidth is greater that the bandwidth of all the signals. This means that the receiver will only see one signal time within the selected resolution bandwidth.

The designer must strive to limit and control the noise spectra of digital signals. Switching noise, typically in the MHz range, is a byproduct of digital circuits switching between logic level "1" (high) and logic level "0" (low). Switching noise will find its way outside of the intended operational environment and cause EMC problems.

In reviewing the spectra of digital pulses, other parametric behaviors are also observed. The spectrum is governed by amplitude (A_f), which is in turn a function of the pulse amplitude, A. Limiting the pulse amplitude has a direct effect on the RF noise created. A slower edge transition time creates a smaller spectrum of RF energy. Wide pulses concentrate energy at lower frequencies than do narrow pulses.

Appendix D - Conversion Tables

Common Suffixes

Suffix	Refers to
dBm	1 milliwatt
dBW	1 watt
dBμW	1 microwatt
dBV	1 volt
dBmV	1 millivolt
dBμV	1 microvolt
dBV/m	1 volt per meter
dBμV/m	1 microvolt per meter
dBA	1 amp
dBμA	1 microamp
dBμA/m	1 microamp per meter

Conversion of dbV, dbmV, and dbμV

dBV	dBmV	dBμV
-120	-60	0
-100	-40	20
-80	-20	40
-60	0	60
-40	20	80
-20	40	100
0	60	120
20	80	140
40	100	160
60	120	180

Frequency --- Wavelength --- Skin Depth

Frequency	λ	$\lambda/2\pi$	Skin Depth
10 Hz	30,000 km	4,800 km	820 mil
60 Hz	5,000 km	800 km	340 mil
100 Hz	3,000 km	480 km	260 mil
400 Hz	750 km	120 km	130 mil
1 kHz	300 km	48 km	82 mil
10 kHz	30 km	4.8 km	26 mil
100 kHz	3 km	480 m	8.2 mil
1 MHz	300 m	48 m	2.6 mil
10 MHz	30 m	4.8 m	0.8 mil
100 MHz	3 m	0.48 m	0.3 mil
1 GHz	30 cm	4.8 cm	0.08 mil
10 GHz	3 cm	4.8 mm	0.03 mil

λ = wavelength

$\lambda/2\pi$ = near field to far field distance convergence (meters)

Frequency–Wavelength Conversion

To convert between frequency and wavelength of a signal, use the following conversion equations.

$f\,(\text{MHz}) = 300 / \lambda\,(\text{MHz})$ $\quad$ $f\,(\text{MHz}) = 984 / \lambda\,(\text{ft})$

$\lambda\,(\text{m}) = 300 / f\,(\text{MHz})$ $\quad$ $\lambda\,(\text{ft}) = 984 / f\,(\text{MHz})$

λ = wavelength, f = frequency

Throughout this book, reference is made to critical frequencies or high-threat clock and periodic signal traces that have a length greater than $\lambda/20$. A summary of miscellaneous frequencies and their respective wavelength distance is shown in the following table, based on the equations above.

Frequency of Interest	$\lambda/20$ Wavelength Distance
10 MHz	1.5 meters (5 feet)
27 MHz	0.56 meters (1.8 feet)
35 MHz	0.43 meters (1.4 feet)
50 MHz	0.33 meters (12 inches)
80 MHz	0.19 meters (7.44 inches)
100 MHz	0.15 meters (5.88 inches)
160 MHz	9.4 cm. (3.72 inches)
200 MHz	7.5 cm (2.95 inches)
400 MHz	3.75 cm (1.48 inches)
600 MHz	2.5 cm (0.98 inches)
1000 MHz	1.5 cm (0.59 inches)

Power and Voltage/Current Ratios

Ratio	V or I in dB	P in dB
10^6	120	60
10^5	100	50
10^4	80	40
10^3	60	30
10^2	40	20
10	20	10
9	19.08	9.54
8	18.06	9.03
7	16.9	8.45
6	15.56	7.78
5	13.98	6.99
4	12.04	6.02
3	9.54	4.77
2	6.020	3.01
1	0	0
10^{-1}	-20	-10
10^{-2}	-40	-20
10^{-3}	-60	-30

dB	Power Ratio	Voltage/Current Ratio
120	10^{12}	10^6
100	10^{10}	10^5
80	10^8	10^4
60	10^6	10^3
40	10^4	10^2
30	10^3	32
20	10^2	10
10	10.0	3.2
6	4.0	2.0
3	2.0	1.4
0	1.0	1.0
-3	0.50	0.71
-6	0.25	0.50
-10	0.10	0.32
-20	10^{-2}	0.10
-30	10^{-3}	0.03
-40	10^{-4}	10^{-2}
-60	10^{-6}	10^{-3}
-80	10^{-8}	10^{-4}
-100	10^{-10}	10^{-5}
-120	10^{-12}	10^{-6}

Conversion of Volt/m to mW/cm² for Linear and dB Scales

V/m	dBμV/m	mW/cm²	dBmW/cm²
1.00×10^{-6}	0	2.67×10^{-16}	-155.8
1.00×10^{-5}	20	2.67×10^{-14}	-135.8
1.00×10^{-4}	40	2.67×10^{-12}	-115.8
1.00×10^{-3}	60	2.67×10^{-10}	-95.8
1.00×10^{-2}	80	2.67×10^{-8}	-75.8
1.00×10^{-1}	100	2.67×10^{-6}	-55.8
1.00	120	2.67×10^{-4}	-35.8
$1.00 \times 10^{+1}$	140	2.67×10^{-2}	-15.8
$1.00 \times 10^{+2}$	160	2.67	-4.2
$1.00 \times 10^{+3}$	180	267	-24.2
$1.00 \times 10^{+6}$	6	1.06×10^{-15}	-149.7
$2.00 \times 10^{+6}$	12	4.24×10^{-15}	-143.7
$6.00 \times 10^{+6}$	15	9.55×10^{-15}	-140.2
$8.00 \times 10^{+6}$	18	1.70×10^{-14}	-137.7

dBμV versus dBm for Z = 50Ω

dBμV	μV	dBm	Power Level
-20	0.1	-127	0.0002 pW
-10	0.316	-117	0.002 pW
0	1.0	-107	0.02 pW
5	1.778	-102	0.063 pW
7	2.239	-100	0.1 pW
10	3.162	-97	0.2 pW
15	5.623	-92	0.632 pW
20	10.0	-87	2.0 pW
30	0.03162	-77	0.02 pW
40	0.10	-67	0.2 pW
50	0.312	-57	2.0 pW
60	1.0	-47	20.0 pW
70	3.162	-37	0.2 μW
80	10.0	-27	2.0 μW
90	31.62	-17	20.0 μW
100	100.0	-7	2000.0 μW
120	1.0V	+13	20 mW

Appendix E – Glossary

Related to Electromagnetic Compatibility (Frequency Domain Terms)

AC Impedance (*Z*). The combination of resistance, capacitive and inductive reactance observed by AC or time varying voltage and current within a transmission line. Impedance restricts the flow of current or RF energypropagation.

Alternating Current (AC). A propagating electromagnetic field that varies with respect to time. This definition is commonly applied to a power source that switches polarity many times per second, such as the power supplied by utility companies. It is usually in a sinusoidal shape but could be a square or triangular wave.

American National Standards Institute (ANSI). An organization that sponsors and supports various standards including EMC test procedures classified under the ANSI C63.x series where "x" represents various test methodologies, requirements and specifications.

Amplitude Modulation (AM). A technique for putting information on a sinusoidal "carrier" signal by varying the amplitude of the wave.

Amplitude. The height or magnitude of a signal measured with respect to a reference, generally 0V.

Anechoic chamber. A test facility shielded from the electromagnetic environment using RF absorbing material. RF energy does not enter or leave this room which is used for compliance testing. Anechoic chambers for measurement of RF energy are different than those used for acoustic purposes commonly found in the recording industry.

Antenna Factor (AF). The ratio of received field strength to the voltage appearing at the terminals of a receiving antenna, or the ratio of the transmitted field strength to its voltage applied to the terminals of a transmitting antenna. Antenna factors are functions of polarity, height above a ground plane and frequency.

Antenna. A device used for transmitting or receiving electromagnetic signals. Antennas are designed to maximize coupling to an electromagnetic field.

Anti-resonance. The opposite of resonance and which generally occurs from the parallel combination of capacitors and inductors at a particular frequency. The frequency at which the impedance of an component or structure is infinite under theoretically perfect conditions.

Attenuation. A general term used to denote a decrease in signal magnitude in a transmission line from one point to another due to losses in the media, measured in units of decibels (dBs).

Bandwidth. A narrow range of frequencies in which a signal falls within specific upper and lower limits. The bandwidth of a signal is classified as being either narrowband or broadband.

Bonding. The process of making a low-impedance electrical connection between two or more metal surfaces, or the permanent joining of metallic parts to form an electrically conductive path that will ensure electrical continuity and the capacity to conduct safely any hazardous current likely to be present on the structure due to a fault condition.

Brownout. A slow varying decrease in the voltage level from the AC mains input sometimes over an extended period of several cycles to hours.

Bulk capacitor. A capacitor with a fairly large capacitance value that recharges power and 0V planes quickly to ensure decoupling capacitors located near switching components are sufficiently recharged prior to their use in minimizing plane bounce and also that a required voltage level is present throughout the entire printed circuit assembly. Bulk capacitors recharge planes. Decoupling capacitors takes charge from these planes.

Bypass capacitor. A capacitor located in a transmission line (series implementation) or between transmission lines (parallel application) that allows an electromagnetic field to be transferred between two locations. A series bypass capacitor keeps the DC bias of signals from being propagated between locations, commonly found in circuits using level translators, or to transfer RF return current across a boundary condition to complete the RF return path and to comply with Ampere's law.

Capacitance. A numeric value that describes the ability of a capacitor to hold charge when a voltage is impressed between the two conductive terminals separated by a dielectric. Capacitance is measured in units of Farads.

Capacitor (*C*). Two metallic conductors separated by a dielectric material. These conductors need not be parallel to each other but can be broadside or in any relationship to each other in 3-dimensional space. The dielectric can be air or any physical/chemical material.

Characteristic impedance (Z_o). The numeric value of a transmission line containing both AC and DC currents that presents both resistance and reactance to a propagating signal. Normally, characteristic impedance is a constant value over a wide range of frequencies when referenced to a transmission line, but may change value in circuits that vary their input or output impedance during operation.

Circuit referencing. The process of providing a common 0V reference for circuits that allows communication between devices. Circuit referencing is the most important reasons for providing a 0V reference. This reference is not intended to carry functional AC power or fault current that may present a safety hazard of electric shock.

Circuit. Multiple devices with source impedance, load impedance and interconnect between that propagates an electromagnetic field between elements. Multiple sources and loads may be part of one circuit where all devices are referenced to the same point, or may use a common signal return conductor. Circuits usually originate in one location and terminate in another.

Coax. A term used to describe a conductor concentric about a central axis. A coax generally take the form of a round transmission line surrounded by a conductive return path (shield), separated by a dielectric which could be air or a physical material. This conductive return path acts as both RF shield (AC currents) on the outer portion and RF return path (DC currents) on the inner half, depending on skin effects, frequency of operation and application of use. A trace or transmission line on a printed circuit board adjacent to a reference planes functions 100% identical to a coax, electromagnetic wise.

Committee on Special International Committee on Radio Interference (CISPR). An international organization concerned with developing standards for detecting, measuring and comparing electromagnetic interference in electric devices.

Common-impedance coupling. Occurs when both source and victim share a transmission line path through a common impedance. Common impedance is generally found within shared conductors resulting in an undesired voltage being developed within the transmission line structure, co-mingling RF return current.

Common–mode current. The component of a signal current that creates both electric and magnetic fields and which do not cancel each other between source and return paths. For example, a circuit with one signal conductor and one 0V reference conductor will have common–mode current as the summation of the total signal current flowing in the same direction on both conductors (in-phase), generally 0V or return. Common-mode currents are established through any type of loss in the transmission line and are the primary source of undesired EMI.

Common–mode interference. Interference that appears between both signal and a common reference, causing the potential of both sides of the transmission path to be changed simultaneously and by the same amount relative to the common reference point.

Common–mode. The instantaneous algebraic average of two signals applied to a balanced circuit, with both signals referred to a common reference traveling in the same direction. Both signals are identical in amplitude and phase, or the potential that exists between neutral and ground that are at different reference levels.

Conducted emissions. The component of propagated RF energy that is transmitted through a transmission line as a propagating electromagnetic wave, generally through a wire or interconnect.

Conducted immunity. The relative ability of a product to withstand electromagnetic energy that penetrates it through external cables, power cords and I/O interconnects.

Conducted susceptibility. Radio frequency energy that couples from outside of the equipment to the inside through interconnect cables, power lines or signal cables.

Containment. A process whereby RF energy is prevented from passing through boundary condition, generally by shielding an enclosure to create a Faraday cage structure or by using a plastic housing with conductive paint or metalized by a special process during manufacturing. Conversely we can also refer to containment as preventing RF energy from entering an enclosure and causing harmful disruption to electrical circuits and components.

Continuous Wave (CW). A propagated waveform with constant amplitude and frequency.

Coupling. The association of two or more circuits or systems in such a way that power or signal information is transferred from one to another.

Crosstalk. Unintended electromagnetic coupling between transmission lines (traces, wires, trace-to-wire, cable assemblies, components and any other electrical elements) subject to electromagnetic field disturbance.

Current (*I*). The total amount of electrons propagating within a transmission line per second with units of Amperes as the result of a voltage difference between two points.

Decibel (dB). A unit of measurement for expressing the ratio between two parametric values using logarithms to the base of 10. Decibels provide a convenient format to express power that range several orders of magnitude between elements. When making reference to voltage (V) or current (I), the decibel has a multiplier of 20 whereas with power (P) the multiplier is 10. Increases or reductions of 6 dB doubles or halves of the power level within the circuit.

Decoupling. A design technique on printed circuit boards using capacitors to prevent switching noise pulses from being injected in the power distribution network by digital components, causing the power and/or 0V reference to bounce beyond voltage operational margins. In addition, switching noise that has been injected into the power distribution network from other components can disturb other components sharing the same power network. Bouncing planes causes signal integrity problems and creation of common-mode current observed as EMI. Decoupling is achieved with capacitor(s) located between power and 0V by ensuring there is a stable voltage level present *prior* to each edge rate transition, and are recharged by both bulk capacitors and planes.

Detector. An electrical circuit that performs detection (extraction of signal or noise from a modulated input) by using a weighting function (extraction of a particular characteristic of the signal or noise being measured).

Dielectric constant. The physical property that determines the electrostatic energy stored per unit volume for unit potential gradient (generally given relative to a vacuum–see Permittivity).

Differential pair. Parallel routed signals exhibiting mutual inductance between both lines, typically 50 to 150 ohms. The signal on each pair is generally equal and opposite in amplitude with regard to the same time reference, and does not require use of a 0V reference.

Differential-mode current. The component of electromagnetic energy that is present on both signal and return path that are equal and opposite of each other. The magnetic fields between these two transmission lines are cancelled if tight coupling occur, thus minimizing development of undesired common-mode current and EMI.

Differential-mode. The aspects of data communication where an electromagnetic signal is propagated from a source to a load along with a return path containing an equal amount of energy in both transmission lines. Single-ended signals between components are propagated in differential-mode.

Direct current (DC). Current produced by a voltage source that does not vary with time.

Dynamic range. The maximum ratio of two signals simultaneously present at the input of a receiver, or analyzer that can be measured to a specific accuracy or range of performance.

Earthing. The connection of a safety ground wire to earth ground at the service entrance of a building, or a ground rod driven deep into soil to prevent hazard of electric shock or electrocution.

Edge rate transition. The rate of change in voltage with respect to time of digital logic signal transitions usually referenced to 0V. Expressed in volts per (nano)second.

Effective Radiated Power (ERP). The relative gain of a transmitting antenna with respect to the maximum directivity of a half-wave dipole, multiplied by the net power accepted by the antenna from the transmitter.

Effective relative permittivity (ε_r'). The actual permittivity experienced by an electromagnetic field propagated through a physical medium relative to free space.

Electric field. A vector of electric field strength, or electric flux density that has significant magnitude. The electric force acts on a unit electric charge independent of the velocity of that charge. Spatial gradients in the field between conductors at different potential have capacitance between them that allows an electric field to propagate.

Electromagnetic compatibility (EMC). The capability of electrical and electronic systems, equipment, and devices to operate in their intended electromagnetic environment within a defined margin of safety, and at design levels or performance without suffering or causing unacceptable degradation as a result of electromagnetic interference.

Electromagnetic interference (EMI). The lack of EMC, since the essence of interference is the lack of compatibility. EMI is the process by which disruptive electromagnetic energy is transmitted from one electronic device to another via radiated or conducted paths (or both). In common usage the term refers particularly to RF signals, but EMI can occur in the frequency range from "anything greater than DC to daylight."

Electromagnetic pulse (EMP). A strong electromagnetic transient such as one created by lightning or a nuclear blast.

Electrostatic discharge (ESD). A transfer of electric charge between bodies of different electrostatic potentials in proximity to each other (air), or through direct contact characterized by rise times less than one nanosecond and total pulse widths on the order of microseconds. This event is observed as a high voltage pulse that may cause damage or loss of functionality to susceptible circuits. Although lightning qualifies as a high voltage pulse, the term ESD is generally applied to events of lesser amperage and more specifically to events that are triggered by human beings and machines operating within an environment of use nearby, not the environment.

Electrostatic field. The element of static electric charge that is not time variant containing a high energy level that will eventually discharge from one electrode to another which is at a lower potential level.

EMI filter. A circuit or device containing components that blocks the flow of certain high–frequency RF and allows the flow of desired RF energy within a specific frequency range This filter may also be used to protect a particular circuit from electromagnetic field disturbance.

Equipment Under Test (EUT). The device or system being evaluated for electromagnetic compatibility compliance, or any other required test to be performed.

Equipotential reference or ground plane. A metallic structure used as a common connection for power and signal referencing. This structure may not be at equipotential levels at RF frequencies owing to its electrically large size and impedance of transmission lines associated with circuit operation.

Far field. A region in space where electromagnetic fields from a radiator appears as a propagating plane wave. This field contains both electric and magnetic field components at right angles to each other. The direction of travel is called a transverse electromagnetic field (TEM) if not disrupted in its path of travel. The ratio or impedance of a propagating wave in the far field is 377 ohms (free space impedance).

Faraday shield. A term referring to conductive shielding used to contain or control an electric field with a metallic barrier. This shield may be located between the primary and secondary windings of a transformer or may completely surround a circuit (or system) to provide electrostatic shielding. No functional earth ground is necessary for the shield to work. (*Note*: A Faraday shield is in reality a Gaussian structure. Gauss's law describes the functional purpose of this shield, while Faraday's law describes the creation of electric fields from time varying magnetic fields. Faraday validated or proved the validity of Gauss's law; hence his name is attributed to this function of shield barrier protection).

Ferrite component. Powered magnetic (permeable) material in various shapes used to absorb conducted interference on wires, cables and harnesses. These devices absorbs magnetic fields due to lossy resistance and increased self-inductance. Ferrite material converts magnetic–flux density field into heat (an exothermic process). One benefit of this process, in contrast to filters that perform by reflecting EMI in their stop-bands, is that ferrites dissipate rather than reflect EMI which otherwise could enhance radiation and disturb other victim components or circuits.

Ferrite material. A combination of metal oxides sintered into a particular ceramic shape with iron as the main ingredient. Ferrites provide two key features: (1) high magnetic permeability that concentrates and reinforces a magnetic field, and (2) high electrical resistivity that limits the amount of electric current flow. Ferrites exhibit low energy losses, are efficient, and are usually used for higher frequency signals (1 MHz to >1 GHz).

Filter. A physical device that blocks propagation of RF current or passes desired frequencies of interest. For communication circuits, a filter suppresses unwanted frequencies and noise, or separates channels from each other.

Frequency domain analysis. The study of electrical signals to determine the characteristics of signal propagation with regard to frequency.

Frequency Modulation (FM). A technique for putting information on a sinusoidal "carrier" signal by varying the frequency of the carrier.

Frequency span. The difference between two frequencies (low to high) on an EMC analyzer.

Gain (antenna). The ratio of power increase delivered to an isotropic (omnidirectional) antenna that is required to develop a given field at a given distance, to the power delivered to that antenna acting as a receiver in the direction of maximum radiation or bore-sight efficiency.

Ground. A term used to describe the terminal of a voltage source that serves as a 0V measurement reference for all voltages in the system. Often the negative terminal of the power source but sometimes may be the positive terminal. The word ground must be prefixed by a descriptor that describes they type of ground system being referenced (e.g., analog, digital, chassis, frame, earth, signal, common, RF, etc.).

Ground loop. A potential interfering condition formed between circuits when interconnected by a conducting element (plane, trace, wire or transmission line) assumed to be at ground or 0V potential where RF return currents travel. Although a ground loop is acceptable for many applications of use, the severity of the problem of currents flowing through the loop depends on the unwanted signals that may be present and those that can cause system malfunction or EMI.

Ground stitch. The process of making a solid connection from a printed circuit board's 0V reference plane to a metallic structure or enclosure for the purposes of providing system wide referencing regardless of grounding methodology utilized.

Grounding. A generic term with as many definitions as there are engineers. This word must be preceded by a descriptor to make sense (i.e., grounding what?).

Grounding methodology. A chosen method for directing RF return currents to their source in an optimal manner, which must be appropriately designed or implemented for an intended application. Three basic methodologies are single-point, multi-point (actually multiple connections to a single point reference) and hybrid.

Harmonics. The component of a signal that is observed in the frequency spectrum at both even and odd integer multiples of a primary frequency.

Hybrid ground. A grounding methodology that combines single-point and multi-point grounding topologies simultaneously, depending on the functionality of the circuit and frequencies present. Hybrid grounding includes parasitic capacitive coupling to an RF return or metal enclosure in addition to inductive connection using a transmission line such as the DC return of a power supply through a wire.

Immunity. The ability of a device, equipment or system to perform without degradation in the presence of an electromagnetic disturbance.

Impedance (*Z*). The resistance to the flow of energy (voltage, current, power, etc.) within a transmission line. Impedance may be resistive, reactive or both.

Inductance (*L*). The property of a conductor that allows for storage of electrical energy in a magnetic field induced by current flowing through a transmission line or physical component. Units of measure-Henry.

Insertion loss (*IL*). The ratio between power received at a load after the insertion of a filter for a given frequency, or how much loss to the propagating signal a filter provides for its intended function.

International Electrotechnical Commission (IEC). An international organization that prepares and publishes international standards (including many EMC standards) for electrical, electronic and related technologies.

Line Conducted Interference (LCI). Refers to RF energy in a power cord or AC mains input cable that enters the AC mains of a facility. Conducted signals do not propagate as radiated fields but as conducted currents.

Line coupling. The electromagnetic coupling between two transmission lines caused by mutual inductance and capacitance between them.

Load capacitance. The capacitance presented to the output of a digital logic device or other signal source. Total load capacitance is usually the sum of both distributed line and input capacitance of all circuits within the transmission line network.

Loop antenna. A loop antenna is sensitive to magnetic fields and is shielded against electric fields. This antenna is in the shape of a coil. A magnetic field component perpendicular to the plane of the loop induces a voltage across the coil that is proportional to frequency.

Magnetic field. A condition in a medium produced by a magnetomotive force such that, when altered in magnitude, a voltage is induced in an electric circuit linked with the flux. Also commonly referred to as the portion of an electromagnetic field that surrounds every time variant current carrying transmission line and is best visualized using the Right Hand Rule.

Modulation. The process in which the characteristics of a carrier is varied in accordance with a modulating wave in both amplitude and time.

Multi-point ground. A method of referencing different circuits together to a common equipotential or reference point in more than one location. In reality, a more accurate definition of multi-point grounding is "*multiple connections to a single-point reference.*"

Near field. A region in space close to a radiator where far field conditions do not yet exist.

Network analyzer. A measurement instrument commonly used during engineering design to characterize physical components and their performance characteristics.

Oscilloscope. An instrument primarily used for viewing the instantaneous value of one or more rapidly varying electrical quantities as a function of time, or of another electrical quantity.

Parasitic capacitance. The capacitive leakage across a discrete component (resistor, inductor, filter, isolation transformer, optical isolator, etc.) that adversely affects high-frequency performance. Parasitic capacitance is also observed between active components (or printed circuit board assembly) to its sheet metal mounting plate or chassis enclosure, or anything with two conductive plates separated by a dielectric, one at potential the other 0V.

Permeability (μ). The extent to which a material can be magnetized, often expressed as the parameter relating magnetic flux density induced by an applied magnetic field. Permeability is not a constant value but varies with electrical frequency at which the measurement is made and the temperature of the environment.

Permittivity (dielectric constant, ε_r). The ratio of incremental change in electric displacement per unit electric field of a material to that of free space. This term is preferred to the words "dielectric constant." Permittivity is not a constant value but varies with several parameters, including electrical frequency at which the measurement is made, temperature of the environment and extent of water absorption in the material carrying a propagating electromagnetic field, among many other physical parameters.

Power/return plane bounce. Digital components, during a logic transition either high or low, requires DC drive current in addition to quiescent current in charge a capacitive load. Inductance in the interconnect to the power/return distribution network will cause a voltage drop to occur per $L(di/dt)$, consuming larger amounts of current. With a larger amount of current consumption the planes will bounce with each respective edge. When a plane bounces, signal integrity problems may result as well as development of common-mode EMI causing harmful disruption to other components located in the vicinity of the disrupting device that expects to see a pure DC power source.

Printed circuit board. A physical structure used to mechanically support transmission lines.

Radiated emissions. The component of RF energy that is transmitted through a medium as an electromagnetic field, usually free space however other modes of field transmission may occur.

Radiated immunity. The relative ability of a product to withstand electromagnetic energy that arrives via free space propagation.

Radiated susceptibility. Undesired propagating electromagnetic fields radiating through free space *into* equipment from external electromagnetic sources both radiated and conducted.

Radio frequency (RF). A frequency at which electromagnetic radiation of energy is used for communication purposes. Radio frequencies are designated as; very low: 3 kHz to 30 kHz, low: 30 kHz to 300 kHz, medium: 300 kHz to 3 MHz, high: 3 MHz to 30 MHz, very high: 30 MHz to 300 MHz, ultrahigh: 300 MHz to 3 GHz, super high: 3 GHz to 30 GHz, and extremely high: 30 GHz to 300 GHz. This energy may be transmitted as a byproduct of an electronic device's undesired mode of operation. RF is transmitted through two basic modes: radiated and conductive.

Reference (ground) loop. A circuit that includes a conducting element (plane, trace or wire) assumed to be at 0V potential (ground). Reference loops must be present for circuits to work to satisfy Ampere's law. Although a reference loop (as a signal image return) is required for functionality purposes, the severity of a problem for RF currents flowing through the loop depends on unwanted signals that may co-mingle and which may cause system malfunction.

Referencing. The process of making an electrical connection between circuits to create a 0V potential that is identical for components in order establish a reference for logic level switching.

Reflections. RF energy that is sent back toward the source as a result of encountering a change in impedance (higher impedance relative to the source transmission line) on which it is traveling, or bouncing off a metallic enclosure back to its source generator as a radiated propagating field.

Relative permittivity (ε_r). The amount of energy stored in a dielectric insulator per unit electric field, hence a measure of capacitance between a pair of conductors in the vicinity of the dielectric insulator compared to the capacitance of the same conductor pair in a vacuum.

Resonance. A condition within a transmission line (traces or planes) where inductive and capacitive reactance are equal, within the limits of equivalent series resistance that may be present in the circuit configuration.

RF ground. Providing a reference for transfer of RF current through a metallic enclosure. What is achieved using an RF ground is capturing an electromagnetic field propagating through space or circulating on the enclosure skin and shunting them to a 0V reference or earth ground, thus removing or shunting out undesired electromagnetic fields that may cause harmful radiated interference.

Safety ground. The process of providing a current path to earth ground to prevent hazard of electric shock through proper connection and routing of a permanent, continuous, low-impedance, adequate fault capacity conductor that runs from a power source to a load.

Shield ground. Providing a 0V reference or electromagnetic field grounding for interconnect cables or chassis housings.

Signal integrity. The engineering discipline that investigates the propagating quality of a signal (actually an electromagnetic field) to ensure that minimal energy loss occurs and that the desired waveform arrives at its destination within a desired period of time without degradation, ringing, reflection and crosstalk.

Single-point ground. A method of referencing circuits together at a single location using a physically long interconnect or transmission line. All signals will thus be referenced to the same 0V potential.

Skin depth. The physical distance to the point inside a conductor at which the electromagnetic field, and hence RF current, is reduced to 37% of the surface value or 9 dB per skin depth.

Skin effect. The effects or how current propagates physically within a transmission line. At DC or low frequency levels current flow travels in the center of the conductor. At higher frequencies, AC (RF) current will migrate to the outer skin of the transmission line. Above a certain frequency depending on the physical size of the transmission line diameter and material composition, signal loss starts to occurs due to excessive amount of the electromagnetic field traveling in this very thin portion of the transmission line.

Spectrum analyzer. An instrument primarily used to display the power distribution of an incoming signal as a function of frequency. This instrument is useful for analyzing characteristics of electrical waveforms by repetitively sweeping through a frequency range of interest and displaying all components of the signal being investigated.

Suppression. The process of reducing or eliminating undesired RF energy without relying on a secondary method such as a metal housing or enclosure. Suppression may include shielding and filtering as well.

Susceptibility. The inability of a device, equipment or system to resist an electromagnetic disturbance. Susceptibility is the lack of immunity.

Time domain analysis. The study of electrical signals in a transmission line with regard to time, or how long it takes for a signal to propagate from one point to another.

Time Domain Reflectometry (TDR). An instrument primarily used to verify the proper functioning of physical components of a network or transmission line with a sequence of time-delayed reflected electrical pulses.

Transducer. A physical device that provides a means where RF energy flows from one or more transmission systems or media to another transmission system or media. The energy transmitted may be of any form (electromagnetic, mechanical or acoustical) and may be of the same or different form in various input and output systems or media. An antenna or probe is another name for transducer.

Transmission line. Any form of conductor used to propagate an electromagnetic field from source to load. A transmission line can be free space or a metallic interconnect such as a wire or printed circuit board trace.

Transverse Electromagnetic (TEM) field. An electromagnetic plane wave where the electric and magnetic fields are perpendicular to each other everywhere and both fields are perpendicular to the direction of propagation.

Velocity of propagation. The speed at which information in the form of an electromagnetic field is transmitted within a conductive medium or dielectric. In free space, the velocity of propagation is the speed of light. In a dielectric medium, the velocity of the transmitted electromagnetic wave is slower (such as approximately 60% the speed of light in FR-4), and is dependent exclusively on the dielectric constant of the material that allows the electromagnetic field to propagate.

Voltage fluctuation. A variation in the voltage level that is above or below a desired range of operation.

Voltage Standing Wave Ratio (VSWR). The ratio of the magnitude of the transverse electric field in a plane of maximum strength to the magnitude at the equivalent point in an adjacent plane of minimum field strength. VSWR is also commonly referred to as the ratio of forward power from a transmitter to reflected power back to the transmitter.

Related to Signal Integrity (Time Domain Terms)**

AC Impedance (*Z*). The combination of resistance, capacitive and inductive reactance observed by AC or time varying voltage and current within a transmission line. Impedance restricts the flow of current or RF energy propagation.

Alternating Current (AC). A propagating electromagnetic field that varies with respect to time. This definition is commonly applied to a power source that switches polarity many times per second, such as the power supplied by utility companies. It is usually in a sinusoidal shape but could be a square or triangular wave.

Amplitude. The height or magnitude of a signal measured with respect to a reference, such as ground or 0V.

Attenuation. Reduction in the amplitude of a signal due to losses in the media through which it is transmitted.

Backporching. A term used to describe reflections that follow a fast rise or fall time signal traveling down a long transmission line that has not been properly terminated. This signal appears on an oscilloscope as a stair-step function generally observed at the source.

Backward crosstalk. Noise injected into a quiet line (victim) that is placed next to an active line (source) as seen at the signal or source driver.

Busbar. A large copper or brass bar used to carry high power supply current onto a printed circuit board or backplane.

Capacitance (*C*). A numeric value that describes the ability of a capacitor to hold charge when a voltage is impressed between the two conductive terminals separated by a dielectric. Capacitance is measured in units of Farads.

Characteristic impedance (Z_o). The numeric value of a transmission line containing both AC and DC currents that presents both resistance and reactance to a propagating signal. Normally, characteristic impedance is a constant value over a wide range of frequencies when referenced to a transmission line, but may change value in circuits that vary their input or output impedance during operation.

Coaxial based transmission line. A printed circuit board trace routed physically adjacent to a reference plane, potential irrelevant. Magnetic flux in the source trace couples with the return flux if closely co-located. When a trace is adjacent to a plane, electromagnetic field performance is identical to that of a regular coax.

Common–mode current. The component of a signal current that creates both electric and magnetic fields and which do not cancel each other between source and return paths. For example, a circuit with one signal conductor and one 0V reference conductor will have common–mode current as the summation of the total signal current flowing in the same direction on both conductors (in-phase), generally 0V or return. Common-mode currents are established through any type of loss in the transmission line, and are the primary source of undesired EMI.

Common–mode rejection ratio. The ratio of common–mode interference voltage at the input of a device to the corresponding interference voltage at the output of the same component. The higher the ratio the better the performance. This ratio describes the capability of the device to reject the effects of a voltage applied simultaneously to both inputs that are not balanced.

Crossover. Intersection of two conductors separated by insulation.

Crosstalk (XTALK). A measure of the electromagnetic coupling from one transmission line to another.

** Some of these terms are extracted from: Glossary–Signal Integrity Terms provided by ICP-2141, *Controlled Impedance Circuit Boards and High Speed Logic Design*, and IPC-D-317A, *Design Guidelines for Electronic Packaging Utilizing High-Speed Techniques*.

Current (*I*). The total amount of electrons propagating within a transmission line per second with units of Amperes as the result of a voltage difference between two points.

DC power distribution. The output voltage and current from a power source that energizes circuits and components without any change in voltage amplitude.

Decoupling. A design technique on printed circuit boards using capacitors to prevent switching noise pulses from being injected in the power distribution network by digital components, causing the power and/or 0V reference to bounce beyond voltage operational margins. Bouncing planes causes signal integrity problems and creation of common-mode current, commonly observed as EMI. Switching noise injected into the power distribution network can disturb other digital and analog logic components sharing the same network. Decoupling is achieved with capacitor(s) located between power and 0V by ensuring there is a stable voltage level present prior to each edge rate transition, and are recharged by bulk capacitors and the planes located adjacent to the capacitor.

Dielectric constant (ε_r). The physical property that determines the electrostatic energy stored per unit volume for unit potential gradient (generally given relative to a vacuum–see Permittivity).

Differential pair. Parallel routed signals exhibiting mutual inductance between both lines, typically 50 to 150 ohms. The signal on each pair is generally equal and opposite in amplitude with regard to the same time reference, and does not require use of a 0V reference.

Direct current (DC). Current produced by a voltage source that does not vary with time.

Edge rate. The rate of change in voltage with respect to time of a logic signal transition. This edge transition is usually expressed as volts per (nano)second.

Edge transition attenuation. The loss in sharpness of a switching edge caused by absorption of the highest frequency component of the transmission line.

Effective relative permittivity (ε_r'). The actual permittivity experienced by an electromagnetic field propagated through a physical medium relative to free space.

FET probe. A high impedance transducer used to measure signal characteristic usually by an oscilloscope without adding capacitive loading, affecting signal quality or operational performance.

Flat conductor. A rectangular transmission line that is wider than it is thick, such as a flat braid. Traces on a printed circuit board are flat conductors.

Forward crosstalk. Noise induced into a quiet line (victim) placed next to an active line (source) as seen at the far end of the quiet line farthest from the signal source.

Ground. A term used to describe the terminal of a voltage source that serves as a measurement reference for all voltages in the system. Often the negative terminal of the power source but sometimes the positive terminal.

Impedance (*Z*). The resistance to the flow of energy (voltage, current, power, etc.) within a transmission line. Impedance may be resistive, reactive or both.

Inductance (*L*). The property of a conductor that allows for storage of electrical energy in a magnetic field induced by current flowing through a transmission line or physical component. Units of measure-Henry.

Input/output Buffer Specification (IBIS). A method for specifying the parameters of a semiconductor device's input and output pins used for computational modeling.

Line coupling. Electromagnetic field transference between two transmission lines caused by mutual inductance and capacitance that exist between the two transmission lines.

Load capacitance. The capacitance seen by the output of a logic circuit or other signal source. Usually the sum of the distributed line and input capacitance of all load circuits.

Logic device. A general term used to describe functional circuits that perform computational functions.

Microstrip. The outer routing layers on a printed circuit board with one side of the transmission lines referenced to a plane regardless of potential. The other side of the transmission line is referenced to free space or covered with a dielectric that separates it from air such as soldermask or conformal coating. Microstrip layers are either the outer layer or embedded one layer deep on a printed circuit board.

Noise budget/noise margin. The allowance for a change in the system's DC and/or AC voltage levels which permits a device to operate within specific limits or margin. There are two primary components for noise budgets-the DC power supply of each integrated circuit and the logic signal AC noise budget.

Overshoot. The effect of an excessive voltage level above or below the power rail or a reference level.

Permeability (μ). The extent to which a material can be magnetized, often expressed as the parameter relating magnetic flux density induced by an applied magnetic field. Permeability is not a constant value but varies with electrical frequency at which the measurement is made and the temperature of the environment.

Permittivity (dielectric constant, ε_r). The ratio of incremental change in electric displacement per unit electric field of a material to that of free space. This term is preferred to the words "dielectric constant." Permittivity is not a constant value but varies with several parameters, including electrical frequency at which the measurement is made, temperature of the environment, and extent of water absorption in the material carrying a propagating electromagnetic field, among many other parameters.

Power distribution. The DC and AC characteristics for defining electrical power to components and circuits that is stable without switching noise injected into the system. Two concerns exist when implementing a stable power distribution network: conductive losses (DC) and dielectric (AC) in addition to plane bounce due to components consuming high levels of current during a switching event (power/return bounce).

Power/return plane bounce. Digital components, during a logic transition either high or low, requires DC drive current in addition to quiescent current to charge a capacitive load. Inductance in the interconnect to the power/return distribution network will cause a voltage drop to occur per *L(di/dt)*, consuming larger amounts of current. With a larger amount of current consumption planes will bounce with each respective edge. When a plane bounces, signal integrity problems may result as well as development of common-mode EMI causing harmful disruption to other components located in the vicinity of the disrupting device that expects to see a pure DC power source. Without an accurate model of the power and return structure which includes package models, accurate simulation or computational analysis is difficult.

Printed circuit board. A physical structure used to mechanically support transmission lines.

Propagation delay. The time required for a signal to travel through a transmission line from source to load, or the time period required for a logic device to perform its desired function from its internal input to output pins.

Pulse. A logic signal that switches from one state to the other in a short period of time. Generally used as a clock signal for logic devices.

Referencing. The process of making an electrical connection between circuits to create a 0V potential that is identical for components in order establish a single reference required for optimal logic level switching.

Reflections. RF energy that is sent back toward the source as a result of encountering a change in impedance (higher impedance relative to the source transmission line) on which it is traveling, or bouncing off a metallic enclosure back to its source generator as a radiated propagating field.

Relative permittivity *(ε_r)*. The amount of energy stored in a dielectric insulator per unit electric field, hence a measure of capacitance between a pair of conductors in the vicinity of the dielectric insulator, compared to the capacitance of the same conductor pair in a vacuum.

Resonance. A condition within a transmission line (traces or planes) where inductive and capacitive reactance are equal, within the limits of equivalent series resistance that may be present in the circuit configuration.

Ringback. The effect of the rising edge of a logic transition meeting or exceeding voltage level requirements then re-crossing the threshold before settling both in the high and low state numerous times or the amount by which a signal rebounds below V_{IL} or above V_{IH} after an overshoot has occurred.

Ringing. The effect within a transmission line that contains overshoot (going past the maximum voltage level of the circuit and below the low voltage reference level) before stabilizing to a quiescent level (i.e., many transitions across the reference level before stabilizing).

Rise time. Time required for a logic signal to switch from a low state (logic 0) to its high state.

Settling behavior. A measure of how long it takes a pulse to stabilize at its maximum amplitude.

Signal integrity (SI). The engineering discipline that deals with the propagating quality of a signal (actually an electromagnetic field) to ensure that minimal energy loss occurs, and that the desired waveform arrives at its destination within a desired period of time without degradation, ringing, reflection and crosstalk.

Signal line. Any conductor or transmission line used to transmit an electromagnetic field between circuits.

Simultaneous Switching Output noise (SSO). SSO noise (also SSN=Simultaneous Switching Noise) is the voltage fluctuation that occurs on an electronic device's power bus due to rapid changes in the current drawn by the component. These power bus fluctuations may also appear on signal output pins referenced to a respective power bus.

Skew. The effect of a signal being delayed with respect to another due to different path lengths, or a delay during a transmission state that may cause timing errors in the design. Skew can be affected by conductor impedance, differing conductor lengths, power supply variations, device tolerances and load capacitance of inputs.

SPICE. (Simulation Program with Integrated Circuit Emphasis). A computational program for analysis of transmission line behavior in an effort to ensure optimal signal integrity in the time domain.

Stripline. Routing layers internal within a printed circuit board with a solid reference plane located on both the top and bottom sides of the transmission line.

Stub. A branch on the main line of a signal net generally used to reach a load that is not on the direct signal path, or a transmission line without any load or termination provided which can cause a reflection to occur.

Switching noise. When component are switching logic states, current is either drawn from, or passed to, the power distribution network. When switching logic states, these currents contains high frequency RF energy created by the inductance of the package leads and board interconnects. This switching noise can cause a signal integrity or EMI event due to loop inductance. A printed circuit board layout must be designed to reduce this loop inductance to the smallest amount possible to reduce switching noise.

Switching transient impedance. The impedance between a decoupling capacitor and associated components. This element is also referred to as power or return bounce. It is the highest frequency component of the circuit.

Threshold violation. Threshold violations are caused when a rising pulse edge does not reach the voltage threshold of the device's input. Weak drivers or poor terminations are often the cause although it can also be created by device drivers with a large rise time versus pulse width time.

Transmission line. Any form of conductor or media used to carry an electromagnetic field from a source to a load.

Undershoot. A condition in which the voltage level does not reach a desired amplitude for both maximum and minimum transition levels.

References

Archambeault, B. *Inductance and Partial Inductance – What's it all mean?* Retrieved from http://web.mst.edu/~jfan/slides/Archambeault3.pdf

Armstrong, K. "Advanced PCB Design and Layout for EMC." *EMC & Compliance Journal*, published online at http://www.cherryclough.com.

Bogatin, E. 2009. *Signal Integrity-Simplified.* Englewood Cliffs, NJ: Prentice Hall.

Cain, J. 2003. "The Effects of ESR and ESL in Digital Decoupling Applications." AVX Corp.

Coombs, C. F. 1996. *Printed Circuits Handbook.* New York: McGraw-Hill.

Dockey, R. W. & R. F. German. 1993. "New Techniques for Reducing Printed Circuit Board Common-Mode Radiation." *Proceeding of the IEEE International Symposium on Electromagnetic Compatibility*, pp. 334–339.

Drewniak, J. L., T. H. Hubing, T. P. Van Doren & D. M. Hockanson. 1995. "Power Bus Decoupling on Multilayer Printed Circuit Boards." *IEEE Transactions on Electromagnetic Compatibility.* Vol. 37(2), pp. 155–166.

Erdin, I. 2003. "Delta-I Noise Suppression Techniques in Printed Circuit Boards, for Clock Frequencies Over 50 MHz." *Proceeding of the IEEE Symposium on Electromagnetic Compatibility.* Boston, MA, pp. 1132-1134.

Gerke, D. & W. Kimmel. 1994. (January 20). "*The Designers Guide to Electromagnetic Compatibility.*" EDN.

German, R. F., H. Ott, & C. R. Paul. 1990. "Effect of an Image Plane on Printed Circuit Board Radiation." IEEE International Symposium on Electromagnetic Compatibility, pp. 284–291.

Hartal, O. 1994. *Electromagnetic Compatibility by Design.* W. Conshohocken, PA: R&B Enterprises.

Hubing, T. H., T. P. Van Doren, & J. L. Drewniak. 1994. "Identifying and Quantifying Printed Circuit Board Inductance." *IEEE International Symposium on Electromagnetic Compatibility*, pp. 205–208.

Huray, P. G. 2010. *Maxwell's Equations.* Hoboken, NJ: John Wiley & Sons/IEEE Press.

Joffe, E. & Lock, K. S. 2010. *Grounds for Grounding-A Circuit-to-System Handbook.* Hoboken, NJ: John Wiley & Sons/IEEE Press.

Johnson, H. W., and M. Graham. 1993. *High Speed Digital Design.* Englewood Cliffs, NJ: Prentice Hall.

King, W. Michael, "*EMCT: Electromagnetic Compatibility Tutorial.*" Module One. Co-branded by IEEE.

King, W. Michael, United States Patent #4,145,674.

Micron Technology Inc. Application Note TN-00-06. Bypass Capacitor Selection for High-Speed Designs.

Mardiguian, M. 1992. *Controlling Radiated Emissions by Design.* New York: Van Nostrand Reinhold.

Mardiguian, M. 2000. *EMI Troubleshooting Tricks.* New York: McGraw-Hill.

Mardiguian, M. 2009. *Electrostatic Discharge-Understand, Simulate, and Fix ESD Problems,* 3rd ed. Hoboken, NJ: John Wiley & Sons/IEEE Press.

Montrose, M. I. 2000. 2nd ed. *Printed Circuit Board Design Techniques for EMC Compliance-A Handbook for Designers.* Hoboken, NJ: John Wiley & Sons/IEEE Press.

Montrose, M. I. 1999. *EMC and the Printed Circuit Board Design-Design, Theory and Layout Made Simple.* Hoboken, NJ: John Wiley & Sons/IEEE Press.

Montrose, M. I. 2005. *Testing for EMC Compliance-Approaches and Techniques.* Hoboken, NJ: John Wiley & Sons/IEEE Press.

Montrose, M. I. 1996. "Analysis on the Effectiveness of Image Planes within a Printed Circuit Board." *Proceedings of the IEEE International Symposium on Electromagnetic Compatibility*, pp. 326–331.

Montrose, M. I. 1999. "Analysis on Trace Area Loop Radiated Emissions from Decoupling Capacitor Placement on Printed Circuit Boards." *Proceedings of the IEEE International Symposium on Electromagnetic Compatibility*, pp. 423–428.

Montrose, M. I. 2007. "Power and Ground Bounce Effects on Component Performance Based on Printed Circuit Board Edge Termination Methodologies." *Proceedings of the IEEE International Symposium on Electromagnetic Compatibility.*

Montrose, M. I. 2012. "Analysis on Decoupling Capacitor Placements Associated with Power and Return Plane Bounce." *Proceedings of the Asia-Pacific International Symposium on Electromagnetic Compatibility*, pp. 425-428.

Morrison, R. 2007. *Grounding and Shielding-Circuits and Interference,* 5th ed. Hoboken, NJ: John Wiley & Sons/IEEE Press.

Novak, I., et. al. 2006. "*Comparison of Power Distribution Network Design Methods: Bypass Capacitor Selection Based on Time Domain and Frequency Domain Performance.*" DesignCon.

Novak, I., et. al. 2008. "*Power Distribution Network Design Methodologies.*" DesignCon.

Ott, H. 1988. *Noise Reduction Techniques in Electronic Systems.* 2nd ed. New York: John Wiley & Sons.

Ott, H. 2009. *Electromagnetic Compatibility Engineering.* Hoboken, New Jersey: John Wiley & Sons.

Paul, C. R. 2006. *Introduction to Electromagnetic Compatibility*, 2nd ed. Hoboken, New Jersey: John Wiley & Sons.

Paul, C. R. 1989. "A Comparison of the Contributions of Common-Mode and Differential-Mode Currents in Radiated Emissions." *IEEE Transactions on Electromagnetic Compatibility.* Vol. 31(2), pp. 189–193.

Paul, C. R. 1992. "Effectiveness of Multiple Decoupling Capacitors." *IEEE Transactions on Electromagnetic Compatibility*, Vol. 34(2), pp. 130–133.

Paul, C. R. 2004. *Electromagnetics for Engineers.* Hoboken, NJ: John Wiley & Sons.

Paul, C.R., K. White, & J. Fessler. 1992. "Effect of Image Plane Dimensions on Radiated Emissions." *Proceeding of the IEEE International Symposium on Electromagnetic Compatibility*, pp. 106–111.

Radu, S., R. E. DuBroff, T. H. Hubing & T. P. Van Doren. 1998. "Designing Power Bus Decoupling for CMOS Devices." *Proceedings of the IEEE International Symposium on Electromagnetic Compatibility*, pp. 175-379.

Ricchuti, V., 2001. "Power-Supply Decoupling on Fully Populated High-Speed Digital PCBs." *IEEE Transactions on Electromagnetic Compatibility*, Vol. 43(4), pp. 671-676.

Roy, T., Smith, L., & Pyrmak, J. 2000. "ESR and ESL of Ceramic Capacitor Applied to Decoupling Applications." In *Proceedings of the 1998 Electrical Performance of Electronic Packaging Conference, West Point, NY.*

Smith, L. D., R. Anderson, D. Forehand, Pelc, T. & Roy, T. 1999. "Power Distribution System Design Methodology and Capacitor Selection for Modern CMOS Technology," *IEEE Transactions on Advanced Packaging*, Vol. 22, No. 3, p 284.

Tsaliovich, A. 1995. *Cable Shielding for Electromagnetic Compatibility.* New York, NY: Van Nostrand Rheinhold

Van Doren, T. P., J. Drewniak & T. H. Hubing. 1992. "Printed Circuit Board Response to the Addition of Decoupling Capacitors." Tech. Rep. TR92-4-007, University of Missouri, Rolla EMC Lab.

Van Doren, T., J. Drewniak, & T. Hubing. 1992 "Printed Circuit Board Response to the Addition of Decoupling Capacitors." Tech. Rep. #TR92-4-007, UMR EMC Lab.
Wikipedia, *The Free Encyclopedia*, published online at http://www.wikipedia.org.
Weir, S. 2006. "Bypass Filter Design Considerations for Modern Digital Systems, A Comparative Evaluation of the Big "V, Multi-pole, and Many Pole Bypass Strategies." DesignCon.
Xu, M., T. Hubing, J. Chen, T. Van Doren, J. Drewniak & R. DuBroff, 2003. "Power-Bus Decoupling With Embedded Capacitance in Printed Circuit Board Design." *IEEE Transactions on Electromagnetic Compatibility*, Vol. 45(1), pp. 22-30.
Zeef, T. & T. Hubing, 1995. "Reducing Power Bus Impedance at Resonance with Lossy Components." *IEEE Transactions on Electromagnetic Compatibility*, Vol. 37 (2), pp. 307-310.

Index

absorption loss, 228
Ampère, Andrè-Marie, 3
Ampère's Law, 9, 11
antenna efficiency, 39
anti-resonant frequency, 115, 130
apertures
  in shield barriers, 235
  multiple, 240
  single, 237
back-drilling on a printed circuit board, 168
balanced circuits, 211
balanced line. *See transmission line*
band-pass filter, 290
band-reject filter, 290
barrier penetrations, 245
board level component shielding, 260
Bode plot, 136
bonding
  definition, 176
bounding surface, 184
braided coax shield, 250
braided shield, 250
bulk capacitor
  definition, 112
  selection of, 165, 167
buried capacitance, 167
bypass capacitor
  calculation of use, 159
  definition, 113
cable shield, 248
  aspects to consider, 259
  grounding, 252
  termination, 256, 258
Canadian Standards Association (CSA), 183
capacitance
  buried, 167
  calculation of, 133
  efficiency, 133
  power and return planes, 135, 138
  relative permittivity, 133
capacitive coupling, 197
capacitive reactance, 69
capacitor brigade, 143
capacitors
  anti-resonant frequency, 130
  anti-resonant response, 134
  Bode plot, 131
  bulk, 112
  bypass, 113
  Class 1 family type, 118
  Class 2 family type, 119
  decoupling, 113
  decoupling guidelines, 171
  definition, 111
  dielectric–type description, 117
  energy storage, 122
  impedance, 123
  lead length inductance, 126
  parallel placement, 129
  physical characteristics, 116
  power and return planes, 132
  self-resonance, 127
  self-resonant frequency, 46, 125, 126
  trace connections, 158
ceramic capacitors, 117
characteristic impedance, 60
Christiaan Huygen, 238
circuit definition
  model of a printed circuit board, 145
  reference, 178
circuit reference definition, 176
circuit(definition), 176
clearance distance, 184
closed loop circuit, 23
closed loop environment, 9
CMRR. *See Common Mode Rejection Ratio*, *See* Common Mode Rejection Ratio
coaxial cable. *See transmission line*
common mode rejection ratio, 35, 211
common-impedance coupling
  inductance, 206
  power and return, 204
common-impedance path
  avoiding, 201
common-mode
  choke, 210, 295
  currents, 27, 32, 77
  filter, 288, 292
  radiation, 35
  rejection, 291
component level shield, 262
compression set, 271
conductance, 64
conductive coating, 277
  implementation concerns, 278
  types of coating, 278
conductivity, 10

copper wire characteristics, 26
Coulomb (unit of measurement), 3
coupled inductors, 96
coupling
  common-impedance, 186
creepage distance, 183
crosstalk, 53
current
  definition, 54
  density distribution, 24
DC bias, 302
DC resistance, 200
decibel (dB), 315
decoupling capacitor
  bulk capacitor selection, 165
  calculation for use, 159
  capacitor bridgade, 143
  definition, 113
  effective range of operation, 121
  implementation, 142
  minimizing loop inductance, 157
  multipole methodology, 139
  placement location, 150
  power and return planes, 134
  radius of operation, 144
  rules-of-thumb, 149
  sharing capacitors, 154
Delta-I noise, 94, 129
dielectric constant, 10, 54
dielectric loss, 61
differential signaling, 298
differential-mode
  choke, 295
  currents, 27, 30, 35, 77
  filter, 289, 293
  radiation, 30
digital-to-analog partitioning, 87
dipole antenna structure, 195
distributed capacitance, 77
distributed line. *See transmission line*
divergence theorem, 11
earthing
  definition, 176
eddy currents, 25, 213, 230
electric
  charge, 10, 14, 21
  current, 21
  dipole, 14
  fields, 10, 36
  shock, 180
electric shock, 182
electrochemical compatibility, 273
electrolytic capacitors, 119
Electromagnetic Interference (EMI), 53
electromotive force, 6, 93
Electrostatic discharge (ESD) ground, 178
embedded microstrip. *See transmission line*
EMF. *See electromotive force*, *See electromotive force*
endface, 227
energy storage of capacitor, 122
energy suppression principles, 40
equipotential reference plane
  definition, 176
equivalent series inductance, 46, 137
equivalent series resistance, 45, 46, 138
ESL. *See equivalent series inductance*
ESR. *See equivalent series resistance*
external inductance, 26, 200
far field, 13
Faraday
  cage, 6, 11, 220
  Law, 9
  law of induction, 6
  shield, 49, 224
Faraday shield, 262
Faraday, Michael, 6
feedthrough capacitor, 303
ferrite bead, 87
ferrite material, 296
fiber optic. *See transmission line*
filters
  AC mains application, 294
  applications of use, 300
  attenuation calculation, 300
  basic configurations, *286*
  capacitive, 284
  common-mode, 288
  common-mode choke, 293, 295
  common-mode rejection, 291
  Curie temperature, 302
  definition, 281
  differential-mode, 289
  differential-mode choke, 295
  environmental cconditions, 301
  environmental conditions, 301
  feedthrough capacitor, 303
  ferrite material, 296
  inductive, 284
  insertion loss, 281
  installation guidelines, 305
  parasitics, 285

resistivity, 302
signal line, 290
three terminal capacitor, 304
flux cancellation, 40, 76, 78
foil shield, 250
Fourier analysis, 319
Fraunhofer region, 13
free space
impedance of, 16
plane field, 16
frequency domain, 2
Fresnel loss, 227
Fresnel region, 13
Fresnel, Augustine-Jean, 238
fundamental concepts of suppression, 40
common-mode RF currents, 41
RF voltages, 41
gaskets, 262
closure pressure, 275
compression set, 271
corrosion, 271
electrochemical effects, 273
environmental aspects, 269
joint unevenness, 271
material composition, 264, 265
mechanical problems, 269
Gauss
divergence theorem, 7
flux theorm, 7
Gauss's law, 9
law of induction, 7
Gauss, Carl Friedrich, 7
Gaussian structure, 224
ground
loop, 207
ground (definition), 176
loop, 197
stitch, 214
strap, 201
symbols, 178
ground loop
definition, 176
ground noise voltage, 186
ground plane discontinuity, 80
ground stich
definition, 176
ground symbols, 178
grounding
chassis, 185
concepts, 179
electric
shock, 180
electric shock, 182
fundamental concepts, 180
ground currents, 184
ground loops, 207
ground noise voltage, 186
hybrid methodology, 197
impedance, 187
misconceptions, 180
multi-point methodology, 194, 196
resonance, 213
signal voltage referencing, 180
signal voltage referencing ground, 184
single-point methodology, 189
various methodologies, 188
voltage referencing, 182
grounding methodologies
definition, 176
ground-noise voltage, 77
Heavyside, Oliver, 8
hidden schematic, 41
high pass filter, 290
history of electromagnetics, 2
Huygen's Principle, 238
hybrid ground, 202
definition, 177
hybrid ground methodology, 197
capacitive coupling, 197
inductive coupling, 197
rack assembly example, 198
image
plane, 73, 75, 90, 100
plane violation, 80
impedance
free space, 16
planes, 196
transmission line, 44
inductance
copper planes, 197
minimizing ground inductance, 203
mounting pads, 156
trace length, 147, 197
inductance (definition), 93
coupled inductors, 96
mutual inductance, 95
mutual partial inductance, 99
partial inductance, 98
self-inductance, 93
self-partial inductance, 99
inductive coupling, 197
inductive reactance, 297

insertion loss, 281
internal inductance, 200
interplane capacitance, 101
isolated plane, 89
isolation in
  Faraday shield implementation, 208
isolation methods
  balanced circuit, 211
  common-mode choke, 210
  optical, 209
  transformer, 208
Kirchhoff
  circuit law, 4
  voltage law, 5
Kirchhoff, Gustav, 4
Kirchhoff's Law, 34
layer jumping, 82
leakage current, 296
lecher lines, 59
Lentz's Law, 49
Lenz, Heinrich, 5
Lenz's law, 5, 95
lighting ground, 177
line filter, 182
longitudinal mode, 32
loop
  impedance, 146
  inductance, 104
  mutual inductance, 105
  structure, 93
loop antenna, 30
loop area, 218
Lorentz force relation, 10
loss resistance, 297
loss tangent, 66
low pass filter, 290
magnetic field, 10, 21
magnetic lines of flux, 21
Maxwell, James Clerk, 8
Maxwell's equations, 9, 19
metal screw inductance, 215
microstrip. *See transmission line*
microstrip topology, 54
mode conversion, 79, 299
  common-mode, 38
  differential-mode, 38
mounting pad inductance, 156
multiple apertures, 240
multiple reflection loss, 232
multi-point grounding
  definition, 177
  ground stitch location, 214
  resonance between ground points, 213
multi-point grounding methodology
  grounding, 196
multi-point grounding methodology, 194
multi-pole decoupling, 139
mutual inductance, 77, 95
mutual partial inductance, 77, 78, 99, 100
noise coupling mechanisms
  electric field, 18
  magnetic field, 18
noise margin upset, 70
Ohm, Georg, 3
Ohm's law, 3
ohmic loss, 61
optical fiber. *See transmission line*
optical isolators, 209
over-damping, 69
parallel capacitors
  effectiveness, 130
parallel resonance, 115
partial inductance, 77, 98, 100
passive component behavior, 41
path of least impedance, 102
permeability, 10, 297, 298, 301
plating of metal, 247
polarization
  slot antenna, 242
power distribution network, 110
Poynting vector, 15
quality factor (*Q*), 115, 120
radiating near field, 13
radio frequency (definition), 55
rail collapse, 53
reactive near field, 12
reference, 175
reference loop (definition), 176
referencing (definition), 177
reflection loss, 226
reflections, 66, 339
relative permittivity, 133
re-reflection loss, 232
resistive loss, 61
resistivity of materials, 200
resistors, 45
resonance
  definition, 114
  parallel, 115, 135
  power and return plane, 135
  series, 114, 135

resonances, 213
RF
  current density distribution, 72
  return current path, 73
RF ground, 178
  definition, 177
Right Hand Rule, 21, 99
ringback, 66
ringing, 69
round conductors, 206
round wire, 201
safety ground, 177, 180
  definition, 177
  electric shock, 181
  green wire, 181
  ground path, 182
  series choke, 182
self-compatibility, 262
self-inductance, 77, 93
self-partial inductance, 99
self-resonant frequency, 46, 127
series resonance, 114, 135, 136
series-point ground methodology
  series and parallel configuration, 189
served shield, 249
shield
  braid, 250
  braided coax, 250
  component level, 262
  foil, 250
  penetrations, 245
  served, 249
  solid, 250
shield ground (definition), 177
shielded cable, 248
shielded compartments, 260
shielding
  effectiveness, 221
  effects, 224
  theory, 222
shielding integrity violation, 245
signal integrity
  definition, 51
signal quality, 53
signal reference, 178
signal voltage referencing ground, 184
simulation software, 70
simultaneous switching noise, 105, 165
single aperture, 237
single-point ground methodology
  definition, 177
  earthing, 193
  ground, 189
  hybrid, 201
  star, 201
  star configuration, 193
skin
  depth, 25, 74, 230
  effect, 11, 24, 74, 187, 229, 230
slot antenna polarization, 242
slot in planes, 82
solid shield, 250
SoS. *See* system-of-systems
SPICE, 52
split plane, 85, 194
SSN. *See* simultaneous switching noise
standoffs, 215
star ground, 201
star quad. *See transmission line*
static fields, 11
stray impedance, 181
stripline. *See transmission line*
stripline topology, 54
swiss cheese syndrome, 80
system-of-systems, 1
telegrapher's equations, 8
TEM. *See transverse electromagnetic mode*
termination impedance, 66
theory of electromagnetics
  closed loop environment, 9
  electric fields, 10
  Faraday's Law, 9
  Gauss's Law, 9
  Lorentz force relation, 10
  magnetic fields, 10
  Maxwell's equations, 9
  static fields, 11
  time-varying currents, 11
through-hole components, 80
time domain (definition), 2
time-varying currents, 11
trace inductance, 147
transformer isolation, 208
transformers, 48
transmission line
  balanced line, 57
  basics, 64
  characteristic impedance, 111
  coaxial based, 83
  coaxial cable, 56
  description, 55
  distributed line, 68
  embedded microstrip, 56

fiber optic, 60
impedance, 64, 66
lecher lines, 59
lossless, 62
lossy, 63
microstrip, 56
optical fiber, 60
reflections, 66
star quad, 58
stripline, 57
termination overview, 70
twin-lead, 58
twisted pair, 57
unbalanced lines, 59
waveguide, 59, 231
transverse electromagnetic mode, 56, 248
twin-lead. *See transmission line*
twisted pair. *See transmission line*
unbalanced lines. *See transmission line*
underdamped, 69
Underwriters Laboratories, 183
velocity of propagation, 68
vias
effect on plane capacitance, 134
Volta, Alessandro, 3
voltage (definition), 54
voltage referencing, 182
wave impedance free space, 223
wave shaping capacitor
calculation of, 162
waveguide. *See transmission line*
wire impedance, 43
X capacitor, 182, 282
Y capacitor, 182, 282

About the Author

Mark I. Montrose is principle consultant of Montrose Compliance Services, Inc., a full service regulatory compliance firm specializing in electromagnetic compatibility (EMC) and industrial product safety. Prior to becoming a consultant, Mark was responsible for regulatory compliance at several high technology companies in the Silicon Valley region of California. He has over thirty five years at time of this book development within applied EMC with expertise as a trainer, consultant, systems designer, product engineer, manufacturing and component engineer in addition to management positions in both Regulatory Compliance and Engineering Services.

Work experience includes design, test and certification of Information Technology along with Industrial, Scientific and Medical Equipment, specializing in the international arena for the European EMC Directive as well as product safety assessment to SEMI and CE requirements for industrial systems.

In addition to his consulting work, Mr. Montrose is an assessed EMC test laboratory to ISO 17025 and accredited by Exemplar Global (*iNARTE*) as an EMC Master Design Engineer and Product Safety Engineer.

He graduated from California Polytechnic State University, San Luis Obispo, California, with a B.S. degree in Electrical Engineering and a B.S. degree in Computer Science. He also has a Master's degree in Engineering Management from the University of Santa Clara in California.

Mr. Montrose is a senior member of the IEEE, past member of the IEEE Board of Directors (2009-2010), long term Board member of the IEEE EMC Society, champion and first president of the IEEE Product Safety Engineering Society, and life member of the American Radio Relay League (ARRL) with the Amateur Extra Class License, K6WJ in addition to being a member of the dB Society.

Mark has authored and published numerous technical papers based on sophisticated research in the field of electromagnetic compatibility and signal integrity related to printed circuit board design and theory. He presents at symposiums and colloquiums in North America, Europe, and Asia, and is a frequent speaker at international events. It is from his work as a professional trainer that he developed a unique method to teach advanced EMC concepts in a simplified manner for those without extensive experience in the field of EMC.

He is an adjunct professor in the United States and Asia, providing printed circuit board design and layout seminars to clients worldwide.

Mark Montrose authored the following best-selling reference/textbooks published by Wiley/IEEE Press prior to self-publishing this book. All books are published under the sponsorship of the IEEE Electromagnetic Compatibility Society with translation into Chinese, Japanese and Korean.

- *Printed Circuit Board Design Techniques for EMC Compliance–A Handbook for Designers*, 1st edition–1995, 2nd edition–2000.
- *EMC and the Printed Circuit Board–Design, Theory and Layout Made Simple*, 1999.
- *Testing for EMC Compliance-Approaches and Technique*. 2004.
- *Electronics Packaging Handbook.* Contributing Author - Chapter 6. IEEE Press/CRC Press. 2000.

9338131R00206

Printed in Great Britain
by Amazon.co.uk, Ltd.,
Marston Gate.